QUANTUM PHYSICS IN AMERICA
1920-1935

This is a volume in the Arno Press collection

THREE CENTURIES OF SCIENCE IN AMERICA

See last pages of this volume for a complete list of titles

QUANTUM PHYSICS IN AMERICA
1920-1935

Katherine Russell Sopka

ARNO PRESS
A New York Times Company
New York • 1980

Editorial Supervision: Steve Bedney

Reprint Edition 1980 by Arno Press Inc.

THREE CENTURIES OF SCIENCE IN AMERICA
ISBN for complete set: 0-405-12525-9
See last pages of this volume for titles.
Manufactured in the United States of America

Library of Congress Cataloging in Publication Data

Sopka, Katherine Russell.
Quantum physics in America, 1920-1935.

(Three centuries of science in America)
Originally presented as the author's thesis, Harvard, 1976.
Bibliography: p.
Includes index.
1. Quantum theory--History. 2. Physics--United States--History. I. Title. II. Series.
QC173.98.S66 1980 530.1'2'0973 79-7997
ISBN 0-405-12585-2

QUANTUM PHYSICS IN AMERICA 1920-1935

A thesis presented

by

Katherine Russell Sopka

to

The Committee on the Degree of Doctor

of Philosophy in Education

in partial fulfillment of the requirements

for the degree of

Doctor of Philosophy

in the subject of

Education

Harvard University

Cambridge, Massachusetts

May, 1976

CONTENTS

Appendices

Preface

by Gerald Holton
Mallinckrodt Professor of Physics
and Professor of the History of Science
Harvard University

Dr. Katherine Sopka's thesis is centered on two historic transformations that took place in science at about the same time in the mid-1920s, one intellectual and the other institutional. The first of these was characterized by Werner Heisenberg; speaking of the Copenhagen group of physicists working on quantum physics, he said:

"We were for a number of years, say, between 1921 or 1922 until 1927, in a state of continuous discussion, and we always saw that we got into trouble, because we got into contradictions and into difficulties. And we just could not resolve these difficulties by rational means. One would argue in favor of the waves, and the other in favor of the quanta. I have heard many discussions between Bohr and Einstein, and Einstein would argue for the light quantum, and Bohr would say, even if you had a definite proof that you had found the light quantum, if you would send me a radio cable telling me of the new research, this cable couldn't arrive here without the waves. So we actually reached a state of despair, even when we had the mathematical scheme--which was to every one of us, to begin with, a kind of miracle....Out of this state of despair finally came this change of mind. All of a sudden we said, well, we simply have to remember that our usual language does not work anymore, that we are in the realm of physics where our words don't mean much....After these infinite discussions, well, they first threw us into a state of almost complete despair, and only afterwards we felt this kind of liberation that things really were consistent." 1

Heisenberg's insistence on the desperate state of morale among physicists prior to the sudden sequence of papers proposing the matrix form of quantum mechanics, the equivalent wave mechanics, and electron spin, is borne out by many others. Wolfgang Pauli declared in 1925:

"Physics is decidedly confused at the moment; in any event, it is much too difficult for me and I wish I...had never heard of it." 2

The American physicist John H. Van Vleck, in his recent (December 1977) address on the occasion of accepting the Nobel Prize--precisely for his work on the theory of magnetism that he launched in the period covered by this thesis--described the era of the early 1920s "as one of increasing disillusion and disappointment, in contrast to the hopes which were so high in the years immediately following 1913." But by 1928, P.A.M. Dirac could confidently write:

"The general theory of quantum mechanics is now almost complete. The underlying physical laws necessary for the mathematical theory of a large part of physics and of all chemistry are thus completely known."

Indeed, in the span of a few years modern physics had come to maturation, with quantum mechanics furnishing the tools for attacking the whole range of puzzles, from spectroscopy and magnetism to physical chemistry and nuclear physics.

Parallel to this transformation there occurred a second one, which in the same convenient short-hand terminology can be called the "coming of age" of physics in the United States of America. To be sure, in earlier years the work of a few outstanding contributors in the U.S., had achieved world renown, and some others, such as the theorist J. W. Gibbs, had been unjustly neglected. But in the early years of the decade of the 1920s, America seemed far from ready to play a major role on the world stage of physics. The atmosphere has been caught in a few lines by Van Vleck:

"The American Physical Society was a comparatively small organization, with only 1400 members in 1921....The papers at that time were sufficiently sparse that parallel sessions [at A.P.S. meetings] were not necessary, and only a small number of the communications were theoretical. Very few physicists in this country were trying to understand the current developments in quantum theory, but there were a few exceptions. One of them was a Case graduate, E. C. Kemble, later on the faculty of Harvard, and I am proud of the fact that I was his first PhD. The problem I worked on was trying to explain the binding energy of the helium atom by a model of crossed orbits which Kemble proposed independently of the great Danish physicist, Niels Bohr, who suggested it a little later. In those days the calculations of the orbits were made by means of classical mechanics, similar to what an astronomer uses in a three-body problem. The Physics Department at Harvard did not have any computing equipment of any sort, and to get the use of a small hand-cranked Monroe desk calculator, I had to go to the business school. I felt very blue when the results of my calculation did not agree with experiment." 3

Van Vleck notes that matters soon began to change--"somehow, in the 1920s, America became interested in physics"[4] --and that one should not overlook the fact that in the middle years of the 1920s American physicists did make important contributions, mainly experimental.

"Still, the discovery of the new fundamental equations of quantum mechanics was primarily a European achievement.

"Then, fairly suddenly, at about the time these basic equations were established and many applications to specific problems were possible, America came of age in physics, for although we did not start the orgy of quantum mechanics, our young theorists joined it promptly."[5]

It was as if they had been waiting in the wings; Carl Eckart, Robert S. Mulliken, Linus Pauling, John Slater, Van Vleck himself, and a rapidly increasing number of others came forward with widely noted contributions on the very stage where Europeans at well-established centers, such as Heisenberg and Pauli, had so recently been near despair.

To be sure, the "coming of age" of the physics profession in America did not mean immediate maturity. Of the total number of papers in quantum mechanics during the late 1920s, only some 10-20% came from the U.S. in a given year, many of them the work of about 50 of the young people who entered physics in the 1920s. But the world had no doubt that a major change had occurred, that after a long period of intellecual and institutional subservience to European contributions and styles the American physicists, in the quantity and quality of their contributions, had joined the front ranks. By the end of the decade, there was in this country in physics--as had begun to be achieved in chemistry a decade earlier--an adequate balance between experimental and theoretical work, adequate provision for undergraduate and graduate training, a much-strengthened professional society, and a spectrum of well-run research publications. The laboratory facilities in several key places were remarkably good, the opportunities for study and colleagueship on the national and international level were excellent, and the interplay between academic and industrial science, "pure" and "applied" research that characterizes the modern U.S. was well launched. The country was well on its way to fulfilling the brash prophecy R. A. Millikan had made in 1919:

"In a very few years we shall be in a new place as a scientific nation and shall see men coming from the ends of the earth to catch the inspiration of our leaders and to share in the results [of] our developments." 6

As Dr. Sopka writes on one of her final pages, even before the major influx of European physicists in the 1930s, "the center of gravity of international activity in physics had shifted westward toward the United States," thereby providing not only a refuge but also a center of intellectual attraction for the immigrants.

We now know the long-term results of the transition in the mid-1920s, the period J. Robert Oppenheimer called the "heroic time." The number of trained physicists in the U.S. is now about 50,000, and their work has helped to shape not only modern science but modern life itself. (One may well remember that the "graduates" of the transformation in the 1920s, and their students, were destined to play a major role in the fight for survival against the war machine of the totalitarian

Axis nations, e.g., in turning their skills and ingenuity to radar, loran, sonar, and other engineering applications.) Using the admittedly somewhat simplistic criterion of the number of Nobel Prize awards as a semi-quantitative indicator of the quality of physics done in the U.S., this field is now second to none when compared with other sciences in the U.S. or with physics contributions abroad. Thus the U.S. share in the total number of Nobel Laureates, starting from 1901 for successive 15-year intervals, was 13%, 31%, 52%, levelling off at 53% in 1961-1976.[7] In terms of absolute numbers, these figures have been markedly above those for all other nations ever since the end of World War II (although if calculated proportionate to the total population, the U.K. outranked the U.S. until the last decade and a half). The sheer size of U.S. journals of physics has become almost unmanageable: binding a single year's run of just one of the many journals published by the American Institute of Physics now requires 19 fat volumes (Physical Review, Sections A, B, C, and D) and 4 more volumes for the brief reports to the Editor (Physical Review Letters).

The role of U. S. scientists in the world literature of physics is correspondingly predominant. U. S. authors in U.S. physics journals in recent years have consistently accounted for about one-third of the published world literature,[8] while one-third of the articles in U.S. journals of physics, as part of these journals' international function, are publications that were submitted by foreign authors. When using as indicator the number of citations (i.e., references) to U.S. literature in the world's science publications, physics again stands out (second only to chemistry): nearly 60% of all the citations found in the physics literature of other countries is to U.S. physics articles. This predominance is only partly due to the sheer size of the U.S. output. For if we normalize by adjusting for the absolute size of the U.S. share of the world's journal literature, the share of citations to U.S. physics publications, found in non-U.S. physics publications, is still over 40% larger than would be expected.[9] Figures such as this indicate, if not quality, at least the degree of global interest in the physics now being done in the U.S.

Against this background it is evident that a host of problems suggest themselves that will test the ability of such relatively young fields as the social study of science to handle the material. Dr. Sopka, who had studied at the Harvard Physics Department at the undergraduate and graduate levels, and whose PhD thesis was undertaken in a joint program of the Graduate School of Arts and Sciences and the Graduate School of Education, has studied the historical and social conditions responsible for the coming of age and the subsequent maturing of physics in America in the period 1920-1935, focussing on the field of quantum physics. It is one of the merits of her study that she also attends to the conceptual development of physics. It is the combination of these elements that has enabled her to do here the essential, inductively oriented and meticulously fine-grained groundwork which future attempts at synthesis, or the testing of sociological theories about the rise of contemporary science, undoubtedly will wish to use.

The questions that motivated her study were of this kind: What was the nature of the "American community" of physicists just before and just after their "fairly sudden" propulsion into the world stage? In terms of a factual narrative, who was where, when, doing what, and why? What were the research programs and styles, and what were the genealogies of the master teachers and major producers? (Here the key roles of Kemble and Van Vleck are shown to be at least as seminal as Oppenheimer's well-known and crucial role.) What were the intellectual responses of different groups of physicists in the U.S. to quantum physics, starting from the almost universal distaste for quantum theory and ending with the triumphal rise of quantum mechanics? Who were the pioneers and who the hold-outs?

What were opportunities for study abroad, or of collaboration with foreign scientists? How were these initiated, funded, nurtured--and what were the fruits? What courses and texts set the pace in curricula, and how did the professional societies and the habits of research publication change? Was the rise of theoretical physics helped or hindered by the older tradition of the predominance of experimental research? What were the respective roles of various modes (symposia, publications, visits by foreign lecturers, curriculum innovation) in the spreading of quantum theory? From the European perspective, which ideas by U. S. scientists were recognized, long neglected, or regarded as simultaneous discoveries? And above all, what were the degrees of freedom or restraint upon innovation that can be related to specifically American institutions? Here the most telling passages concern the role of grants made by the Rockefeller Foundation, and the freedom given to young researchers to pursue a line of investigation with which their older patrons may have had little sympathy or even firsthand experience

In addition to this array of materials, Dr. Sopka has also provided identifications and bibliographies that amount to a Who's Who of theoretical physics in the U.S. during the period chosen; a calendar of the visits of foreign physicists; and cross-correlations of important information mined from the great storehouses assembled by the history of physics projects of the American Institute of Physics and of the Archive for the History of Quantum Physics. These she supplemented by her own extensive interviews and correspondence with many of the most important contributors to quantum physics.

Finally, to touch only briefly on the kinds of research problems which Dr. Sopka's study will now permit us to pursue with far greater certainty and ease:

To what degree can one trace the present U. S. superiority in basic scientific research to differences between American and European university systems (as Joseph Ben-David has held)?

Was the role of European visitors and immigrants to the U.S. of first importance, as Stanley Coben seems to believe, or is it not more correct to hold with Van Vleck who wrote:

"You immediately ask whether the importation of distinguished European physicists was not the prime factor, but the point I want to make is that most of them had not arrived at the time we had begun to reach critical size....I would say that these distinguished Europeans were responsible not for giving us maturity, but rather for carrying us still further to pre-eminence, at least at an earlier date than otherwise. Lord Hailsham to the contrary, American physics is not, in my opinion, parasitical." 10

Was the intellectual transition in the mid-1920s in the U.S. the response to an earlier "crisis" involving a sharp break with the past? Or was it not rather the culmination of trends that had been underway for a number of years, as Dr. Sopka proposes?

And was not perhaps the most prototypical American ingredient in the maturing process the tendency to put one's money on the younger people, as R. A. Millikan often put it, and to permit young researchers to pursue their ideas with a remarkable degree of independence? This chraacteristic--so different from the system by which the young _Docent_ usually has to work on the problem sanctioned by the _Ordinarius_ --would help to explain why the older physicists in the U.S., while overwhelmingly experimental and initially antagonistic to quantum theory, nevertheless allowed and often encouraged the younger ones to work on the theoretical problems at the center of the new excitement.

I have little doubt that studying this volume will suggest to the reader answers to some of these current research problems, and open the way to new studies which can draw upon the material here provided.

Preface: Notes and References

1. "Discussions with Heisenberg," in Owen Gingerich, ed., The Nature of Scientific Discovery, Washington, DC: Smithsonian Institution Press, 1975, pp. 567-569.

2. Quoted in R. de L. Kronig, "The Turning Point," in M. Fierz and V. F. Weisskopf, eds., Theoretical Physics in the Twentieth Century, New York: Wiley Interscience Publishers, Inc., 1960, p. 22.

3. J. H. Van Vleck, "American Physics Comes of Age," Physics Today, v. 17 (June 1964), p. 22.

4. Ibid., p. 25.

5. Ibid., p. 24.

6. Quoted in Daniel J. Kevles, The Physicists: The History of a Scientific Community in Modern America, New York: A. A. Knopf, 1978, p. 169.

7. Calculated from data in Science Indicators 1976: Report of the National Science Board, Washington, DC: National Science Foundation, 1977, pp. 194-196. When this criterion is applied to the sciences within the U. S., physics ranks significantly higher than chemistry, but barely higher than physiology and medicine.

8. Ibid., p. 11. A world-wide sample of 2400 influential science journals was used.

9. Calculated from ibid., p. 14.

10. Van Vleck, "American Physics Comes of Age," p. 24.

Introduction

The period 1920-1935 saw the coming of age of the American physics community and the firm establishment of quantum theory in modern physics. These two developments were more than merely accidentally simultaneous. Not only did the great upheaval associated with the advent of quantum mechanics affect the way in which the study and practice of physics evolved in the United States but also there was significant participation by American physicists in the accompanying international activity. The latter fact is frequently overlooked by authors who view the development of quantum physics as an almost exclusively European enterprise.

The study which follows examines the factual aspects of the propagation of quantum physics among institutions and individuals in the United States at the same time as the contributions of particular American physicists to quantum physics are delineated. Characteristics which appear to differentiate the American scene from its European counterpart are stressed and explanations presented in terms of social, economic, political, intellectual and philosophical tenor of the period.

The work is divided chronologically into four main subsections:

1) the period before 1920 that provided the base from which the subsequent American activity in quantum physics emerged;

2) the period 1920-1925 when the old quantum theory was being pursued generally with only limited success, and serious American participation was becoming evident;

3) the period 1926-1929 which saw the quantum-mechanical revolution involving chemists and mathematicians as well as physicsts and also saw theoretical physics attain a new professional status in America; and,

4) the period 1930-1935 that were years of confirmation and application for quantum mechanics especially in the areas of quantum chemistry and the theory of magnetism where American contributions stand high.

The final Summary section contrasts the situation prevailing in the international physics community by the mid-thirties with that of fifteen years earlier. By 1935 the pressing questions, the locations of the centers of activity, the educational and institutional patterns within the world of physics had all altered in the direction of increased American participation and stature.

The timeliness of this study becomes apparent when we consider that almost half a century has elapsed since the birth of quantum mechanics and the attainment of professional status by American theoretical physicsts. This is long enough for some perspective to emerge on what were significant contributions, yet short enough to still be able to confer with active participants in the developments being considered.

The scope of this study differs from those by other authors who have dealt with this period insofar as it involves the internal history of modern physics, the sociological place of Americans within the international community of physicists and the growth of scientific activity in America. In addition, through focusing on one particular discipline, there is opportunity to consider the role of informal as well as formal educational institutions in the United States.

The implication of the phrase "coming of age," as applied to physics in America during the period on which this study focuses, is a double one. It signifies, on the one hand, that the quality, quantity and diversity of the scientific output of the members of the physics community is sufficient to rank high on the international scale and, on the other hand, that the profession as a whole has attained a viable status within its own country providing its membership with social recognition and a means of livelihood.[1]

As a result of the investigation that I have carried out I am led to the following conclusion: Physics in America came of age in the late 1920's due to the conjunction of two separate circumstances; social conditions in America were favorable in terms of the apprecia-tion of the value of pure science accompanied by adequate institution-al and financial support and, most importantly, a sufficiently large number of bright, young Americans were attracted to the study of physics by the excitement of the rapidly developing quantum concept to overcome America's earlier deficiency in theoretical areas.

While it is true that the activity of the members of the physics profession in America was stimulated by and oriented about the quantum developments pioneered by European physicists, the quantity and quality of the physics being done in America by the 1930's were of sufficiently high caliber that the Europeans who came to the United States then were no longer "missionaries" bringing physics to America as in the past, nor were they simply refugees fleeing impossible political conditions at home. They came as colleagues entering an hospitable, respectable enterprise which held great promise despite the severe economic conditions associated with the Depression.

The resources that have been used in this study are uniquely available at this time and comprise a fortunate blend of items in the public and private domains. Public information was gleaned from contemporary sources such as research and popular scientific journals, conference proceedings, records of funding agencies, textbooks, university catalogues, departmental reports, etc. Later publications, such as biographies, Festschrift volumes, posthumously published "Collected Works of ..." and recent studies related to the history of quantum physics have been examined with an eye to discerning the conceptions, and misconceptions, that have begun to be embedded in the public domain. In addition, I have also made use of unpublished materials, such as departmental histories, personal letters and notebooks of the period which have been gathered in recent years by the Archive for the History of Quantum Physics and the Center for the History of Physics of the American Institute of Physics. These provide valuable insight into the private aspects of the scientific enterprise.

Another useful resource for this study has been the interviews with many of the early quantum physicists which have been made in recent years under the auspices of the Archive for the History of Quantum Physics and the Center for the History of Physics. Such interviews have the advantages that they help to fill in many gaps in the written records and involve perspective developed over the intervening years. Although obviously subject to the fallibility of human memory for physical and psychological reasons, they often bring to the surface unexpected correlations among people and events which can then be documented from contemporary records. Moreover, in my particular case, I have been greatly aided by personal contact with several scientists who were active professionally during the period 1920-1935. They have graciously given of their time to help illuminate obscure points in existing records as well as to provide helpful suggestions and encouragement. Most notable among these has been Professor John H. VanVleck who provided the original impetus for my undertaking this study and has maintained a close interest in its progress. The identities of others will become clear in the subsequent chapters when their communications are specifically cited. The Bibliography contains a list of all their names. At this time, however, I wish to acknowledge my special indebtedness to Professor Edwin C. Kemble, America's first quantum physicst, who has not only unstintingly shared with me his recollections, books and correspondence but also has greatly helped to improve the early drafts of these pages by his thoughtful criticisms.

To Professor Gerald Holton I am particularly grateful for his wise, kind and patient guidance as teacher and thesis supervisor during my recent years of graduate study as well as for the inspiration of his own work which was responsible for my entering the field of history of science.

Finally, I wish to express my gratitude to the Radcliffe Institute for its encouragement of mature women scholars and for its assistance to me in meeting the expenses associated with the completion of this thesis.

Note

1 This definition of maturity, or coming of age, combines the views expressed by John H. VanVleck in "American Physics Comes of Age" which appeared in Physics Today, June 1964, pp 21 - 26 and by Everett Mendelsohn in Social Science 119: The Social Context of Science, lectures at Harvard University, Fall Semester, 1969 - 70.

Abbreviations Used for Institutions and Publications

Institutions:

AAAS	American Association for the Advancement of Science
ACS	American Chemical Society
AHQP	Archive for History of Quantum Physics
AIP	American Institute of Physics
AMS	American Mathematical Society
APS	American Physical Society
BAAS	British Association for the Advancement of Science
CHP	Center for History of Physics
MIT	Massachusetts Institute of Technology
NAS	National Academy of Sciences
NRC	National Research Council

Publications:

Amer. Jour. Phys.	American Journal of Physics
Ann. d. Phys.	Annalen der Physik
Astrophys. Jour.	Astrophysical Journal
Bell Sys. Tech. Jour.	Bell System Technical Journal
Bull. Amer. Phys. Soc.	Bulletin of the American Physical Society
Chem. Rev.	Chemical Reviews
Dict. Amer. Biog.	Dictionary of American Biography
Dict. Sci. Biog.	Dictionary of Scientific Biography
Helv. Phys. Acta	Helvetica Physica Acta
Hist. Stud. Phys. Sci.	Historical Studies in the Physical Sciences

Internat. Jour. Qu. Chem.	International Journal of Quantum Chemistry
Jour. Amer. Chem. Soc.	Journal of the American Chemical Society
Jour. Chem. Phys.	Journal of Chemical Physics
Jour. Chem. Soc.	Journal of the Chemical Society (London)
Jour. Frank. Inst.	Journal of the Franklin Institute
Jour. Opt. Soc. Amer. J.O.S.A.	Journal of the Optical Society of America
Jour. Pure and App. Chem.	Journal of Pure and Applied Chemistry
MIT Jour. Math. Phys.	Massachusetts Institute of Technology Journal of Mathematics and Physics
N.A.S. Biog. Mem.	National Academy of Sciences Biographical Memoirs
Naturwiss.	Naturwissenschaften
Phil. of Sci.	Philosophy of Science
Phys. Rev.	Physical Review
Phys. Zeit.	Physikalische Zeitschrift
Proc. Camb. Phil. Soc.	Proceedings of the Cambridge Philosophical Society
Proc. N.A.S.	Proceedings of the National Academy of Sciences
Proc. Roy. Soc.	Proceedings of the Royal Society
Rev. Mod. Phys.	Reviews of Modern Physics
Rev. Sci. Inst. R.S.I.	Review of Scientific Instruments
Sci. Mon.	The Scientific Monthly
SHQP	Sources for History of Quantum Physics
Verh. Deut. Phys. Ges.	Verhandlungen der Deutschen Physikalischen Gesellschaft
Zeit. f. Phys.	Zeitschrift für Physik

Chapter 1

Physics in America Before 1920

Chapter 1

Physics in America Before 1920

An assessment of the state of a discipline, such as physics, in a particular period and country requires consideration of the following kinds of questions: How many practitioners were there and under what conditions did they operate? What were the intellectual characteristics of their output? How did they, and their work, relate to the international activity in the field?

When such questions are investigated in the case of physics in America before 1920 it is expedient to recognize three periods of time corresponding to the markedly different stages of development. Although these periods merged into and overlapped with each other to such an extent that the use of strict boundary dates would be meaningless, we can, nevertheless, assign approximate dates corresponding to years when a new trend was clearly visible or a definite level of achievement had been attained.

The earliest of the three periods to be discussed extended from colonial times to about 1870. During those years there were a very few individuals carrying on research in physics. They operated in practical isolation from compatriots and from the European world of physics, usually on an amateur or avocational basis without institutional support. Then, during the last three decades of the nineteenth century, the picture brightened for physics in America in terms of the quantity and quality of the intellectual output of individuals, coupled with the development of a concerted effort to improve the conditions under which physicists operated, thus making possible the emergence of a physics profession. By 1900 practically all the essential groundwork for that purpose had been accomplished and the first two decades of the twentieth century were to witness the strengthening of many newly initiated American enterprises in physics at the same time that the discipline of physics itself was coming into its "modern" period with startling innovations being made especially in the theoretical areas of atomic structure, quanta and relativity.

For the purpose of studying quantum physics in America it is obvious that our attention will focus primarily on the last of the periods listed above but, in order to appreciate the development of the physics profession in America, it will be necessary to briefly examine the American activity in physics before 1900.[1]

Part I: Physics in America Before 1900

The Rise of Physics circa 1870

As we have indicated above, the last third of the nineteenth century marked the real beginning of professional physics in America. While it should not be overlooked that, before 1870, such figures as Benjamin Franklin and Joseph Henry conducted significant electrical investigations and the Drapers, John William and his son Henry, made worthwhile spectroscopic studies, all of these individuals operated in essentially total isolation from others engaged in the study of physics. There were too few Americans during those early years devoted to the study of physics to be mutually supportive and the European activity in physics was too remote to have immediate influence. In addition, for all of those named, their principal commitments were elsewhere.

Franklin was occupied with affairs of state; Henry was burdened, first with teaching duties[2] and then with administrative work associated with the founding of the Smithsonian Institution, the American Association for the Advancement of Science and the National Academy of Science; the Drapers were members of a medical school faculty.

What then differentiated the study and practice of physics in America during the last three decades of the nineteenth century from that of earlier times? There were a number of interrelated developments. For example, we find that, although the number of young men entering the field of physics was still small, it was increasing and, most importantly, at least three of them promptly attained international recognition. In addition, the growing number of American physicists began to enjoy more intellectual intercourse among themselves and with the European physics community. The latter was accomplished through study and travel abroad by Americans and visits to America by European physicists. Very likely it was a direct result of these international contacts that there emerged in America a spirit of self-criticism concerning the conditions under which physics was pursued in America during this period. Fortunately, self-criticism carries within itself the seeds for improvement, for once shortcomings are recognized steps can be taken toward their elimination. This is precisely what took place in the American physics community during the closing decades of the nineteenth century. Working within the social and cultural framework of the period, a few, farsighted individual physicists took those essential steps required for the emergence of a physics profession in America that could, less than fifty years later, take its place as a coequal within the larger, international community of physicists.

Gibbs, Michelson and Rowland

We shall begin a closer examination of the salient details of the developments that took place with a discussion of those three Americans whose contributions to physics attracted the admiration of the European scientific community. Their productivity contributed heavily to the rise of physics in America which began in the last three decades of the nineteenth century, for, not only did their accomplishments stand out as bright spots against a generally dismal background, but their very presence inspired and encouraged their fellow countrymen of lesser distinction in the world of physics.[3] Moreover, an examination of their careers will serve to illuminate the content and the social context of physics in America as the nineteenth century came to an end. The three, who came from diverse backgrounds and worked in diverse areas of physics, were Josiah Willard Gibbs, Albert A. Michelson and Henry A. Rowland.

Gibbs, one of the founders of physical chemistry, famous for his memoir "On the Equilibrium of Heterogeneous Substances," his phase rule and for the beauty and power of his formulation of Classical Statistical Mechanics, was "a man whose achievements command the reverent admiration of the scientists of the whole world."[4]

Gibbs studied in Europe at the Universities of Paris, Berlin and Heidelberg after receiving his doctorate in physics from Yale University in 1863.[5] He returned to Yale in 1871 as Professor of Mathematical Physics, a post he retained until his death in 1903. During the more than thirty years that Gibbs spent at Yale he worked in practical isolation from his American colleagues. On the other hand, although he never returned to Europe after becoming a productive physicists, Gibbs' work was well known and admired by such figures as Maxwell, Ostwald and LeChatelier. Gibbs' classic monograph on Thermodynamics, "On the Equilibrium of Heterogeneous Substances", became particularly well known in Europe due to a German translation by Ostwald and a French translation by LeChatelier. These translations supplemented the circulation of his work that Gibbs himself achieved through sending reprints of his publications to many of the leading scientists in England and on the Continent.[6]

Rowland's debut into the world of physics was a curious one in that it came in Europe with the backing of Maxwell before Rowland himself had ever set foot in Europe.[7] Rowland graduated from Rensselaer Polytechnic Institute in 1870 and subsequently taught there while engaging in research and independent study in the field of electricity and magnetism. When he was rebuffed in his attempt in 1873 to get some of his results on magnetic permeability published in the <u>American Journal of Science</u> Rowland sent his paper directly to Maxwell who recognized its worth and immediately arranged for its publication in the <u>Philosophical Magazine</u>.[8]

Rowland's later studies also included heat and light, but the two aspects of Rowland's endeavors which had the greatest impact on the development of physics in America were his production of curved diffraction gratings ruled with more lines and to a higher degree of precision than had been previously achieved and his promotion of the ideal of scientific research for its own sake. Rowland's gratings became prized possessions of scientists at home and abroad.[9] Without doubt their availability was a significant factor in the widespread study of spectroscopy that occupied many American physicists in subsequent years.

Rowland, the least retiring of these three distinguished American physicists, became the acknowledged leader of the physics community in America and a relatively frequent visitor to Europe. He visited European scientific centers extensively before taking on his duties as professor of physics at the newly established Johns Hopkins University in 1876 and returned to Europe again later on the occasions of the International Electrical Congresses held in Paris in the 1880's.

Michelson came to the study of physics by way of the United States Naval Academy from which he graduated in 1873 and to which he returned two years later as instructor in physics and chemistry.[10] He was already engaged in measuring the speed of light when he went to Europe in 1880 for a two year period of study in Berlin and Paris. These investigations culminated in 1887 in the famous Michelson-Morley Experiment conducted at Western Reserve College. Aside from the theoretical significance of this experiment it was important in that it involved the use of a new tool of great precision--the interferometer--invented by Michelson. It was in this area, the development of experimental tools and techniques, that Michelson's great genius lay. The echelon spectroscope was another of his inventions.

Michelson spent a few years at Clark University before joining the faculty of the new University of Chicago in 1893. He was the only one of this early triumvirate to live on and witness the wonders of physics in the early decades of the twentieth century. Rowland died in 1901, Gibbs in 1903, but Michelson lived until 1931 and was responsible for much of the prestige that the University of Chicago enjoyed in physics.

These three figures, Gibbs, Michelson and Rowland, present an interesting contrast and balance. Gibbs and Michelson represented dedication and high achievement in theory and experiment respectively. Neither of them was greatly concerned about matters outside the focus of his own research. Rowland, on the other hand, had a strong feeling for both the experimental and theoretical aspects of physics[11] and, in addition, sought to advance the level of all the physics that was being done in America.

Criteria for Scientific "Maturity"

With the works of these three men America made its real entry into the world of physics. It would be a mistake, however, to conclude that physics was flourishing in America in the closing decades of the ninenteenth century. It was not and the successes of these three men were all the more remarkable because of the general conditions that did prevail. Recalling our earlier statement defining our conception of scientific maturity, or "coming of age," as a combination of quality, quantity and diversity of output together with a satisfactory professional environment, we have noted already the excellent quality and diversity represented in the work of Gibbs, Michelson and Rowland. With regard to quantity, their individual outputs were certainly praiseworthy, but these men were outstanding all the more because so few of their American colleagues in physics were accomplishing anything of a comparable nature. The extent to which this phenomenon can be ascribed to adverse environmental conditions is uncertain. It is, however, appropriate at this time to turn our attention to a consideration of those conditions, the recognition of their inadequacies and the steps taken toward amelioration.

Before focusing critical attention on the details of the environment in which physics was pursued in nineteenth century America it will be worthwhile to consider what are the attributes of a professionally mature scientific environment. Everett Mendelsohn has set forth[12] characteristics that are so consistently found in societies where science has attained a mature status that they almost seem to be prerequisites for attaining maturity. Most notable among these attributes are: the existence of an organized professional society, the availability of journals where members can publish the results of their research, suitable physical facilities for conducting research, positions paying salaries sufficient to allow individuals to devote most of their time to scientific endeavor thereby graduating from an amateur or avocational status, recognition by peers in the form of prizes, medals or other honors such as invitations to prestigious conferences, and last, but by no means least, the availability of adequate educational opportunity for young aspirants into the profession.

Another characteristic, particularly applicable to physics and common among the more scientifically advanced European nations in recent times, is the existence of clearly recognizable centers of theoretical as well as experimental activity, together with some established leaders to serve as master-teachers, models, leaders and inspiration--roles such as Lorentz, Maxwell and Poincaré, for example, played in late nineteenth century European countries.

Criticism of the American Scene

If we consider, in the light of such criteria, the circumstances under which Gibbs, Michelson and Rowland entered and operated in the world of physics, the deficiencies of the American scene became apparent. In the 1870's, when all three were beginning their careers, America had no professional society or publication devoted exclusively to physics. Positions paying adequate salaries and providing time and facilities for research were largely non-existent. Physicists looked to Europe for training and recognition. On the positive side, however, was the fact that America was a young and vigorous nation already displaying respect, appreciation and financial support for the practical applications of scientific principles. In addition, there were men in America knowledgeable enough about physics to see what was needed and optimistic enough about American potentialities to speak out and work for improved conditions. Before describing their chief accomplishments it will deepen our insight into their perspective to consider a few of the speakers and the occasions on which individuals sought to motivate others to aid the advancement of science.

One of the earliest examples of such an occasion took place in 1872 and an Englishman was used as spokesman. It was in that year that a group of about two dozen Americans headed by Joseph Henry persuaded John Tyndall to come to America and deliver a series of popular lectures over a period of several months specifically for the purpose of stirring up appreciation and financial support for pure science in America.[13] Tyndall's approach was primarily to beguile his audiences with the beauty of physical phenomena and the exhilaration of comprehending them, but, toward the end of his presentation, he gently chided the American public for its adulation of the applications of science while failing to appreciate and support those men who made the original scientific investigations that are later exploited as applications. In particular, he urged that potential American contributors to science be relieved of excess teaching and administrative duties.

Among the Americans who raised their voices in criticism of the domestic scientific environment were F. W. Clarke, Simon Newcomb and H. A. Rowland.[14] Their particular concerns included the lack of appreciation for the distinction between science and technology, the inadequacies of the scientific and mathematical training available to students, the need for vigorous scientific societies and suitable journals and the paucity of positions providing research facilities and time for their use along with adequate recompense in terms of salary and prestige.

The fact that such flaws characterized the American scientific scene at that time is wholly understandable when one recognizes the fact that America in the nineteenth century was still a pioneer country with little time for the study of such an abstract and sophisticated subject as physics. As long as the collective energies of the nation are concentrated on the expansion and development of habitable land and the acquisition of means for comfortable living, practical achievement will rate a higher priority for support and greater prestige will be accorded those responsible for its progress.

Furthermore, scientific progress which might be esteemed for its practical benefactions was not always held in such high regard for its more speculative aspects, especially when they seem to threaten religious beliefs. Many American educational institutions had close ties to particular religious denominations and under such conditions scientific and mathematical courses were frequently considered to be of less importance than courses in theology and classics.[15]

The Record of Progress Achieved

It was possible, nevertheless, for considerable progress to be made before the close of the nineteenth century toward improving the conditions under which American physicists studied and worked. Our attention will now focus on the milestones reached in some specific areas.

During the last quarter of the nineteenth century a remarkable number of achievements were made for physics in America in the areas of improved educational opportunity at home, the development of laboratories for research and instructional purposes, the establishment of journals specifically for the publication of research in physics and the founding of a professional society for physicists. That so much could have been accomplished in so short a time attests to the acumen, the dedication and the energy of the growing number of Americans who entered the world of physics as well as to the prevailing social conditions which made such progress possible.

1) Educational Opportunities

In the middle of the nineteenth century any young American seeking advanced education, especially in the sciences, knew that he must go to Europe to find it and the German universities were particularly attractive for that purpose.[16] For the young aspiring physicist there were the attractions of great men under whom to study and established laboratories in which to be trained.[17] In some cases the American visitors remained long enough to complete the work required for a Ph.D. but, at that time, the degree had not yet attained the prerequisite status it later came to have.[18] As early as 1861 it had become possible to receive a Ph.D. in physics at an American institution,[19] but that did not obviate the need for European study as we have seen in the case of J. Willard Gibbs. Practically every member of the American physics profession before 1900 went to Europe for graduate or post-graduate study.[20] In fact it was so much the rule, rather than the exception, that when E. H. Hall wrote his account of the history of the Harvard University Physics Department[21] he made particular note of the fact that Professor John Trowbridge had never studied abroad.

The first major step toward providing a genuine American counterpart to German university training was made with the founding of the Johns Hopkins University in 1876. It was unique among American institutions in that, during its early years, it provided only graduate level instruction and was strongly research oriented. The faculty, themselves trained for the most part in Germany, adopted the format of large lectures for general discussion, small seminars for specialized topics and, above all, laboratories for experimentation. The influence of the Johns Hopkins University soon spread to other institutions, such as Columbia, Harvard, Princeton and Yale which already had programs of graduate level instruction in physics, as well as to the University of Chicago and Clark University which were to establish new programs in 1892 and 1889 respectively.

An appreciation of the growth of facilities for pursuing the study of physics at American institutions may be gained by noting that, in addition to the completely new Johns Hopkins and Clark Universities, the Jefferson Physical Laboratory was opened at Harvard University in 1884, the Ryerson Physical Laboratory was opened at the University of Chicago in 1894 and Columbia's physics laboratory grew from six rooms to four floors with the opening of the "Physics Building," in 1896.[22]

The emphasis placed by American universities on the building of physics laboratories did not stem from German models alone. By the latter part of the nineteenth century laboratory training had become a hallmark of the European scientific tradition.[23] The Cavendish Laboratory was opened in 1874 and attracted many American visitors. It should be noted however, that the idea of laboratory work as part of an American student's training in physics was introduced by E. C. Pickering at the Massachusetts Institute of Technology and by John Trowbridge at Harvard University in the early 1870's.[24]

Most of these early institutional advances were made possible through the generosity of wealthy American individuals. This was a phenomenon which began in the late nineteenth century and developed in subsequent years with significant impact on the evolution of the study of physics in America. The fortunes of Jonas Gilman Clark, Johns Hopkins and John D. Rockefeller, Sr. made possible Clark University,[25] the Johns Hopkins University and the University of Chicago.

While much emphasis was being placed on laboratory activity in physics it is worth noting that, of the few American authors writing textbooks before 1900, there were three who wrote works of a mathematical-theoretical nature which were, at least in part, in response to the fact that American college students are:[26]

"...as a rule unsuffiently prepared in the departments of mathematics necessary in approaching the subject of mathematical physics."

J. W. Gibbs wrote a short pamphlet Elements of Vector Analysis Arranged for the Use of Students in Physics;[27] B. O. Peirce, of Harvard University, wrote Elements of the Theory of the Newtonian Potential Function;[28] and A. G. Webster, of Clark University was the author of an ambitious volume The Theory of Electricity and Magnetism.[29]

Before leaving this topic of the development of educational opportunity it is appropriate to discuss the awarding of doctorates in physics by American institution prior to 1900. Between 1861, when the first Ph.D. in physics was granted by Yale University, and 1900 there were approximately 75 other doctoral degrees earned by students of physics at American universities.[30] The Johns Hopkins University, as might be expected, led with 35 degrees awarded; Yale University was next with 16; Cornell University, third with 11; Harvard University, fourth with 6; Princeton University, fifth with 3; and, the University of Chicago, sixth with 2. In addition, Brown, Stanford and Western Reserve Universities awarded one apiece.

2) Research Publications

The growing amount of research in physics being done by Americans at the developing academic institutions increased the need for adequate publication channels. Until the middle of the 1890's American physicists who wished to have the results of their research published in this country had to submit their papers to journals of a general scientific nature such as: American Journal of Science ("Silliman's Journal") which was founded in 1818; Journal of the Franklin Institute, founded in 1826; Proceedings of the American Academy of Arts and Sciences, founded in 1848; and Transactions of the Connecticut Academy of Arts and Sciences, founded in 1874. Then, in 1894, two physicists at Cornell University, Edward L. Nichols and Ernest Merritt, founded The Physical Review as "A Journal of Experimental and Theoretical Physics," according to its title page. In 1895 the Astrophysical Journal, "An International Review of Spectroscopy and Astronomical Physics," was founded with astronomers George E. Hale and James E. Keeler as editors but with a substantial number of physicists, such as J. S. Ames, A. A. Michelson and H. A. Rowland, serving as assistant or associate editors.

3) Professional Societies

The first step toward establishing a professional society for physicists was taken in 1882 when Section B was established with the American Association for the Advancement of Science. This section was, and still is, devoted to physics. During the annual meeting of the Association, held in a different location each year during the summer months, an active program was presented by Section B, including an address by the Chairman who was also designated a Vice-President of the Association. Election to this office provided one of the first honors that the American physics community could bestow on one of its members. T. C. Mendenhall, H. A. Rowland and John Trowbridge were the first to be so designated.[31]

Section B served as the only organization of American physicists until the founding of the American Physical Society in 1899 and for many years, while the Society was firmly establishing itself, the two held joint meetings during the annual meeting of the Association.

By 1899 there was a group of American physicists who felt that more frequent meetings with colleagues would be desirable. A committee was formed, headed by A. G. Webster of Clark University, which sent out a letter of invitation to a meeting to be held at Columbia University on May 13 for the purpose of organizing a Physical Society that would meet four times a year. Most of the approximately forty physicists who participated in this meeting came from east coast academic institutions north of Washington, D. C., but there was one each from the Universities of Chicago and Michigan, Lake Erie College and Dalhousie University of Canada. H. A. Rowland was elected President, A. A. Michelson Vice-president, Ernest Merritt Secretary and William Hallock, of Columbia University, Treasurer.[32]

International Contacts

One final facet of the activity in physics that took place during the closing decades of the nineteenth century merits our attention: the relationship of the American physics community to the European physics community. We have already commented upon the visits to Europe by American physicists and upon the European recognition accorded the work of Gibbs, Michelson and Rowland. What remains to be considered are the visits of European physicists to America and instances of European honors attained by American physicists.

Several European physicists came to the United States for varying periods of time and for various purposes.[33] We have already mentioned the visit which John Tyndall made in 1872-1873 to deliver a series of popular lectures. In 1884 William Thomson, later Lord Kelvin, delivered twenty lectures at the Johns Hopkins University where his audience was limited to professional physicists and consisted of about twenty-one regular attendees. The lectures, which were later written up for publication,[34] were delivered informally and involved the participation of Kelvin's "21 coefficients" as they became known.[35] A third emminent British physicist to lecture in America was Sir J. J. Thomson who made his first trip to the United States in 1896 to attend the sesquicentennial celebration of Princeton University and to deliver four lectures on the electrical conductivity of gases.[36] In addition to these lecturers several other European physicists had opportunity to visit the United States in connection with large scale scientific meetings. Following the meetings of the British Association for the Advancement of Science that were held in Montreal in 1884 and in Toronto in 1897 some British scientists crossed the border to visit the United States before returning home.[37] The Electrical Congress of 1893, held in Chicago in connection with the World Exposition attracted, among the official delegates, Hermann von Helmholtz who was elected honorary president of the meeting.[38]

All accounts stress the cordial relations between the Europeans and their hosts that were associated with these visits. The Europeans were treated as honored guests and the Americans were pleased to sponsor occasions of intellectual discussion on this side of the Atlantic while, at the same time, showing off their growing institutions.

Turning now to the topic of recognition of American productivity in physics by the international physics community, we find that this can be assessed in a general way by noting the incidence of papers in foreign journals of physics written by American authors. In the final decade of the nineteenth century there were some thirty American physicists whose articles were accepted for publication in such prestigious foreign journals as Annalen der Physik und Chemie, Annalen der Physik and Chemie Beiblätter and the Philosophical Magazine.[39] Practically all of the work reported was of an experimental nature. Viewed from the perspective of the total literature of physics during that decade, the American contribution was but a small percentage. Viewed from the perspective of a young nation with a scarcely established physics profession, however, the picture represents considerable achievement.

More specific and higher recognition came to a smaller number of American physicists. Gibbs, Michelson and Rowland were named honorary members of foreign scientific societies. In addition, the Copley Medal of the Royal Society was awarded to Gibbs, the University of Paris Medal to Michelson and the Matteucci Medal to Rowland.[40] Furthermore, equipment sent by Michelson to the Paris Exposition of 1900 was awarded a Grand Prix.[41] During the same year of 1900 J. S. Ames and Carl Barus participated in the Congrès International de Physique held under the auspices of La Societé Francaise de Physique[42] and Robert W. Wood, then an assistant professor at the University of Wisconsin, was invited to address The Royal Society.[43] W. D. Bancroft and E. L. Nichols, both of Cornell University, were invited to contribute to the Festschrift prepared to honor Lorentz on the occasion of the 25th anniversary of his doctorate.[44] In short, by 1900 several American physicists enjoyed international professional status but, with the exception of Gibbs, it was always for experimental achievement.

Summary

Looking back over the accomplishments in the study and practice of physics in America during the final decades of the 19th century, it becomes clear that, although the criticisms voiced earlier remained valid to a considerable degree, there were strong grounds for pride in what had already been accomplished and for optimism about the future prospects of physics in America. Although the number of practicing physicists was small and they worked for the most part in isolation, some excellent work had been accomplished and accorded international recognition. Innovations had been made in the very important areas of publication, professional society establishment, development of programs and facilities for graduate instruction and financial support. While the status of these endeavors at the turn of the century may seem puny by latter-day standards, their great importance lies in the fact that the fundamental groundwork had been accomplished[45] and the next twenty years were to witness considerable progress toward the attainment of a fully mature status for physics in America.

Part II: The Growth of Physics in America 1900-1920

The study and practice of physics in America during the first two decades of the twentieth century evolved on a course that was determined by a variety of independent circumstances. On the one hand, the American physics profession continued to develop within the prevailing social context and in the direction of increased emphasis on experimental activity. On the other hand, American physicists were forced to begin to cope with the innovations of the relativity and quantum theories arising within the discipline itself in Europe. In addition, the circumstances surrounding the World War that engulfed Europe between 1914 and 1918 had its particular impact on the development of physics in America. It brought an interruption to the normal scientific activity on both sides of the Atlantic but it also provided the American physics profession with a new visibility and cohesiveness which would signficantly affect its postwar development. Specific details associated with each of these influences will be incorporated into the discussion which follows. The discussion itself will focus on: 1) the pattern of growth experienced by the American physics profession; 2) the productivity and general outlook of the members of the profession; and, 3) their specific response to quantum theory.[46]

The Pattern of Growth

All of the activities concerned with the emergence of a physics profession in America that had been initiated as the nineteenth century drew to a close continued and gained momentum in the new century. We shall now examine the progress that was achieved in the areas of professional society, journal publication, laboratory building, Ph.D. production and participation within the international community of physicists in sufficient detail to be able to appreciate the status of physics in America by 1920.

1) Professional Societies and Research Publications

Looking first at the American Physical Society, we find that the membership by 1920 had risen to approximately 1,300 and was divided into three categories: regular members, associate members and honorary members from foreign nations.[47] New York, Washington, D. C. and Chicago had become established as regular meeting places for the Society. A Pacific Coast Division was formed in 1917 with its own Local Secretary. Contacts with other scientific disciplines were maintained through the expedient of scheduling a Physical Society meeting to be held at the same time and in the same city as the annual meeting of the American Association for the Advancement of Science in late December. Programs for individual meetings were in the hands of local committees who sometimes arranged symposia on topics of wide interest to the membership to supplement the usual contributed papers dealing with individual research. Election to the Presidency of the Society became one of the honors the profession could bestow on its esteemed members.

Ties between the American Physical Society and the Physical Review were established after the Society stopped publishing its own Bulletin in 1902. In 1903 the Physical Review began to carry the designation "Conducted with the cooperation of the American Physical Society" and, henceforth, published all official notices of the Society's meetings. In 1913 the Society assumed responsibility for the publication of the Physical Review and a new series of journal numbers was begun.

Two other journals of importance to the physics profession were inaugurated a few years later: Proceedings of the National Academy of Science in 1915 and the Journal of the Optical Society of America in 1917. Both of these publications carried many articles by physicists in subsequent years.

2) Laboratory Facilities

Universities continued to improve their facilities for laboratory instruction and faculty research. New Physics Laboratories were opened at Cornell in 1906, at Princeton in 1909, at Illinois in 1909 and at Yale in 1913.[48] Other institutions sponsoring laboratory facilities for work of interest to the physics profession, inaugurated during these years, were the National Bureau of Standards (1901) and the Carnegie Institution (1902), both located in Washington, D. C. It was during this period also, that industries began to recognize the importance of research facilities for their own growth and development.

By 1920 research laboratories had been established, for example, by the American Telephone and Telegraph Company, the General Electric Company, the Eastman Kodak Company and the Westinghouse Electric and Manufacturing Company.[49]

These developments provided a broadening of the employment horizons for young physicists beyond academic institutions. In 1913, when the membership of the American Physical Society stood at about 350, some 70 percent of the members were associated with academic institutions, while 10 percent were employed by private and governmental agencies and about 6 percent had found position in industries.[50]

3) Educational Opportunities

During this period the number of Americans pursuing graduate study in physics increased markedly and although the statement made in 1904 that[51]:

"No American need any longer come to this Germany or any other country for higher education. In my judgement the United States offers today facilities for collegiate, academical and postgraduate studies equal in quantity and quality to those offered by any country in the Old World."

is, without doubt, an exaggeration, the fact is that, after 1900, very few Americans went abroad for their doctoral studies. In the first twenty years of the new century about 400 doctoral degrees in physics were awarded by some twenty American institutions--more than five times the number of degrees and twice the number of institutions of the forty preceding years. Chicago, Cornell and The Johns Hopkins Universities were each responsible for more than fifty; while Columbia, Harvard and Yale Universities produced twenty-five or more, apiece, California, Illinois, Princeton and Wisconsin each added twelve or more.[52]

One factor in this remarkable growth rate was the circumstance that, by this time, most of the American physicists who had obtained their doctoral degrees between 1875 and 1900 at either American or European universities had found positions on the faculties of the growing number of American universities offering sound graduate level instruction. The increase in the number of state universities that developed significant programs in physics is one of the striking characteristics of this period. It is notable that, while all of the Ph.D. granting institutions before 1900, as listed on pages 1.14 and 1.15 were private and, for the most part, located in the eastern third of the nation, the period 1900-1920 saw as newcomers to this class the Universities of California, Michigan, Minnesota and Wisconsin. It will be seen in subsequent chapters that all of these universities became sites of quantum theoretical activity in the years 1920-1935. In addition, Illinois, Indiana and Iowa State showed up in the ranks of universities granting Ph.D.'s in physics prior to 1920. Virtually without exception, the dissertations accepted by these institutions contained experimental results--a point that we shall return to later in discussing the outlook of the physics profession in America.

American institutions of higher learning, especially those state supported universities growing up around the turn of the century, were very much a product of the American social pattern of the times.[53] They became multipurpose institutions, borrowing some academic traditions from Europe, but adding much that was uniquely American. From Germany came the reverence for scholarly research, ("Wissenschaft") and the concept of the doctor of philosophy degree; from England, the idea of a "liberal education."[54] To these were added elements arising from the democratic and pragmatic nature of American society: a great diversity of courses was offered; large enrollments were provided for; practical advisory services were made available to the public. In turn, the funding of these institutions was primarily the responsibility of the individual states, although, in some cases, wealthy residents of the state (such as Mrs. Phoebe Hearst of California[55]) took the initiative of providing additional revenues.

The resulting phenomenon of many sizeable universities spread throughout the country had its impact on the conditions under which the American physics community was to come to maturity. On the one hand, the multiplicity of universities, each of which involved local pride and responsibility, made for widespread employment opportunity and encouraged competitive excellence; but, on the other hand, the relatively large number of universities, coupled with the geographic size of the continent, made for real isolation of the individual endeavors. In the early decades of the twentieth century physics departments employing as many as half a dozen faculty members were the exception rather than the rule.

With the increased opportunity for graduate level education at home, the trend of study after 1900 for young American physicists was more in the direction of a year or two of postdoctoral work abroad for those who could manage it. Some were still drawn to the great German universities but an increasing number spent their time abroad at the Cavendish Laboratory.[56]

4) International Contacts

While proportionately fewer Americans were going abroad to study physics, the flow of European lecturers to the United States increased in the years from 1900-1914.[57] These visits were made possible, in many cases, by two phenomena associated with American universities: the endowed lectureship and the summer school. The Silliman Lectures[58] were begun at Yale University in 1903 with J. J. Thomson then making his second visit to America.[59] Columbia University was able to attract H. A. Lorentz, Max Planck and Wilhelm Wien through the Ernest Kempton Adams Fund. Among the Europeans coming to lecture at Summer Sessions were Ludwig Boltzmann and Ernest Rutherford who visited the University of California at Berkeley in 1905 and 1906, respectively.[60] The University of Chicago Summer School had Max Born and F. A. Lindemann as lecturers in 1912 and 1913.[61] As will be seen in subsequent chapters, endowed lectureships and summer school lecturing continued and grew in importance as channels for European influence on the development of the physics profession in America.

Some European physicists made more extended visits to American universitites and a few emigrated permanently to America during this pre-1920 period. In such instances they were in a position to significantly affect the development of the physics departments with which they were associated. A particular case in point was the importation to Princeton University of J. H. Jeans who served as professor of applied mathematics from 1905 to 1909 and O. W. Richardson who arrived in 1906 and remained with the physics department until 1913. They were both members of the Princeton faculty during the years when the Palmer Physical Laboratory was being planned and built. Jeans wrote two books during his Princeton tenure and Richardson's work on electron emission strongly influenced the research interests of a number of young graduate students among whom were K. T. Compton and C. J. Davisson.[62]

What can we say of the relations between these visitors from Europe and their American hosts? In particular, what was the scientific level on which they approached each other? Looking into the account of the International Congress of Arts and Sciences (held in conjunction with the Universal Exposition at St. Louis in 1904) written by Hugo Munsterberg,[63] we find it described as an epoch making event where "the Old and the New Worlds stood on equal levels and (where) Europe really became acquainted with the scientific life of these United States," following years of condescension toward America. Munsterberg claimed that a marked change followed immediately after this Congress, but this seems to have been an exaggeration in the light of the actual state of science in America at that time and of remarks made by physicists some twenty years later which we shall have occasion to quote in our next chapter.

Commentaries on their American travels by European scientists[64] politely stress the warmth of their welcome, their admiration for the new buildings that were rising on campuses and the enthusiasm for the pursuit of science evidenced by the physicists of America. With few exceptions, in those days, the Europeans were the teachers, the Americans the learners. America remained "mission country" long after John Tyndall first came as an "apostle of science" in 1872.[65]

There were, however, some instances when American physicists were honored on an international level. Of these, the most prestigious was the awarding of the Nobel Prize in physics to A. A. Michelson in 1907. This was the first time an American was so honored since the founding of the award in 1901 and the only time an American physicist received this award in the period before 1920.[66] Michelson also was awarded the Copley Medal of the Royal Society that same year and in 1911 he lectured as an exchange professor at the University of Göttingen.[67]

No Americans were present at the first Solvay conference in 1911, but R. W. Wood took part in the second, held in 1913 and delivered an address entitled "Rayonnement de Resonance et Spectres de Resonance."[68] By this time Wood had become one of the best known American physicists among the Europeans. He made two extended, pre-war, visits to Europe in 1910-1911 and 1913-1914. In the course of these he combined research, lecturing and travel with his family. During the earlier trip, he collaborated with J. Franck and H. Rubens in laboratory investigations at the University of Berlin and was invited to lecture at many scientific gatherings including those associated with the Nobel awards of 1910. Wood's second European visit was undertaken at the invitation of Sir Oliver Lodge to address the annual meeting of the British Association for the Advancement of Science and receive an honorary degree from the University of Birmingham along with H. A. Lorentz and Madame Curie who were being similarly honored at that time. Wood, a man of independent spirit and private financial means, found little difficulty in arranging to be absent from the Johns Hopkins University whenever he wished.[69]

Meanwhile, R. A. Millikan, who was more closely tied to his research and teaching duties at the University of Chicago and was less affluent than Wood as well as being less well known at the time, did manage a six month European sojourn in mid-1912 during which he was cordially received by members of the physics communities of England and Germany. He was twice invited to address the German Physical Society and present reports on the research he had been carrying out in Chicago. Before returning home Millikan attended the meeting of the British Association for the Advancement of Science, held that year in Dundee, where he also spoke on his research.[70]

The balance of international honors and recognition in the pre-war period remained predominantly on the side of the Europeans, but American experimental physicists were beginning to take their place within the international physics community.

5) Effects of the World War

Before closing our discussion of the social aspects of the American activity in physics during the early decades of the twentieth century we must take cognizance of how that activity was altered by the war that engulfed Europe in 1914.[71]

The outbreak of hostilities in Europe affected not only the scientists of the nations initially involved in the conflict but also those on this side of the Atlantic well in advance of our entry into the war in 1917. Some American scientists who were familiar with the political developments in Europe through their travels there in the years just before 1914 were convinced of the righteousness of the Allied cause and of the need to be prepared to come to the assistance of England and France in, at least, an advisory, if not military, capacity in order to insure their ultimate victory. To this end the National Research Council was established under the auspices of the National Academy of Science and carried the endorsement of President Woodrow Wilson. The Council's function was to make available the scientific resources of America for the defense of this country and the assistance of those nations which were soon to become our wartime allies.

As America's involvement in the European conflict intensified many American physicists interrupted their normal professional activities and eagerly applied their expertise to war-related problems such as those associated with submarine and airplane detection, optical signalling and radio communication.

While it can scarcely be claimed that the work of American (or European) physicists played a decisive role in the outcome of World War I, the physicists of the Allied Nations did cooperate wholeheartedly in the war effort.

For the American physics community, the wartime activity of its members resulted in an increased cohesiveness among themselves and a new visibility among members of governmental, military and industrial circles in the United States and among their scientific colleagues in England and France. In particular, the strong experimental competence that was characteristic of American physicists commanded widespread respect.

Meanwhile, the National Research Council began taking steps toward becoming a permanent institution even before the war had come to an end. Among its later peacetime activities were two which had special significance for the American physics profession: its fellowship program and its serial publication, <u>Bulletin of the National Research Council</u>. We shall have more to say about both of these in our next chapter where the postwar situation will be described in detail.

Summarizing, now, the pattern of growth of the American physics profession and the social aspects of the environment in which the individual American physicists pursued their careers in the first two decades of the twentieth century, we can make the following observations: Domestically, all of the institutional advances initiated in the closing decades of the nineteenth century grew in numbers and strength. Society membership, journal publication, educational and research facilities and professional employment opportunity all improved markedly. Fewer students went to Europe to study at the predoctoral level but a year or more of postgraduate work at one of the European laboratories was highly prized. Further contact with the European scientific community was achieved through visits to America by outstanding physicists from England and the Continent. These forms of intercourse served to make Americans aware of the state of physics in Europe and accelerated the rate of rise of the general level of physics in this country. Above that general level some American figures clearly stood out and achieved cordial recognition by their European colleagues. While World War I interrupted the scientific careers of individual physicists of all ages, it also brought increased attention to the potential role which physicists could play in society at large.

If we say that the birth of American physics took place in the late nineteenth century, then it is appropriate to consider the years 1900-1920 as bringing its adolescence.

The Output and Outlook of American Physicists

An examination of the volumes of Physical Review and other scientific journals carrying contributions from American physicists during the early decades of the twentieth century shows that American activity in physics was almost exclusively experimental and, in a number of instances , it was of very high quality.

Excellent spectroscopic investigations, for example, were carried on in several laboratories using high precision gratings that were made in America. The ruling engine developed at the Johns Hopkins University by Rowland continued to be used after his death in 1901 by his successor R. W. Wood, himself a skilled experimenter with a particular interest in resonance radiation. In addition, at the University of Chicago Michelson built a ruling engine capable of producing large gratings of high resolving power. At Harvard University Theodore Lyman developed techniques for probing the ultra-violet region of the spectrum which led to his discovery of the series of hydrogen lines corresponding to transitions to the n = 1 state, the so-called Lyman Series.[72]

During this period Harvard's Jefferson Physical Laboratory was also the site of P. W. Bridgman's early work on high pressures that later, led to his being awarded a Nobel Prize in Physics, and of an active program in X-Ray study headed by William Duane.

At Princeton University K. T. Compton, who had taken his doctorate there in 1912 with O. W. Richardson writing a thesis on the photoelectric effect,[73] returned to join the faculty in 1915 and institute a broad research program on electron impacts. Significant studies on electron impact phenomena were also begun about 1918 at the National Bureau of Standards where the war had brought together John T. Tate, Paul D. Foote and Fred L. Mohler.[74] At the University of Michigan Harrison Randall, after a European sojourn studying spectroscopy at Tübingen with Paschen, was developing a laboratory to concentrate on the study of the infra-red region of the spectrum.[75] At Dartmouth College the pressure of light[76] was successfully measured by E. F. Nichols and G. F. Hull.

The American experimental investigations of this period which attracted the greatest attention and admiration were those carried on at the University of Chicago by R. A. Millikan who developed his oil drop method for determining the fundamental unit of electrical charge and made photoelectric studies which confirmed the photoelectric equation postulated by Einstein in 1905 and provided a value for the Planck constant "h" to a greater degree of accuracy than was previously possible. These were the "works in progress" about which Millikan lectured during his European visit of 1912, mentioned earlier. The importance of his final results were formally recognized and rewarded in 1923 when Millikan was given the Nobel Prize in Physics for these achievements.[77]

While excellent experimental studies such as those cited above were being carried out by American physicists theoretical activity was virtually at a standstill. The theoretical work of J. W. Gibbs, looked at in retrospect, appears as a "supernova" in the field of physics in America. He appeared suddenly, produced brilliantly and left little immediate after effect. When Gibbs died in 1903 he left no group of younger theoreticians to carry on in the creative spirit which had characterized his work.[78] It appears that during Gibbs' tenure at Yale University some dozen theses in physics and mathematics were inspired by their authors' contact with Gibbs as a teacher but, as a thesis supervisor, Gibbs remained aloof from his students, neither suggesting topics originally nor keeping in close touch once his approval had been obtained for a topic proposed by the student. He showed his appreciation, however, of the student's work when it was completed, all the more so because it was entirely the student's product.[79]

The expectation that other, younger American physicists who were mathematically competent might have become creatively productive in theoretical areas did not materialize. Of B. O. Peirce, whom we have mentioned earlier, it was said:[80]

"Though profoundly versed in scientific theory he was not a theorizer. His constant effort was to add to our certainties of knowledge."

This opinion is borne out by inspection of the twenty-one papers that he wrote during the final decade of his life.[81] Fewer than half a dozen were of a mathematical or theoretical nature. In Peirce's final years he was predominantly occupied with teaching duties and experimentation. A. G. Webster continued to write textbooks of mathematical physics that were well received at home and abroad and to plead the cause of increased mathematical training for physics students but his own research was primarily experimental.[82] The subsequent generation of mathematical physicists which included such figures as H. A. Bumstead, Max Mason and Leigh Page were practioners rather than innovators, competent teachers and text book writers and, for the most part, concerned with electromagnetic theory.[83]

The fact that American physicists of this period were so much more strongly oriented toward experimental rather than theoretical activity stems from a number of roots. Practical, applied physics continued to be held in high esteem in a rapidly developing society. The electrical pursuits of M. Pupin and the acoustical studies of W. C. Sabine, both of very admirable quality, are cases in point.[84] In addition, the generation of physicists which formed the established profession of the early twentieth century had received their training at a time when classical physics was enjoying its most triumphant period following the introduction of the Maxwell electromagnetic theory. In many circles[85] it was felt that theoretical physics was a completed area of study with new discoveries likely to emerge only from laboratory investigations. As we have already noted, the building of laboratories was of prime concern for institutions wishing to have programs in physics. Concomitant with these facts was a situation with regard to mathematical training which left something to be desired. Although by about 1910 good courses in mathematical physics were offered at practically all the major graduate physics departments active at that time, these courses were, as a rule, concerned solely with the application of mathematical techniques to the well established areas of mechanics, acoustics and electromagnetism, i.e. classical physics. What was lacking to American physics students was first hand contact with mature, creative mathematicians and theoretical physicists such as some of their European counterparts enjoyed.[86] The American mathematics community was itself reaching maturity during this period but there was little interest among its members in collaborating with physicists.[87] A. G. Webster, a staunch advocate of more mathematical training for physicists was highly critical of the way mathematics departments taught those topics particularly needed by students of physics.[88] Doctoral candidates, lacking the tools for mathematical physics, usually chose their thesis topic in their first year of graduate work and were kept fully occupied by the necessity of building their own apparatus--as did all experimental physicists at that time.

Nevertheless, valid as such criticisms of mathematical training may have been, it seems that the dominant factor in explaining the American failure to progress as rapidly in the theoretical areas of physics as in experimental was the widespread attitude on the part of the established professional physicists of America that physics was, first and foremost, an experimental science in which mathematical or theoretical activity was, at most, of an ancillary character.[89] We have already cited views on the preeminence of experimentation that were attributed to B. O. Peirce. Almost identical sentiments were ascribed to H. A. Bumstead to explain why Bumstead's research output was primarily experimental, despite the fact that he had done his doctoral work with Gibbs:[90]

"Perhaps the explanation lies in the feeling he often expressed that experimentation is the real business of the physicist and in a certain distrust of speculative ideas which he frequently manifested. ...He believed in holding close to the facts revealed by experimentation, and he often stressed the point that all physical laws are the result of experimental discoveries and that no mathematical formulation can contain more than is involved in its premises."

Strong statements exalting the value of experimentation also came from other physicists, such as Henry Crew who, three times during his Presidential Address to the American Physical Society in 1909, paid tribute to a score of experimenters. They, he maintained, were responsible for the progress of modern science.[91]

The attitude of the American physics community of the early twentieth century was clearly not one which encouraged purely theoretical pursuit.[92] By about 1915, however, as we shall soon see, considerable attention was becoming focused on theoretical considerations related to atomic structure. Then came the interruption of the World War which, in the words of Harold Webb, "had a very dampening effect on physics research in the U.S., expecially from 1915 to 1920. During this period we received few foreign publications and from 1917 to 1919 were ourselves involved."[93] Thus, with few exceptions, significant involvement of Americans in the development of quantum physics was delayed until 1920, the initial date of the period of the principal focus of this study.

The attitude on the part of respected faculty members regarding the primacy of experimentation meant that graduate students were encouraged to begin work on experimental problems almost as soon as they entered graduate school and frequently they never did take as much mathematical training as might have been desirable.[94] Not only was there a bias favoring experimental work but also an attitude that theoretical work alone would not be sufficient to earn a doctoral degree. David L. Webster received a Ph.D. in physics from Harvard University in 1913 for a thesis whose title, as listed by Marckworth and Harvard University,[95] was: "I, On an electromagnetic theory of gravitation and II, On the existence and properties of the ether," seemingly an entirely theoretical thesis. When Webster's thesis is consulted, however, it is found to contain two additional sections: "The theory of the scattering of Röntgen Radiation" and "The absorption of light in Clorine as dependent on the Pressure."[96] The last named section was experimental. In an interview granted in 1964[97] Webster described his experiences as a graduate student who worked four years on this experimental problem with great difficulty while really feeling much more at home pursuing theoretical topics. Professor Webster has more recently provided the following account of the attitude which he encountered during his years at Harvard University Physics Department which extended from 1906, when he entered Harvard College as a freshman, to 1917, when his position as an Instructor was interrupted by military service:[98]

"It [the "equation," good physics = experimental physics] was a very common belief at Harvard when I was a student there, and at that time I did not know of any university in this country where I could expect to find doubts about it. To me that was a source of regret. I had studied not only the experimental courses, but also some excellent theoretical physics with Sabine, G. W. Pierce and B. O. Peirce, and had read books by Planck, Lorentz and others; and I believed that the great advances in physics came through theories tested by experiment: theories first, then the tests to decide between them. Maxwell's electromagnetic theory of light, for example, came before Hertz's experimental test of it. And the Michelson-Morley experiment and the few others that led to relativity were started as tests of earlier theories. So I wanted to train to be a theoretician, rather than an experimenter. But I was told, plainly and emphatically, that Harvard would never give a Ph.D. degreee for any theoretical thesis.

My good friend and classmate, Robert H. Kent, tried to break that rule. He wrote a thesis on an application of the virial theorem to a thermodynamic problem. It was excellent work, but the rule held; and

Kent's discovery was soon made independently in England by Lennard-Jones, who won his election to the Royal Society largely for this work. Kent went into the U.S. Army, in which he rose to be the most eminent ballistician in Aberdeen Proving Ground; and for his work there some forty or fifty years later, Harvard awarded him an honorary doctorate."

Further evidence of the attitude that doctoral theses must contain experimental results is found in the fact that, although A. G. Webster was one of the most outspoken advocates of increased appreciation of mathematics by physicists, during his tenure as professor of physics at Clark University, 1892-1923, not one theoretical doctorate was awarded.[99]

Certainly conditions favorable to the pursuit of theoretical physics did not prevail in early twentieth century America. There were no established practising theoreticians to encourage and guide young aspirants to such activity; strong mathematical training was not required for all graduate students of physics; and, general speaking, experimental results were expected from successful, productive physicists.

Another aspect of the attitude of the American physics community toward experimentation was that some spokesmen not only held that good physics = experimental physics, but also were ready to equate laboratory activity with character training and the quest for truth. As early as 1886 H. A. Rowland had set a precedent for such statements in a Commemoration Day Address at the Johns Hopkins University saying:[100]

"This, then, is the use of the laboratory in general education, to train the mind in right modes of thought by constantly bringing it in contact with absolute truth and to give it a pleasant and profitable method of excercise which will call all its powers of reason and imagination into play."

He argued further that students "...must try experiment after experiment and work problem after problem until they become men of action and not of theory." A. G. Webster spoke in a similar vein in his address entitled "The Physical Laboratory and its Contribution to Civilization"[101] saying that the pursuit of accuracy and objectivity in the laboratory will provide "an education in morals which is hard to equal in any other part of education."

The outlook that the pursuit of science was infused with idealism, optimism and the potential for benefiting all of humanity was a recurrent theme among other American physicists, such as R. A. Millikan and Michael Pupin.[102] Both these men were of strong religious persuasion. Their outlook on physics was a combination of American optimistic belief in successful progress through a rational, scientific approach and a belief that science was not only compatible with religion but was entitled to a reverence of its own. They seem not to have been encumbered by any strong philosophical bias that would greatly hinder their acceptance of new ideas in physics on a purely pragmatic basis once they were to become convinced that the experimental evidence was incontrovertible.[103]

During these years the American physics community was not without its moments of introspection and self-criticism. On such occasions it was pointed out that America was still far from having reached its full potential intellectually. One of these critics was A. G. Webster whose devotion to the cause of advancing the scientific endeavor in America made him seize every opportunity to speak out on the subject. In 1908 he wrote an article entitled "America's Intellectual Product"[104] which should have dispelled any feelings of self-satisfaction and complacency that might have been overtaking his colleagues. Another who shared Webster's views that American scholarship left much to be desired was F. C. Brown, who had joined the faculty of the State University of Iowa after taking his doctorate at Princeton with Richardson in 1908. In 1914 Brown made the suggestion[105] that what was needed to accelerate the growth of pure science in America was the establishment of a physical institute independent of the existing facilities associated with universities and industries--both of these he held suspect because of their lack of single-minded devotion to the pursuit of basic scholarship. We mention Brown's proposal in two contexts: as an indication of a spirit of self-criticism by some American scientists and because his specific proposal related to one of the American deficiencies that we mentioned earlier and to which we will return in discussing the post-war period when the establishment of such an institute could have taken place, but did not.

Summarizing now the output and the outlook of the American physics community during the first two decades of the twentieth century we can point to many instances of the production of experimental work of excellent quality, accompanied by a strong faith in the validity of a basically empirical approach. While productivity was low in theoretical physics there was a growing recognition that a mathematical-theoretical perspective was essential to attaining the full potential of the study of physics. Further, there was every reason to expect that America would one day stand on equal footing with the older European nations in the production of first-rate science if our intellectual, social and financial resources were advantageously developed. America was a young nation, freely drawing on the cultural resources of Europe while developing its own characteristic institutions and way of life, preparatory to assuming a fully participating status in the international scientific community.

The American Response to Quantum Theory

1) A Turning Point in 1913

The year 1913 was a decisive one for quantum theory in America. That was the year when American physicists began public discussion of those aspects of quantum theory that had been developing since Planck's original postulate in 1900. It was also the year when Bohr's theory of the hydrogen atom was announced. Early forms of quantum theory were already being discussed in the United States when Bohr's papers appeared. Taken together, they made it impossible for Americans to remain uninformed and unopinionated about the new ideas any longer.

Of course, American physicists were not ignorant of the existence of quantum theoretical ideas in the preceding dozen years. They did read Annalen der Physik. In addition, European visitors spoke of it, but it was not strongly advocated. Lorentz mentioned Planck's hypothesis during his lectures at Columbia University in the spring of 1906,[106] but only briefly and very tentatively. Planck, himself, when he came in 1909, discussed quantum ideas during only one of his eight lectures which ranged over several theoretical topics.[107] Of the Americans who went abroad for study during this period, one, Harold Webb, had this to say[108]:

"When I was in Europe in 1908-1909 I attended the lectures of Larmor in Cambridge and of Planck and Nernst in Berlin. I do not recall any discussion of quantum theory in those lectures nor among the other students."

There seems to have been few advocates of quantum theory before 1910.

By the academic year 1909-1910 Planck's ideas had entered the curriculum of at least two[109] American educational institutions. The 1908-1909 catalogue of the Massachusetts Institute of Technology was the first to mention Planck in the description of a course called "Radiation" to be given the following year by H. M. Goodwin. The Annual Register of the University of Chicago for 1909-1910 announced a new course "Heat Radiation" to be given by R. A. Millikan. It was described as a combined lecture and seminar course based on the material in Planck's Wärme Strahlung (sic.).[110] Karl Darrow, a member of the University of Chicago class of 1911, took that course as an undergraduate. According to his recent recollection[112] quantum theory was not "taught" in the course--the material in Wärmestrahlung, and in relevant papers, as they appeared in the literature, was read and discussed by the members of the class.

There was little noticeable impact by quantum theory on the research output of American physicists in the years before 1913. Granted that, as we have already discussed, American physicists were almost exclusively experimentalists during this period, it is still surprising to note, for example, that when Millikan published his first paper on the photoelectric effect in 1907, he seems to have been completely unaware of the photoelectric equation postulated by Einstein two years earlier.

In the opening paragraph of Millikan's paper[112] we find the statement:

"Two views have been advanced regarding the mechanism of the discharge of corpuscles from metals under the influence of ultraviolet light."

This is followed by a summary of the views of Thomson and Lenard. Nowhere in the paper is there any mention of Einstein.

It is not clear just when Millikan began to think of the photoelectric effect in the context of quantum theory but he is listed as having delivered a paper at the American Physical Society meeting that was held in Washington, D. C. in December of 1911 with the title "The Velocity-distribution Curves of Electrons Liberated by Different Sources of Ultraviolet Light and the Bearing of these Curves on the Planck-Einstein Theory."[113] An abstract of this presentation, later published in the Physical Review,[114] shows that Millikan's findings at that time, based on the use of a spark type source of illumination, were at variance with, what he called, "the Planck-Einstein light unit theory." He later came to realize that these results were spurious.[115]

The photoelectric papers of Compton and Richardson, published about 1912, which we have already mentioned, discussed the Einstein equation but gave preference to an alternate, non-quantal, treatment of the photoelectric effect proposed by Richardson.

The earliest research paper dealing expressly with Planck's work that I have found in American journals was one by E. Buckingham and J. H. Dellinger, entitled "Calculation of the Constant c_2 in Planck's Equation by an Extension of Paschen's Method of Equal Ordinates."[116]

This paper was not concerned in any way with a critique of the quantum theory postulates but merely with the validity of the equation from the experimental point of view. At the time Buckingham and Dellinger were associated with a program at the National Bureau of Standards on thermal radiation standards which extended for many years.[117]

In general, it can be stated that, prior to 1914, any mention of quantum theory in American research journals was rare. When it did occur it was in the context of experimental confirmation. The acceptability of the postulates involved had not yet become a matter for discussion in the American physics community.

The fact that American physicists were not involved in quantum theory discussion until about 1913 is really not surprising. Armin Hermann has shown[118] that, until 1911, quantum theory remained the property, as it were, of the German school of physicists. A significant breakthrough of the cultural barrier took place on the occasion of the first Solvay Conference of 1911 when physicists from Denmark, England, France and Holland met with German physicists to discuss "La Theorie du Rayonnement et les Quanta."[119] It was as a result of these discussions that Poincaré became convinced not only of the usefulness, but indeed, of the inevitability of quantum concepts. An account of Poincaré's newly acquired views were included in his Dernières Pensées,[120] written shortly before his death in 1912. Poincaré's acceptance of quantum theory was a very influential factor in bringing about its acceptance in British circles.[121] In the light of these factual aspects of the dissemination of quantum thinking in Europe it becomes quite understandable that there was little concern with quantum theory in America before 1913.

In contrast to the scant attention paid to quantum theory earlier, the year 1913 heralded an American awakening to its existence. On at least three occasions within a year there was public scrutiny of quantum ideas. At none of these was the presentaion one of enthusiastic advocacy. Rather, they all emphasized the difficulties encountered. The aim of the speakers was to be informative, not persuasive.

The first of these occasions occurred on the last day of 1912 at the Case School of Applied Science in Cleveland, Ohio where the American Physical Society and the American Association for the Advancement of Science were holding their annual joint meeting. R. A. Millikan, the retiring Vice-President and Chairman of Section B of the Association, delivered an address which he called "Unitary Theories in Physics."[122]

In the published text of this talk[123] we find that Millikan emphasized that there was no single quantum theory of radiation, but five different varieties set forth by Planck, Einstein, Thomson and Bragg. In addition to sketching the essential features of each, Millikan presented them in order of their increasing violence to classical concepts. On such a scale Einstein's rated fifth as "the most concentrated form of the quantum hypothesis." Millikan acknowledged that it was the most successful in terms of explaining the largest number of experimental facts, then he asked:

"Why not adopt it? Simply because no one has thus far seen any way of reconciling such a theory with the facts of diffraction and interference, so completely in harmony in every particular with the old theory of ether waves." (emphasis Millikan's)

There seems little doubt that Millikan's choice of topic for this address was prompted by discussions of quantum theory that he encountered during the six months he had recently spent in Europe. In describing his trip Millikan recounted[124] that, in addition to enjoying many personal contacts with Planck, he attended lectures by Planck on Wärmestrahlung in which quantized absorption of radiation was repudiated. He was also present at a discussion of the application of quantum theory to the specific heats of gases that took place during the meeting of the British Association for the Advancement of Science that was held in Dundee. Following that, he visited with W. H. Bragg with whom he "sat up long after midnight talking over the existing status of physics." Now that quantum theory was being so widely discussed in Europe[125] Millikan apparently found it timely to present an exposition of it to his American colleagues.

While Millikan's presentation was, as was appropriate to the occasion, entirely descriptive and non-mathematical, W. Wien gave a much more detailed and mathematical treatment of the current state of quantum theory when he came to Columbia University as Foreign Lecturer in the Spring of 1913. He stressed how unsatisfactory many aspects of quantum theory were and how much remained to be resolved and he invited American physicists to take up the issues with the following words taken from his final lecture:[126]

> "Bisher ist die Quantentheorie made in Germany. Ich würde mich freuen, wenn meine Vorlesungen dazu beitragen würden, die Aufmerksamkeit der amerikanischen Physiker auf diese Theories zu lenken und wenn die bekannte amerikanische Energie und der amerikanische Unternehmungsgeist darauf gerichtet würde, uns bei dem weiteren Ausbau der Theorie zu Hilfe zu kommen "

The following November a group of American physicists did undertake to present a symposium discussion of quantum theory one afternoon during the regular meeting of the American Physical Society that was held at the Ryerson Physical Laboratory.[127] The local committee in charge of arranging this event consisted of R. A. Millikan, A. A. Michelson and Henry Crew. The speakers and their topics were as follows:

C. E. Mendenhall - The Quantum Theory and Radiation

R. A. Millikan - The Quantum Theory and the Photoelectric Effect

Max Mason - The Quantum Theory and Statistical Mechanics

Jacob Kunz - The Quantum Theory and Atomic Structure

A. C. Lunn - The Quantum Theory and Specific Heats

Unfortunately, no resumés of these proceedings seem to have been printed so we have no way of knowing what information was presented, what was the attitude of the speakers toward quantum theory and what was the tenor of the discussion with the audience. The only clue to what took place lies in a remark found in the Minutes of the meeting which said: "The presentation of each topic was followed by a twenty-five minute general discussion." A little arithmetic shows that those discussion periods would have consumed more than two hours of the afternoon session; hence, it can be assumed that each speaker's presentation must have been limited to ten or fifteen minutes at most--scarcely long enough for more than a cursory glance at the topic listed yet sufficient to make everyone in the audience aware of the widening impact of quantum ideas inthe physics of the day.

2) Theories of Atomic Structure: Bohr, Lewis, Parson and Langmuir

The introduction of quantum theory into the topic of atomic stucture that took place in 1913 with the publication of Bohr's theory of the hydrogen atom was a milestone to which Americans were very soon alert, but not necessarily receptive.

One American physicist who reacted favorably to these new ideas was R. T. Birge. He has related[128] that he not only found Bohr's theory very exciting but became something of a "missionary" in the cause of its dissemination. He recalled giving a Sigma Xi lecture on it at the University of Syracuse, where he was an instructor at the time, very soon after its publication. In the course of trying to explain Bohr's ideas Birge hit upon the notion of drawing an energy level diagram. To the best of his knowledge he was the first person to do this. Certainly there were no such diagrams in Bohr's papers in the *Philosophical Magazine* and it was only many years later that energy level diagrams became standard pedagogical devices appearing in virtually all textbooks that discuss atomic theory.

Birge recalled recently that, as a student of spectroscopy writing a doctoral thesis on the band spectrum of nitrogen:[129]

"I was always interested in atomic structure and when Bohr's first paper appeared in 1913, I recognized that it gave the first, at least, hopeful clue to the understanding of that complex matter. I do not know why everyone else did not immediately agree, *except* for those who had pet theories of their own on the matter, like G. N. Lewis."

We shall discuss later the competition Lewis' theory presented to the acceptance of Bohr's theory, but it must be noted that there were American physicists who opposed Bohr's ideas on grounds other than the chemical considerations that motivated Lewis. One such physicist was D. L. Webster, who has stated:[130]

"I disbelieved it [Bohr's theory], because I could not see how Bohr's atoms could collide with each other without spoiling the game entirely; and as for its quantitative accuracy, I felt sure that any quantitative theory for those lines [in the spectrum of atomic hydrogen] would have to depend... on the atomic constants--for dimensional reasons--and that the numerical coefficient was [so] simple a combination of small numbers and π as to be incredible without Bohr's postulates."

Opposition to Bohr's theory was also raised by G. S. Fulcher[131] on the basis of recent publications by Stark which included his discovery of the splitting of spectral lines in an electric field, "The Stark Effect," and a volume entitled <u>Elektrische Spektralanalyse Chemischer Atome</u>.[132] It seems that Stark erroneously ascribed certain lines in the helium spectrum to doubly ionized helium, leading Fulcher to the conclusion that the electrons responsible for all spectral lines must reside in the nuclei of atoms. Fulcher bolstered his arguments with citations of recent statements by Stark that were unreceptive to Bohr's atomic theory and to quantum theory in general.

This paper by Fulcher is of more interest than it may seem in isolation. It appeared shortly after Fulcher had participated in a symposium discussion of "Spectroscopic Evidence Regarding Atomic Structure" that was held during the 74th meeting of the American Physical Society in November of 1914 in Chicago, arranged by the same local committee of Millikan, Michelson and Crew that had been responsible for the Quantum Theory symposium a year earlier.[133] The paper by Fulcher, just discussed, no doubt reflected the views he presented at this symposium. The views of the other panelists are not known at this time, but the very fact that Bohr's name did not appear in the title of any of the presentations would indicate that there was no rush on the part of American physicists to embrace Bohr's theory of atomic structure.

Discussion of atomic structure aroused considerable interest in American scientific circles which included chemists as well as physicists.[134] The former found Bohr's ideas of little or no value in undertanding the problems of valence that were their main concern.[135] Furthermore, there were popular among American chemists, home-grown theories that seemed much more promising.

As early as 1902[136] G. N. Lewis had conceived an atomic model consisting of concentric cubes with electrons located in stationary positions at the vertices, the number of electrons in the outermost cube being a measure of the valence of the atom. Lewis postponed developing this theory until 1916,[137] by which time he had added the shared pair theory of chemical bonding. This proved of great value to chemists, especially in later studies of the dynamics of chemical reactions.[138]

Lewis' theory was taken up by Irving Langmuir in 1919[139] and developed into what became known as the Lewis-Langmuir theory of atomic structure. The principal innovations introduced by Langmuir were: 1) the electrons no longer had to be static but could "rotate, revolve or oscillate about definite positions in the atom;" 2) electrons were arranged in concentric shells, rather than cubes, of radius 1, 2, 3,... and area 1, 4, 9, 16,...; 3) each shell consisted of cells of equal area with two, non-repelling electrons per cell; and, 4) groups of eight electrons formed stable octets.[140] With these ideas Langmuir believed he could successfully explain the arrangement of elements in the Periodic Table, the chemical valences of elements in organic and inorganic compounds, as well as such physical properties as boiling points, magnetism and electrical conductivity. At the time when Langmuir's theory was introduced, and lectured on widely by Langmuir himself, the Bohr theory had done little other than explain the spectrum of hydrogen and ionized helium.

Another theory of atomic structure was set forth in America in 1915 by A. L. Parson, a young Englishman, who came to study in the chemistry departments at Harvard University and later, at the University of California.[141] Parson's atom had a positive nucleus surrounded by rings of negative electric charge that rotated about the nucleus in definitely oriented planes. With these ideas Parson set out to explain both valence and magnetic properties of atoms in an eighty page exposition that was published by the Smithsonian Institution[142] and attracted considerable attention in American scientific circles.[143]

The topic of atomic structure was of sufficient interest to warrant an all-day symposium on "The Structure of Matter" at a combined meeting of Sections B and C of the American Association for the Advancement of Science, The American Chemical Society and the American Physical Society held on December 27, 1916 at Columbia University. There were eight invited speakers, four physicists and four chemists:"[144]

R. A. Millikan - Radiation and Atomic Structure

G. N. Lewis - The Static Atom

R. W. Wood - Molecular Resonance and Atomic Structure

W. J. Humphreys - The Relation of Magnetism to the Structure of of the Atom

A. P. Wills - The Relation of Magnetism to Molecular Structure

W. D. Harkins - The Evolution of the Elements as Related to the Structure of the Nuclei of Atoms

I. Langmuir - The Structure of Solids and Liquids and the Nature of Inter-atomic Forces

L. W. Jones - Electromerism

Millikan and Lewis were the featured speakers on this occasion. Millikan, delivering his Address as President of the American Physical Society, spoke first and came out firmly in favor of Bohr's theory of atomic structure showing its reasonableness in the light of work done after Bohr's original hypothesis in 1913. Millikan appeared to have become convinced of the validity of Bohr's theoretical ideas only after they were supported by the work of Moseley correlating the X-Ray spectra of elements with atomic numbers and the work by Einstein and deHaas in German and by S. J. Barnett[145] in the United States indicating the existence of permanent, non-radiating electronic orbits. Millikan also argued, however, that Bohr's theoretical determination of Rydberg's constant to within 0.1 percent constituted "most extraordinary justification of the theory of non-radiating electronic orbits."

When Lewis' turn to speak came he noted the wide acceptance by physicists of Bohr's theory of the atom and the lucid arguments for it just presented by Millikan. But, he maintained, the great stability of isomers made it "inconceivable that electrons which have any part in determining the structure of such a molecule [that of an isomer] could possess proper motion, whether orbital or chaotic, of any appreciable amplitude." He argued, further, that Coulomb's Law could not be valid in the realm of atomic dimensions and that Bohr's theory held internal contradiction since it used Coulomb's Law to hold the orbiting electron bound to the nucleus, yet ruled out the electron's effect on any other nearby charged particle. Lewis offered an alternative formulation for the interaction of charged particles which would reduce to the familiar law of Coulomb for large distances and which, he believed, would be able to account for the frequency of spectral lines and Einstein's photoelectric equation, as well. Lewis was clearly trying to meet the physicists on their own ground.

All the other speakers at this symposium stressed the incomplete, tentative nature of each of the theories of atomic structure being discussed. In general, the physicists were partial, on the basis of magnetic properties, to electric charge in orbital motion, but not necessarily localized as in Bohr's model, while the chemists stressed that any valid model of atomic structure must provide for a satisfactory bonding mechanism among atoms of apparently great stability. The conference had little relevance to quantum theoretical considerations, as such. It did, however, serve to illuminate the strong American interest in a variety of mechanically formulated models of atomic structure, which interest would serve to inhibit American participation in developing a quantum theoretical approach to atomic structure based primarily on spectroscopic evidence and involving mathematical sophistication, such as that which was evolving in Europe in the hands of Wilson and Sommerfeld and even in Japan in the hands of Ishiwara.[146]

Several of the characteristics of the American scientific endeavor in the early twentieth century that we have already discussed were contributing factors to this particular situation involving theories of atomic structure. For example, the lack of an established theoretical tradition made possible a kind of "democracy" in which almost any chemist or physicist, established experimentalist or young graduate student could propose, and find an audience for, his own special theory. The very multiplicity of atomic theories on the scene in America around 1916, each with its own relevance to some particular phenomena, tended to dilute the attention paid to any single one.

Further, the strongly empirical outlook of the American physics community of that period tended to make them trust experimental evidence more than mathematical theory. Hence, a theory applicable only to spectroscopic data would not seem nearly as persuasive as the broadly conceived one of Lewis and Langmuir, for example, which embraced all sorts of physical and chemical phenomena. Then too, the distraction from theoretical considerations brought on by the exigencies of war work, plus the scientific isolation imposed by the lack of scientific journals from abroad,[147] kept American physicists from even being aware of how Bohr's early ideas were being developed abroad. On the other hand, the widespread activity by American physicists in studying all portions of the electromagnetic spectrum would inevitably lead them to serious consideration of the quantum theory that was already showing great promise for unravelling the details of atomic, molecular and X-Ray spectra.

3) Doubts and Reservations

In general, however, the American physics community approached quantum theory with great reservation. As we shall see later on in this discussion, the reaction of many of those American physicists, who concerned themselves with theoretical discussion at all, was to seek ways of avoiding the use of quantum notions by suggesting alternate, non-quantal means of attaining the results that were being advertised as the "successes" of quantum theory. The number of American physicists who sought to cope constructively with quantum theory was small indeed. Even they were not really "enthusiasts" for quantum theory but were convinced that modifications must be sought for classical physics which would permit accommodation of phenomena which appeared to require a quantum hypothesis.[148]

The attitude of many members of the American physics community, circa 1915, toward quantum theory is reflected in the following quotation:[149]

"I presume that all of us would agree that the quantum theory is quite distasteful. In working with the theory we have no definite mechanical picture to guide us nor have we any definite clear cut principle as a basis of operations--Physicists everywhere have been making strenuous efforts to find a method of escape from the theory. If such a method could be found I presume that we should all breathe a sign of relief and sleep better thereafter."

Looking now at specific responses by members of the American physics community to those aspects of quantum theory not directly related to atomic structure, it will be significant to consider the dissemination of quantum theory on this side of the Atlantic by means of printed expositions and university lecture courses, the discussion of quantum theory in papers appearing in American research journals, and to note the instances where American research contributed to increasing the domain of quantum theory.

The dissemination in America of information about quantum theory was furthered by two English language publications concerning it. These publications were of value to established physicists for their own reference and for use as reading material by students of physics being introduced to the topic at the increasing number of universities where discussion of quantum theory was incorporated into courses generally bearing the title "Radiation."

In 1914 Morton Masius, a faculty member at the Worcester Polytechnic Institute, published, with the author's approval, an English translation of the second edition of Planck's Wärmestrahlung[150] in order that English speaking physicists might have "the opportunity to study the ideas set forth by Planck without the difficulties that frequently arise in attempting to follow a new and somewhat difficult line of reasoning in a foreign language."[151] In addition to making the translation, Masius prepared as an Appendix, a list of literature references together with some commentary of his own summarizing the state of the acceptance of Planck's quantum theory within the physics profession at large and the experimental evidence favoring it.

Another useful introduction to the current state of quantum theory was J. H. Jeans' Report on Radiation and the Quantum Theory,[152] prepared under the auspices of the Physical Society of London and written soon after his public espousal of quantum theory at the meeting of the British Association for the Advancement of Science that was held in Birmingham in September of 1913 and, which we have mentioned earlier.

The impact of quantum theory on American university physics curricula increased after 1913 so that, by about 1915, practically all the universities with graduate physics programs had either recently introduced or expanded their discussion of it for their advanced students.[153]

A case in point occurred at Harvard University where G. W. Pierce had been giving a course entitled "Radiation" during alternate years for more than a decade. By 1914-1915 he was including quantum discussion and adopted Masius' book for his students' reading. E. C. Kemble, a student in Pierce's course during 1914-1915, recalls that:[154]

> "Pierce, a man of the older generation, gave a good course that did justice to Planck's work but, in view of the inconsistency between it and Maxwell's theory, continued for years to search for a purely classical explanation of the phenomena on which quantum theory was based."

We shall show shortly that such an approach was shared by other American physicists in addition to Pierce.

The American awakening to quantum theory in 1913 did little to alter the pattern of research activity of most of the physicists in the United States. Gradually, however, papers began to appear in Physical Review that related more profoundly to quantum theory than the reporting of additional experimental data on the photoelectric effect, for example. 1914 was the first year in which articles involving critical evaluation of quantum theory were printed in Physical Review and for the next several years practically all such articles were concerned with presenting escape routes from the radical assumptions associated with it. A few examples will serve to illustrate this point.

G. N. Lewis and E. Q. Adams published two articles[155] generally titled "Notes on Quantum Theory." The first carried the subtitle "A Theory of Ultimate Rational Units; Numerical Relations between Elementary Charge, Wirkungsquantum, Constant of Stefan's Law" and purported to show that "h" is merely the square of the charge on the electron multiplied by a suitable coefficient, the coefficient turning out to be

$$\frac{1}{c}\sqrt[3]{\frac{8\pi^5}{15}}.$$

The second part of Lewis and Adams' presentation, subtitled "The Distribution of Thermal Energy," contained an attempt to account for the black body radiation and heat capacity phenomena not by abandoning the equipartition of energy theorem, but by considering the probable interactions among neighboring, individual components of the system under consideration.

R. C. Tolman, in an article "The Specific Heats of Solids and the Principle of Similitude"[156] derived a formula for the specific heat of a solid which gave as good agreement with data taken at low temperatures as did the quantum relation derived by Debye. Tolman noted, with apparent pride, that with his method it was "<u>not</u> necessary to make any use of the various forms of quantum theory." Tolman's derivation was based on a rule which he had recently invented, called "The Principle of Similitude," a scheme analogous to Lewis and Adams' theory of rational units.

When we recall that Lewis and Tolman, both physical chemists, had been active in introducing relativity theory to America as early as 1908[157] it seems curious that they should both have sought so eagerly to avoid quantum theory. The key to their opposition seems to lie in their belief in a physical universe which was basically simple and rational. The tentative and untidy state of quantum theory, as it stood about 1914, seemed insufficiently advantageous over classical physics to warrant its acceptance, at least before all possible alternatives were considered.

A. H. Compton was another American who sought, in 1915, to avoid the use of quantum theory in explaining the dependence of specific heats on temperature. He suggested[158] that, below a certain temperature, neighboring atoms in a substance become "agglomerated," i.e., they form aggregates. A. H. Compton, brother of the previously mentioned K. T. Compton, was at that time a graduate student at Princeton University. We shall see in the next chapter that later work by this same A. H. Compton was to prove of crucial importance in establishing the validity of localized quanta. Of course, at the time of his 1915 paper, Compton could not have had any idea of what lay ahead for him and his work.

On the other hand it is still more curious to note that, in this pre-1920 period there were two American physicists whose current work constituted definite correlation with, if not indeed proof of, the validity of quantum theory, yet they could not bring themselves to accept it. The two were William Duane and R. A. Millikan.

The X-Ray studies mentioned earlier that were going on at Harvard University under the guidance of William Duane came, in 1915, to yield the result that the maximum frequency, ν, of X-Rays produced by electrons impinging on a target was related to the voltage, V, which had accelerated the electrons, in a very simple way, namely $h\nu_{max}=Ve$, where "h" is Planck's constant and "e" is the charge of the electron.[159] I have been unable to determine when this relation became generally recognized as involving the conversion of the kinetic energy of single electrons into individual quanta of energy, a kind of inverse photoelectric effect. In October of 1915 Duane recognized this possibility but instead of seizing upon it, chose to present an analysis of it which did not use that interpretation and, further, proceeded to deduce Planck's original radiation formula "from frequency relations and not from any law according to which radiant energy can be produced or exist only in quanta hν."[160] Duane did not claim that such a course was clearly preferable but his paper seems to be another example of an American physicist seeking to show that a "success" of quantum theory can be arrived at from a non-quantal hypothesis. It was also another example of an American experimental physicist turning his hand to theoretical discussion.

In 1916 Millikan presented a full treatment of his photoelectric studies.[161] After saying that his experiments were in exact agreement with Einstein's photoelectric equation, he added: "Nevertheless the physical theory which gave rise to it seems to me to be wholly untenable."[162] One might ask how Millikan could be so negative in this instance toward quantum theory when at the symposium described above he had come out strongly in favor of the Bohr atomic theory. The clue seems to lie in Millikan's abhorrence of a quantum of energy that is localized in space. The Bohr theory made no statement about the nature of radiation and for that very fact Millikan lauded it, saying:[163]

> "It has been objected that the Bohr theory is not a radiation theory because it gives no picture of the mechanism of the production of the frequency. This is true, and therein lies its strength, just as the strength of the first and second laws of thermodynamics lies in the fact that they are true irrespective of mechanism. The Bohr theory is a theory of atomic structure; it is not a theory of radiation, for it merely states that energy relations must exist when radiation, whatever its mechanism, takes place. As a theory of atomic structure, however, it is thus far a tremendous success. The radiation problem is still the most illusive and the most fascinating problem in modern physics."

It must be recognized that Millikan was not alone in his reluctance to accept a quantum theory of the photoelectric effect. As has been pointed out by Roger Stuewer,[164] in the years 1910-1913 H. A. Lorentz, J. J. Thomson, A. Sommerfeld and O. W. Richardson had all sought ways of explaining this effect without discrete light quanta.

As a matter of fact, in calling attention to the numerous examples we have cited of American reluctance to accept quantum theory, we do not mean to imply that such a response was uniquely American. Classical physics exerted a strong appeal for physicists of every nationality by virtue of its own beautiful inner structure and the broad scope of the phenomena which it could explain. In contrast, quantum theory at that time appeared confused and useful in only a relatively small number of instances.

4) <u>E. C. Kemble: His Training and Early Work</u>

Against a background such as that described above where so many of the established figures of physics and chemistry in America were eager to diminish the scope of applicability of quantum theory the efforts of Edwin E. Kemble, around 1916-1917, stand out all the more by contrast. In the course of developing his doctoral thesis at Harvard University Kemble had the insight to realize that, by means of quantum theory, certain characteristics of the band spectra of diatomic molecules could be satisfactorily explained for the first time.

Kemble was one of the first Americans since Gibbs to work creatively in theoretical physics and, in contrast to Gibbs who left no group of younger theoreticians to carry on in an established tradition, Kemble was to have considerable influence among the young physicists who came into quantum theory in America during the 1920's.

Although the World War was raging in Europe at the time when Kemble's early papers were published in America, notice was soon taken of them by European physicists. Kemble was the only American physicist to be mentioned by Niels Bohr in his 1918 Danish Academy paper "On the Quantum Theory of Line Spectra."[165] Later, when Arnold Sommerfeld wrote his Atombau und Spektrallinien,[166] a work which was to become the "Bible" for students of quantum theory in the early 1920's, he also took note of Kemble's work.

In view of the fact that Kemble's activity in physics represented such a departure from the pattern exhibited by his American contemporaries, it seems appropriate at this point to present some of the details which characterized his training and the circumstances under which he began his career as a theoretical physicist, before describing his initial work in this field.

Kemble took his undergraduate training at the Case Institute of Technology where he received his B.S. degree in 1911. Two details of his undergraduate experience are noteworthy: he came in contact with Dayton C. Miller, one of America's noted experimental physicists, and he developed mathematical skills beyond those normally associated with undergraduate training in America at the time. As a senior, Kemble wrote a thesis which involved a mathematical analysis of the operation of a device, invented by Miller, for analyzing sound waves.

After two years of teaching at the Carnegie Institute of Technology, 1911-1913, Kemble entered Harvard University as a graduate student in the physics department. He has remarked that he decided early on a career in theoretical physics and deliberately took an extra year of mathematical work before beginning to think about a thesis. In his opinion, however, this was not sufficient to give him a mathematical background equal to that of the European mathematical physicists of his generation.

At that time the Harvard physics department had an active colloquium program with weekly presentations, sometimes by faculty members, sometimes by graduate students, on topics of current interest. According to the Harvard University Gazette, Kemble gave several colloquium talks between 1915 and 1917 on such topics as "The Stark Effect," "The Variation of Specific Heats of Solids with Temperature" and "The New Theory of Pyro-electricity," all of which indicate Kemble's theoretical interests.

On February 26, 1916 a colloquium was presented by another graduate student, James B. Brinsmade,[167] entitled "Bjerrum's Hypothesis Regarding Absorption Bands in Gases" and it was while listening to this presentation that Kemble conceived the idea which was to become his thesis. Within the following year he worked out the details with the approval and guidance of Percy W. Bridgman.[168]

Briefly stated, Kemble's analysis was as follows: stimulated by Bjerrum's suggestion[169] that the observed infra-red absorption band spectra of diatomic gases could be explained by applying Planck's ideas of quantization to the vibrational and rotational modes of motion of dumbell-shaped molecules, and by an observation of E. von Bahr that[170] a faint absorption band was found in the spectrum of carbon monoxide at a wavelength of about half that of the principal band, Kemble proposed that if the amplitude of vibration of the atoms in a diatomic molecule is large, their motion would no longer be simple harmonic motion. The molecule would then be an anharmonic oscillator and higher frequency "overtones" would be expected to occur.

Kemble included in his theoretical study of the motion of diatomic molecules an analysis of the distribution of angular velocities that could be expected among the molecules and of the consequent distortion that would occur at high rotational speeds. Concluding that the vibrational frequency of the atoms decreased with increasing angular velocity of the molecule, Kemble was able to explain the observed asymmetrical shape of all diatomic band spectra. The absence of a centrally located line was interpreted as evidence that molecules can never have zero angular velocity.

Kemble predicted that hydrogen chloride and hydrogen bromide would show "overtone" frequencies similar to that already observed for carbon monoxide and, with the aid of equipment constructed by J. B. Brinsmade, he found that faint harmonics did indeed occur at frequencies close to the predicted values.

Another aspect of the band spectra that greatly interested Kemble was the relative intensities of the various lines within each band, since he felt that those intensities were important clues to the internal behavior of the molecules. The problem of determining accurate values for intensities was not easily solved experimentally, but one innovation in experimental procedure for gathering such data that Kemble introduced was the use of gas-filled tubes of various lengths through which the absorption spectra could be photographed. In this way he hoped to gain insight into the actual intensities associated with individual molecules.

Throughout his discussion Kemble assumed, after the manner of Planck, that the absorbed frequencies were the same as the actual internal molecular frequencies. The newer concept, introduced by Bohr for the hydrogen atom, that observed spectral frequencies are equal to changes of energy content divided by Planck's constant "h," had not yet entered into his analysis.

Although Kemble's thesis, which was entitled "Studies in the Application of the Quantum Hypothesis to the Kinetic Theory of Gases and to the Theory of their Infra-red Absorption Bands" was never published as a whole, parts of it did appear in the literature.[171]

Kemble received his degree in June, 1917. He then left Harvard for wartime work at the Curtiss Aeroplane Company and a brief period of teaching at Williams College. In the Fall of 1919 he returned to Harvard University as a member of the faculty in the physics department. By this time he had become familiar with Bohr's work and realized that the use of Bohr's assumptions would account for the systematic deviation of the faint "harmonics" from the values calculated from the frequencies of the corresponding "fundamental" bands. A paper carrying out the revised considerations was published as "The Bohr Theory and the Approximate Harmonics in the Infra-red Spectra of Diatomic Gases."[172]

In addition to resuming his own research activity during the years immediately following the end of the war, as a faculty member at Harvard Kemble inaugurated two courses dealing with quantum theory; one, a lecture course, was titled "Quantum Theory with Applications to the Infra-red, Photoelectric Phenomena and Specific Heats," the other, a research course, "Quantum Theory and Infra-red Spectra."[173] So far as has been ascertained there was no other university in America at that time offering any comparable courses. Millikan's presentations of quantum theory at the University of Chicago, for example, were those of an experimentalist trying to bring the current literature on quantum theory to the attention of students.[174] On the other hand Kemble was a theoretician, himself engaged in research related to quantum theory, teaching the new quantum physics to a new generation of graduate students. Aside from the intrinsic value of Kemble's early work, it holds special significance of a ground-breaking nature for theoretical physics in America and as a prelude to his further work in the 1920's as a researcher and mentor of young physicists entering the field of quantum theory. We shall have more to say about these activities in the next chapter.

5) Changing Attitudes Toward Theoretical Physics

The closing years of the second decade of the twentieth century saw significant changes in the outlook of the American physics profession which had, up to that time, maintained a strongly empirical, non-mathematical approach to science. Theoretical physics began to be held in higher esteem. In part, this was due to the new visibility attained by theoretical physics with the announcement, in 1919, of the experimental verification of Einstein's theoretically predicted gravitational effect on light.[175] Although a discussion of the reception accorded relativity by Americans is beyond the scope of this thesis, it is appropriate to cite some comments made in connection with it which indicate that American physicists were coming to realize that they could no longer confine themselves to seeking explanations for physical phenomena based entirely on mechanical models. Such a change in attitude was certainly to be important in paving the way for the coming of the quantum-mechanical revolution in the next decade.

J. S. Ames chose "Einstein's Law of Gravitation" as the topic to be presented for his Presidential Address to the American Physical Society on December 30, 1919. Near the conclusion of this talk he remarked:[176]

"There is not the slightest indication of a mechanism to explain gravitation, meaning by that a picture in terms of our senses. In fact, what we have learned has been to realize that our desire to use such mechanisms is futile."

The following year, 1920, H. A. Bumstead gave a lecture on the history of physics to the Yale Gamma Alpha Scientific Fraternity which included the following comment:[177]

"...the more one knows of the history and recent developments of physics the more sincere and ardent is one's admiration for the individuality and brilliant originality of Einstein's genius. It does not seem probable at present that his discoveries will have as great an effect upon immediate future of physics as some of the others [quantum theory] which I have discussed. But the ultimate result of his work upon the <u>methods</u> used in the theoretical side of physical science may well prove to be revolutionary; and it seems highly probable that it will change to some extent our philosophical views of the nature of the external world and of our relation to it."

As hinted at in the last quotation, we find that there was increasingly general agreement among the established physicists in America about this time that quantum theory was not only unavoidable but, with its myriad of unsolved problems, was the area of physics that held the most promise for exciting new developments.

In 1918 the <u>American Journal of Science</u> celebrated the 100th anniversary of its founding by Benjamin Silliman with a special issue of commemorative papers on various branches of science. The one on physics,[178] by Leigh Page contained the following:

"The three mentioned [Radiation, Photoelectric Effect and Structure of the Atom] are only the most clearly defined of a growing group of phenomena in which the quantum manifests itself. Its significance and the alteration in our fundamental conceptions to which it seems to be leading is for the future to make clear. That it presents the most important and interesting problem as yet unsolved few physicists would deny."

H. A. Bumstead, in the lecture cited previously, provided the following assessment:

"At present no one doubts that most of our fundamental ideas in mechanics and electrodynamics must be revised in the light of the quantum theory, which, however, is itself still in a very immature state. The problem thus arising of bringing together under one system apparently discrepant bodies of phenomena is an exceedingly difficult one, and we may have to wait for another Newton to solve it. But it possesses the greatest fascination for all theoretical physicists; they are able to congratulate themselves upon the possession of an unsolved problem of the first magnitude and of great difficulty and they know that as long as its lasts, life will not be dull for them."

Although both Bumstead and Page were themselves more oriented toward theoretical phyiscs than were most of the members of the American physics profession at that time, one detects in their comments an attitude of aloofness toward quantum theory. Bumstead died later in the same year in which his comments were made and so did not live to see the quantum revolution to come. As we shall see in subsequent chapters of this thesis, Page did follow the development of quantum theory and gave courses in it during its pre-quantum mechanical stage, but all the while he continued to seek ways of making quantum phenomena compatible with classical physics. His efforts in this direction tended to keep him out of the mainstream of quantum physics as it developed in the 1920's.

The American physics community may have been convinced by 1920 that quantum physics was here to stay but very few were ready to cope with it.

Summary

During the years from 1900 to 1920 the study and practice of physics in America gained notably in scope and in strength, yet there remained inadequacies to be overcome before a fully mature status within the international physics community could be attained.

Among the assets to be counted by the American physics profession in the year 1920 were the following:

a vigorous professional society with a membership of about 1,300 which held regularly scheduled meetings for the presentation by members of their research results and for informative panel discussion of topics of current interest;

channels for the publication of research results which included a journal sponsored by the professional society as well as other scientific publications of a more general nature;

graduate education of sufficient quality and quantity to preclude the necessity for students' going abroad for doctoral training, but at the same time prepared them to take advantage of such foreign study if they wished;

numerous laboratories where high caliber experimental work was carried on;

a leadership within the profession characterized by an optimistic belief in progress through the conscious, rational pursuit of the scientific endeavor; and a new visibility of the physics profession and a cohesiveness among its members which resulted from their wartime activities.

Among the drawbacks to the generally improving picture there remained:

considerable isolation among the members of the domestic physics community due to inevitable geographic dispersal;

remoteness from the European physics community recently worsened by the breakdown of scientific communications during the war years;

so much emphasis on the primacy of experimentation that speculative innovation of a theoretical nature was not generally encouraged and mathematical training was frequently inadequate; and

a theoretical outlook firmly wedded to classical physics and relying on mechanical models rather than mathematical sophistication.

The last named drawbacks accounted for the reaction by most of the American physics profession to quantum physics. Late (1913) in coming to serious consideration of it, the initial response of many was either to actively seek escape routes from it or to remain passively skeptical of it. An isolated exception to these common responses came from E. C. Kemble who wrote the first American Ph.D thesis involving quantum theory and later pioneered in increasing the quantity and quality of the curricular approach to the new physics while extending his own research in the area of the theory of molecular spectra.

The postwar period 1919-1920, however, saw the beginning of the realization in America of the importance of theoretical physics in general and the acknowledgement that the development of a satisfactory quantum theory would be the most significant and challenging problem for the immediate future of physics.

Chapter 1: Notes and References

1.1 The development of science in America during the nineteenth century has engaged the attention of a number of authors whose writings are included in the bibliographic section of this thesis. They have been freely drawn upon in composing the section that follows. In particular, I acknowledge the following which are specifically concerned with physics.
D. J. Kevles, "The Study of Physics in America 1865-1916" Ph.D. Thesis, Princeton University, 1964.
________ "On the Flaws of American Physics: A Social and Institutional Analysis" in Nineteenth-Century American Science: A Reappraisal, George H. Daniels, ed. (Evanston: Northwestern University Press, 1972) pp. 133-151.
Nathan Reingold, "The Rise of Physics" in Science in Nineteenth Century America: A Documentary History, N. Reingold, ed. (New York: Hill and Wang, 1964) pp. 251-322.

1.2 The fourteen years that Joseph Henry spent at the College of New Jersey (later called Princeton University) were the closest approximation to an opportunity to pursue the study of physics professionally available to an American in the first half of the nineteenth century. Information on Henry and his career has been obtained from
Thomas Coulson, Joseph Henry: His Life and Work (Princeton: Princeton University Press, 1950).
Nathan Reingold, "Henry, Joseph" Dictionary of Scientific Biography Vol.VI, pp. 277-281.

1.3 For example, in his autobiography From Immigrant to Inventor (New York: Charles Scribner's Sons, 1923) Michael Pupin recounts his particular admiration for Gibbs and Rowland.

1.4 So wrote Arthur Haas in his Preface to A Commentary on the Scientific Writings of J. Willard Gibbs, Volume II: Theoretical Physics, Arthur Haas, ed. (New Haven: Yale University Press, 1936).

1.5 Gibbs' thesis was more of an engineering nature than physics. It was entitled "The Form of the Teeth of Wheels in Spur Gearing" according the Dissertations in Physics, M. Lois Marckworth, compiler (Stanford: Stanford University Press, 1961).
Information on Gibbs and his career has been obtained from
H. A. Bumstead and R. G. VanName, eds., The Scientific Papers of J. Willard Gibbs (London: Longmans, Green and Co., 1906).
M. J. Klein "Gibbs, Josiah Willard" Dictionary of Scientific Biography, Volume V, pp. 386-393.
L. P. Wheeler Josiah Willard Gibbs: The History of a Great Mind (New Haven: Yale University Press, 1951, re. 1952, 1962; reprinted by Archon Books, 1970).

1.6 This work of Gibbs was originally published in the Transactions of the Connecticut Academy of Arts and Sciences 3 (1876) pp. 108-248; (1878) pp. 343-524. An Abstract appeared in the American Journal of Science 16 (1878) pp. 441-458. A detailed listing of Gibbs' mailing list for reprints is found as Appendix IV of the Wheeler biography cited above.

1.7 Information on Rowland and his career has been obtained from The Physical Papers of Henry Augustus Rowland (Baltimore: The Johns Hopkins University Press, 1902) and
T. C. Mendenhall "Henry A. Rowland" National Academy of Science Biographical Memoirs 5, pp. 115-134.

1.8 The American Journal of Science, founded in 1818 by Benjamin Silliman of Yale College, was edited by James Dana at the time when Rowland's paper was rejected. Considerable detail on this episode is found in Reingold "The Rise of Physics". Rowland's paper "On Magnetic Permeability, and the Maximum of Magnetism of Iron, Steel and Nickel" appeared in Phil Mag. 46 (1873) pp. 140-159.

1.9 In 1885 Sir William Thomson (later Lord Kelvin) was presented with one of Rowland's gratings as a token of appreciation from the American physicists who had compised his audience during a series of lectures that he had delivered at the Johns Hopkins University the year before. His letter of acknowledgement appears in Silvanus P. Thompson The Life of William Thomson, Baron Kelvin of Largs, Vol. II, p. 838 (London: Macmillan and Company, Ltd. 1910). Rowland had demonstrated the use of his gratings to European scientists when he was abroad in 1882. Two letters from John Trowbridge to D. C. Gilman, reprinted in Reingold "The Rise of Physics" pp. 270-274, describe the admiration of it exhibited by the Europeans. (Trowbridge was with Rowland during much of that trip abroad.)

1.10 Information on Michelson and his career has been obtained from Bernard Jaffe, Michelson and the Speed of Light (Garden City, New York: Doubleday and Company, Inc., 1960).
Albert Abraham Michelson: The Man Who Taught a World to Measure (China Lake, California: Michelson Museum Publication No. 3, 1970).
Dorothy Michelson Livingston, The Master of Light: A Biography of Albert A. Michelson (New York: Charles Scribner's Sons, 1973).

1.11 "To the terms Experimental and Mathematical Physics I object since all physics must be mathematical and experimental," from a letter written by Rowland to D. C. Gilman, April 20, 1876 and reprinted in Reingold "The Rise of Physics" p. 270.

1.12 Social Science 119: The Social Context of Science, lectures at Harvard University, Fall Semester, 1969-1970. Professor Mendelsohn equates "maturity" with professionalization.

1.13 K. J. Sopka, "An Apostle of Science Visits America--John Tyndall's Journey of 1872-1873" The Physics Teacher 10 (1972) pp. 369-375. It is noteworthy that Tyndall was so eager to assist the development of science in America that he generously donated the proceeds of his lecture fees for a fellowship program to allow young Americans to go to Europe to study physics. Michael Pupin and William Duane were two of the early beneficiaries of these monies.

1.14 F. W. Clarke, "American Colleges versus American Science" Popular Science Monthly 9 (1876) pp. 467-479.
Simon Newcomb, "Exact Science in America" North American Review 119 (1876) 286-308.
Both of these articles may also be found in Science in America: Historical Selections John C. Burnham, ed. (New York: Holt, Reinhart and Winston, 1971).
H. A. Rowland "A Plea for Pure Science" Science 2 (1883) pp. 242-250 and "The Highest Aim of the Physicist" Science 10 (1899) pp. 825-833.
The first of these articles by Rowland was his address as vice-president and chairman of Section B of the American Association for the Advancement of Science, the second was his presidential address to the American Physical Society. The texts of both are found in The Physical Papers of Henry Augustus Rowland.

1.15 F. W. Clarke, in the article cited above, was especially outspoken on the teaching of science in American religious colleges. Both he and Rowland, in "A Plea for Pure Science," were appalled by the burgeoning number of small colleges of dubious scholastic merit that were being established throughout the United States.

1.16 John S. Brubacher and Willis Rudy, Higher Education in Transition-An American History 1636-1956 (New York: Harper and Row, 1958), Part III, "The Rise of Universities in Nineteenth Century America," passim.
These authors cite figures indicating that about 10,000 American students attended German universities between 1850 and 1900. The numbers fell off sharply after 1900, by which time there were fifteen major graduate schools in the United States.

1.17 Samuel Sheldon "Why Our Science Students Go to Germany" Atlantic Monthly 63 (April 1889) pp. 463-466, provides a contemporary view of the relative merits of German and American institutions. In addition to the lure of great masters and established laboratories, Sheldon maintained that a student could study for three years at a German university at the same cost that would pay for only two years at Harvard University or the Johns Hopkins University, for example; in addition, the student who had studied abroad could expect to obtain a better position upon his return to the United States. (According to American Men of Science, 1st edition, 1906, Sheldon had received his own doctorate at the University of Würzburg in 1888 and was then Professor of Physics and Electrical Engineering at Brooklyn Polytechnic Institute.)

1.18 Neither Michelson nor Rowland, for example, formally earned Ph.D. degrees. Rowland was awarded one, "Honoris Causa," by Johns Hopkins University in 1880, according to the list of honors which appears in The Physical Papers of Henry Augustus Rowland.

1.19 Yale awarded the first American Ph.D. in physics in 1861 to A. W. Wright for a thesis that was really in mathematical astronomy. It carried the title "Having giving the velocity and the direction of motion of a meteor on entering the atmosphere of the earth, to determine its orbit about the sun taking into account the attraction of both these bodies," according to Marckworth, Dissertations in Physics.

1.20 In addition to Gibbs, Michelson and Rowland, the following few examples may be cited: J. S. Ames, R. A. Millikan, B. O. Peirce, A. G. Webster and R. W. Wood. (As the number of individual physicists being mentioned increases it is appropriate to call the reader's attention to the Biographical Notes that will be found in Appendix I.)

1.21 E. H. Hall, "Physics" in Development of Harvard University 1869-1929, S. E. Morison, ed. (Cambridge: Harvard University Press, 1930) pp. 277-291. Trowbridge did, however, go to Europe as a physicist on at least one occasion as seen from his letter to Gilman which was cited in Footnote 1.9.

1.22 Information on Columbia University was obtained from Frederick P. Keppel, ed., A History of Columbia University 1754-1904 (New York: Columbia University Press, The Macmillan Co. Agts., 1904) and from the University Catalogues of 1896-1897 and 1897-1898. In 1892 Columbia established a Faculty of Pure Science specifically to encourage scientific research; Michael Pupin was one of the first appointees to this Faculty.

1.23 Everett Mendelsohn "The Emergence of Science as a Profession in Nineteenth-Century Europe" in The Management of Science, Karl Hill, ed. (Boston: Beacon Press, 1964) pp. 3-48

1.24 Cited by E. H. Hall in "Physics" and by A. G. Webster "The Physical Laboratory and Its Contribution to Civilization" Popular Science Monthly 84 (1914) pp. 105-117.

1.25 According to Brubacher and Rudy, Higher Education, Clark University never reached the potential that was expected of it due to a disagreement between Clark and University President, G. Stanley Hall, whom Clark had recruited from the Johns Hopkins University. This led to the withholding of anticipated financial backing and the shattering of faculty morale. The significance of this for our study of the development of physics in America lies in the fact that it was very probably a strong factor in Michelson's leaving Clark in 1892 to go to the University of Chicago.

1.26 Quotation from Preface to A. G. Webster volume subsequently cited in Note 1.29.

1.27 This carries the designation "not published" (New Haven: Printed by Tuttle, Morehouse and Taylor, 1881-1884).

1.28 Boston: Ginn and Company, 1st edition 1886. By 1902, when this work had reached its third edition, it had grown from 178 pages to 490. Peirce published his Short Table of Integrals for the first time in 1899; this work proved invaluable to many generations of students; a third edition of it was published in 1929, sixteen years after Peirce's death.

1.29 London: Macmillan and Company, Ltd., 1897.

1.30 The data in this section are based on information found in Marckworth Dissertations in Physics. This is the only one of the reference works relating to Ph.D. production (see Bibliography for others) which readily allows for assembling this kind of data. It should be noted, however, that the Marckworth volume does have some serious omissions; for example, neither R. A. Millikan, Columbia 1894, nor Harrison Randall, Michigan 1902, are included. Since my main interest in these data was to identify the leading Ph.D. producing institutions and to determine the order of magnitude of their output, I have not tried to improve significantly on the Marckworth listings.

1.31 Proceedings of the American Association for the Advancement of Science were the source of information for the earliest meetings of Section B. More complete, narrative accounts are found in The Physical Review for the years from 1893 on.

1.32 For about two years after its founding the American Physical Society published its own Bulletin which contained Minutes of the meetings and printed versions of some of the papers that were read at those meetings. Additional information on the early activities of the American Physical Society is found in Ernest Merritt "Early Days of the Physical Society" Review of Scientific Instruments 5 (1934) pp. 143-148.
Frederick Bedell "What Led to the Founding of the American Physical Society" Physical Review 75 (1949) pp. 1601-1604.
Gordon F. Hull "Experimental Discoveries Announced at the Meeting of the American Physical Society Fifty Years Ago" Scientific Monthly 77 (July 1953) pp. 13-18.
________ "The Early Years of the American Physical Soceity" an unpublished, invited address before the American Physical Society at the semi-centennial meeting held in Cambridge, Mass., June, 1949; Xerox copy of manuscript obtained from the Gordon Ferrie Hull Papers located in the Archives of Dartmouth College, Hanover, New Hampshire. Additional details on the American Physical Society will be found in Appendix IV.

1.33 A summary listing of the occasions when European physicists visited the United States and of the instances when some took up permanent residence in this country will be found in Appendices II and III.

1.34 Baltimore Lectures on Molecular Dynamics and the Wave Theory of Light (London: C. J. Clay and Sons, 1904) contained, in addition to the original lectures, 12 appendices written by Kelvin in the years subsequent to 1884. Although this publication did not appear until some twenty years after the delivery of the lectures, there were available in the meantime copies of a "papyrographed" edition of stenographic notes made by A. S. Hathaway, a member of Kelvin's audience, and copyrighted by the Johns Hopkins University.

1.35 In addition to the details of Kelvin's visit found in Silvanus P. Thompson The Life of William Thomson, Baron Kelvin of Largs Vol. II, pp. 810-839, the following commentary was provided by Lord Rayleigh who attended some of the lectures which took place immediately after the Montreal meeting of the British Association for the Advancement of Science:
"Thomson had collected half the physicists in America so that we had almost daily discussion. His lectures were quite in the usual Thomsonian style, a sort of thinking out loud in an enthusiastic, incoherent manner."
Letter written to his mother, October 19, 1884, quoted by his son Robert J. Strutt, Fourth Baron Rayleigh, John William Strutt Third Baron Rayleigh (London: Edward Arnold and Co., 1924) p.146.

1.36 The lectures were published as The Discharge of Electricity Through Gases (New York: Charles Scribner's Sons, 1898) Thomson's own account of his visit to America are found in his autobiography Recollections and Reflections (London: G. Bell and Sons Ltd., 1936) pp. 164-180.

1.37 For example, Lord Kelvin travelled extensively in the United States in 1897 (see Thompson The Life of William Thomson, Vol. II, pp. 1001-1004). It is worth noting that although only three Americans (Rowland, Michelson and Woolcott Gibbs) participated in the 1884 Montreal meeting, about a dozen American physicists and astronomers took part in the Toronto meeting, thirteen years later. (Reports of the 54th and 67th Meetings of the BAAS, London: John Murray, 1885 and 1898).

1.38 H. A. Rowland was President of this Congress and Joseph Henry was posthumously honored during the meeting through the naming of the "henry" as the unit of induction following a motion initiated by delegates from France and Great Britain, according to Physical Review 1 (1893-1894) pp. 226-229.

1.39 Annalen der Physik Register zu Band 51-69 (1894-1899); Annalen der Physik und Chemie Beiblätter Register 16-30; and, selected volumes of Philosophical Magazine (1890-1900).

1.40 Information from the biographical sources cited in Footnotes 1.4, 1.7 and 1.10.

1.41 R. A. Millikan described his being sent to Paris to set up Michelson's equipment which included an interferometer and an echelon grating spectrometer in his Autobiography (New York: Prentice Hall, 1950).

1.42 Rapports Presentés au Congrès International de Physique (Paris: Gauthier-Villars, 1900) Ames discussed the mechanical equivalent of heat, Barus discussed pyrometry.

1.43 Wood presented the results of his newly developed technique for photographing sound waves. William Seabrook Doctor Wood: Modern Wizard of the Laboratory (New York: Harcourt, Brace and Company, 1941) p. 105.

1.44 Bancroft wrote on reaction velocities and solubilities, Nichols wrote on thermocouples. Recueil de Travaux offers par les auteurs à H. A. Lorentz (La Haye, M. Nijhoff, 1900).

1.45 J. H. Van Vleck, in his inaugural address as Lorentz Professor at the University of Leiden, March 4, 1960, tied the scientific development in America to the closing of the frontier of the nation, maintaining: "...as a rule scientific discovery and research do not flourish in a frontier country, which of necessity has other problems to meet, and that it is only when the frontier is closed that the road to progress in pure science is fully opened."

1.46 Stanley Goldberg, in Chapter 5 of "The Early Response to Einstein's Special Theory of Relativity, 1905-1911"(unpublished thesis for Ph.D. in Education, Harvard University, 1968) has explored the American response to relativity theory. We note that quantum theory had a much greater impact on the average practicing physicist than did relativity, for two reasons: the uncertainties involved in its tortuous evolution gave rise to more prolonged controversy and the number of phenomena which quantum theory eventually encompassed brought it closer to the working areas of a large number of physicists.

1.47 The terms "regular" and "associate" members, first used in 1905, were later changed to "Members" and "Fellows" after 1920. Svante Arrhenius was the first to be elected a foreign honorary member of the American Physical Society. This took place at the meeting of the Society that was held in conjunction with the International Electrical Congress in St. Louis, 1904, and in which Arrhenius participated. (Physical Review 19 (1904) 298)

1.48 According to E. Merritt, "Early Days of the Physical Society," there were meetings of the Society held in connection with each of these events.

1.49 From G. E. Hale "Introduction" to The New World of Science: Its Development during the War, Robert M. Yerkes, editor (New York: The Century Company, 1920).

1.50 These percentages are based on the addresses provided for individuals listed as members of the American Physical Society in Physical Review, n.s. 2 (1913) 507-532. The nature of the employment of the remaining 14 percent could not be determined from their addresses.

1.51 "American and German Universities" by Henry W. Diederich, identified as Consul, Bremen, in Science 20 (1904) 157.

1.52 Marckworth Dissertations in Physics.

1.53 "The American State University," Brubacher and Rudy, Higher Education, 139-170; William W. Brickman and Stanley Lehrer, editors, A Century of Higher Education: Classical Citadel to Collegiate Colossus (New York: Society for the Advancement of Education, 1962) passim.

1.54 John T. Merz, A History of European Scientific Thought in the Nineteenth Century (London: William Blackwood and Sons, 1904-1912; reprinted New York: Dover Publications, Inc. 1965), Chapter II: "The Scientific Spirit in Germany", Chapter III: "The Scientific Spirit in England," pp. 158-301.
Friedrick Paulsen, The German Universities: Their Character and Historic Development, Edward D. Perry, trans. (New York: MacMillan and Company, 1895). This work was prepared as an accompaniment to the German Educational Exhibit at the Columbian Exhibition held in Chicago, 1893. It contains an Introduction by Nicolas Murray Butler which is an exposition of American educational philosophy.
D. S. L. Cardwell, The Organization of Science in England: A Retrospect (London: William Heinemann, Ltd., 1959).

1.55 See R. T. Birge, History of the Physics Department Volume I, page designated II(8); this work was issued privately in mimeographed form by the University of California, Berkeley. As noted by Professor Birge, on (V)25, Ludwig Boltzmann was so impressed by the role of Mrs. Hearst that he was moved to write: "Wer ist Mrs. Herst? Es ist nicht ganz leicht, das einen Europäer klar zu machen. Am nächsten würde man der Wahrheit kommen, wenn man sagte sie ist die Universität Berkeley." in his essay "Reise einen deutschen Professors ins Eldorado", to be found in L. Boltzmann, Populäre Schriften (Leipzig: J. A. Barth, 1905).

1.56 According to A History of the Cavendish Laboratory 1871-1910 (London: Longmans, Green and Company, 1910), more than twenty-five Americans came there for a year or more. Among these were E. P. Adams, H. A. Bumstead, William Duane and Theodore Lyman. Thomson, commenting upon these visitors to Cavendish in Recollections and Reflections, noted the wide distribution of institutions from which they came and at which they were later employed.

1.57 See Appendix II

1.58 The Silliman Lectures are not associated with either Benjamin Silliman Sr. or Jr. Rather, they were, and still are, made possible through the establishment of the Silliman Memorial Lectureship in 1901 by Augustus Ely Silliman in memory of his mother, Mrs. Hepsa Ely Silliman. It was stipulated that these lectures should "illustrate the presence and wisdom of God as manifested in the Natural and Moral World." Over the years a number of illustrious physicists, such as Ernest Rutherford, Walter Nernst, Niels Bohr and O. W. Richardson, have delivered these lectures. (Information obtained from Judith A. Schiff, Chief Research Archivists, Yale University Library.)

1.59 Details of this visit may be found in Thomson Recollections and Reflections, 181-189.

1.60 R. T. Birge _History of the Physics Department_, Vol. 1, p. (V)25.

1.61 _Annual Register_, University of Chicago, 1912, 1913.

1.62 Archive for History of Quantum Physics, Interview with Otto Stuhlman, Jr., 8 June 1962;
Allen Shenstone "Sixty Years of Palmer Laboratory," address delivered at the dedication of the Jadwin Physics Building, and ________ "Princeton Physics: A Remarkable History" _Princeton Alumni Weekly_, February 24, 1961, pp. 6-12, 20.
Jean's books were entitled _An Elementary Treatise on Theoretical Mechanics_ (Boston: Ginn and Company, 1906) and _The Mathematical Theory of Electricity and Magnetism_ (Cambridge: The University Press, 1907).

1.63 "The Results of the Congress" _Proceedings of the Congress of Arts and Sciences, Universal Exposition_, St. Louis, 1904, ed. by Howard J. Rogers (Boston: Houghton, Mifflin and Company 1905) Volume I. Munsterberg, Harvard Professor of Psychology, was one of the vice-presidents of the Congress.

1.64 See, for example, L. Boltzmann "Reise eines deutschen Professors in Eldorado," and J. J. Thomson _Recollections and Reflections_, An account of Lorentz' 1906 American visit, written by his daughter who accompanied him on the trip, is found in "Reminiscences" _H. A. Lorentz: Impressions of his Life and Work_, ed. by G. L. deHaas-Lorentz (Amsterdam: North Holland Publishing Company, 1957).

1.65 K. J. Sopka "An Apostle of Science Visits America."

1.66 Michelson's award was given, not for his celebrated work with Morley attempting to detect the presence of the ether, but for his contributions to metrology and spectrology through the use of his interferometer and echelon grating. See _Nobel Lectures--Physics 1901-1921_ (Amsterdam: Elsevier Publishing Company, 1967) pp. 157-180.

It is curious to note that the eminent American historian, S. E. Morison fails to recognize this early Nobel award to Michelson. Writing in _The Oxford History of the American People_ (New York: Oxford University Press, 1965) p. 913, Morison says "Americans received no Nobel Awards for physics, chemistry or medicine prior to 1923." Was this indeed an oversight on Morison's part or was Michelson not considered an "American" due to his European birth, despite the fact that he was brought to the United States at the age of two? In all likelihood it is rather an indication of Morison's lack of concern with the historical development of science since he also fails to mention the Nobel Prize in chemistry that was awarded to his own Harvard colleague, T. W. Richards in 1914. (I am indebted to Gerald Holton for calling Morison's statement to my attention.)

The awarding of the Nobel Prize to Michelson was viewed by some as honoring the entire nation as well as Michelson as an individual. See, for example, the announcement of plans for a dinner in Michelson's honor to be held in Chicago upon his return from Europe "by colleagues and friends...and those who appreciate the honor done to this country by the confirming on him of the Copley Medal of the Royal Society and the Nobel Prize in physics" Science 27 (1908) 37. Popular Science Mon. 72 (1908) 285 carried the statement "The award of one Nobel Prize in science to a citizen of the United States, even though his birthplace was in Germany, is a recognition as great as this country may properly claim." (Michelson's birthplace, Strzelno, Poland, was under Prussian domination at the time when his parents decided to emmigrate to America, in 1855. See Livingston The Master of Light, p. 13.)

1.67 Livingston, The Master of Light, pp. 232-262.

1.68 La Structure de la Matière; Rapports et Discussions du Conseil de Physique Tenu a Bruxelles du 27 Au 31 Octobre 1913 Sous les Auspices de L'institut International de Physique Solvay (Paris: Gauthier-Villars et Cie, 1921) pp. 303-324.

1.69 Information on Wood summarized here was gathered from Seabrook, Dr. Wood, and Report of the 83rd Meeting of the British Association for the Advancement of Science, Birmingham, September 10-17, 1913 (London: John Murray, 1914).

1.70 Information on Millikan summarized here was gathered from his Autobiography, Chapter 8, pp. 87-99, and from Report of the 82nd Meeting of the British Association for the Advancement of Science, Dundee, September 4-11, 1912 (London: John Murray, 1913)

1.71 First person accounts of activities of individual scientists and the organization of the National Research Council are found in Millikan Autobiography, Pupin From Immigrant to Inventor, and Seabrook Dr. Wood, as well as in a chapter by Millikan "Contributions of Physical Science" in Yerkes, The New World of Science: Its Development During the War; reprinted in Burnham, Science in America pp. 303-313. In addition, this topic has been treated by D. J. Kelves in "George Ellery Hale, the First World War, and the Advancement of Science in America" Isis 59 (1968 Winter) 427-437 and by I. B. Cohen "American Physicists at War: from the First World War to 1942" American Journal of Physics 13 (1945) 333-346.

1.72 Lyman's results were published in Nature 93 (1914) 241 soon after the publication of Bohr's theory of the hydrogen atom. From present day perspective one might expect Lyman's results would have been viewed as powerful support for the acceptance of that theory. I have been unable to find instances of such thinking in the writings of Bohr or others. The contemporary view seems to have been expressed by Millikan who wrote: "Lyman's discovery...is not to be regarded as a success of the Bohr atom, but merely as a proof of the power of the series relationships to predict the location of new spectral lines," {"Radiation and Atomic Structure," Presidential Address to the American Physical Society, 1916, published in Physical Review 10 (1917) 149-205.}

1.73 Compton's thesis was published in Phil. Mag. 23 (1912) 579-593; in addition, he was coauthor with Richardson for two subsequent papers entitled "The Photoelectric Effect," ibid. 24 (1912) 575-594, 26 (1913) 549-567. In these, experimental work done by Compton was used to test a photoelectric theory developed by Richardson which was more complicated than the one proposed by Einstein but did not require a quantum nature for the light.

1.74 Rexmond C. Cochrane Measures for Progress: A History of the National Bureau of Standards (Washington, D.C.: Department of Commerce, 1966) p. 248.

1.75 David M. Dennison "Physics and the Department of Physics Since 1900" Research--Definitions and Reflections: Essays on the Occasion of the University of Michigan Sesquicentennial (Ann Arbor: University of Michigan Press, 1967) 120-136.

1.76 Many years later Hull wrote an account of this collaboration, "Reminiscences of a Scientific Comradeship," American Physics Teacher 4 (1936) 61-65. Their work was probably the earliest instance of results being obtained independently and almost simultaneously in Europe and in America. The Russian physicist, P. Lebedev, had made a preliminary announcement of his investigations on the pressure of light at the Paris Exposition of 1900 (Rapports, Tome II, 133-140) and gave a full presentation in Annalen der Physik 6 (1901) 433-458. Nichols and Hull first made their work known at the 1901 meeting of the American Physical Society held in Denver and submitted a paper on it to Physical Review 13 (1901) 307-320 and to the Bulletin of the American Physical Society 2 (1902) 25-26. {G. F. Hull "Experimental Discoveries Announced at the Meeting of the American Physical Society Fifty Years Ago" The Scientific Monthly 77 (1953) 13-18.} Lebedev's results predated those of Nichols and Hull but the latter proved to be the more precise.

1.77 Nobel Lectures-Physics 1922-1941 (Amsterdam: Elsevier Publishing Company, 1965) pp. 49-69.

1.78 Gibbs' classic exposition Elementary Principles in Statistical Mechanics developed with especial reference to the Rational Foundation of Theremodynamics (New York: Charles Scribners Sons, 1902) was issued just one year before Gibbs' death. While Gibbs was busy preparing the manuscript for this work he gave his approval, but no assistance, to the writing of Vector Analysis, "a textbook for the use of students of mathematics and physics founded upon the lectures of J. Willard Gibbs" (New York: Charles Scribner's Sons, 1901). This work was undertaken by Edwin B. Wilson who had studied with Gibbs while taking his doctorate in mathematics at Yale in 1901.

1.79 Wheeler Josiah Willard Gibbs, pp. 171, 172.

1.80 N.A.S. Biog. Mem. Vol. 8, 457.

1.81 B. O. Peirce Mathematical and Physical Papers 1903-1913 (Cambridge: Harvard University Press, 1926).

1.82 In 1904 the first edition of Webster's book Dynamics of Particles and of Rigid, Elastic and Fluid Bodies (New York: G. E. Stechert) was published. A second edition was issued in 1912 and reprinted in 1922. A third edition was published in 1925 by B. G. Teubner of Leipzig. Webster's final work Partial Differential Equations of Mathematical Physics was issued posthumously S. J. Plimpton, ed. (New York: G. G. Stechert, 1927; second edition 1933 and 1950; republished by Dover Publications, 1955).
A complete Bibliography of Webster's writings accompanies J. S. Ames' article "Arthur Gordon Webster" N.A.S. Biog. Mem. Vol. 18 377-347.

1.83 The term "theoretical physics" was rarely used in America during this period, and rightly so. Instead, "mathematical physics" was the usual designation for non-laboratory activity. The distinction was a suitable one in most cases since, properly speaking, the term "theoretical physics" should mean that the practitioner is actively engaged in adding to the body of theoretical understanding, whereas a "mathematical physicist" could be entirely engaged in using mathematical techniques in areas where the essential theory has already been established. In the early years of the twentieth century such areas were dynamics and electromagnetism. D. F. Comstock was probably unique in that he held appointments at the Massachusetts Institute of Technology ranging from Instructor of Theoretical Physics up to Associate Professor of Theoretical Physics in the years between 1905 and 1917. He was not, however, a practising theorist and left MIT to enter the business world. It must also be recognized that the emergence in Europe of theoretical physics as a distinct discipline was of relatively recent origin, having come in the late nineteenth century. A discussion of this topic by Russell McCormmach may be found in the Editor's Foreword to Historical Studies in the Physical Sciences Vol. 3 (Philadelphia: University of Pennsylvania Press 1971). McCormmach points out that "Planck was supreme among the first generation of theorists in establishing theoretical physics as a highly prestigious subspecialty." Planck had received his doctorate in 1879. Ten years later he was called to Berlin to head a new Theoretical Physics Institute. (I am indebted to Keith Nier of the Harvard History of Science Department for calling this reference to my attention.)

1.84 Pupin's work was important for the extension of long distance telephone lines. (See From Immigrant to Inventor) Sabine's work demonstrated that not only could the acoustical properties of existing structures be improved, but also that a building (e.g., Boston Symphony Hall) could be designed so that it would have favorable acoustical properties. (See William D. Orcutt Wallace Clement Sabine: A Study in Achievement (Norwood, Massachusetts: The Plimpton Press, 1933, privately printed).

1.85 The University of Chicago Catalogue, 1898-1899 contains a statement, (attributed by some to Michelson--see discussion in Physics Today April, 1968, p. 56; August, 1968, p. 9; and January, 1969, p. 9) to the effect that the future truths of physics should be sought in the sixth decimal place. The notion that physics was essentially "finished" also appeared elsewhere in the 1890's. Newton Henry Black, who received his B. A. degree in physics from Harvard University in 1896, related to his freshman physics class at Radcliffe College many years later that he had been told this when he was an undergraduate. (The author was a member of that Radcliffe class.)

1.86 Contrast, for example, the experiences which Max Born descibes his having at Göttingen in the chapter entitled "How I became a Physicist," Max Born, My Life and My Views (New York: Charles Scribner's Sons, 1968).

1.87 A summary of the development of mathematics in America may be found in G. D. Birkhoff "Fifty Years of American Mathematics," American Mathematical Society, Semicentennial Addresses, 2 (1938) 270-315, reprinted in George David Birkhoff Collected Mathematical Papers, Vol. III, 606-651 (New York: American Mathematical Society, 1950). As late as 1933 Birkhoff was urging that American mathematicians begin to take an interest in problems arising from the real world of physics.

1.88 See, for example, the Prefaces to his books, Theory of Electricity and Magnetism and Dynamics of Particles, and his Presidential Address to the American Physical Society "Some Practical Aspects of the Relations Between Physics and Mathematics" Physical Review 18 (1904) 297-318.

1.89 This attitude toward mathematical activity was apparently not confined to American physicists. In his Presidential Address to the British Association for the Advancement of Science meeting in Winnepeg, Canada in 1909, J. J. Thomson took occasion to chide those members of the physics community who not only discouraged the use of mathematics in physics but even seemed to consider that ignorance of mathematics was almost a virtue. Report of the 79th Meeting of the British Association for the Advancement of Science (London: John Murray, 1910) p. 6.

1.90 Leigh Page "Henry Andrews Bumstead" N. A. S. Biog. Mem. Vol. 13 pp. 105-124. Quoted passage occurs on page 120.

1.91 "The Debt of Physics to Metaphysics," Physical Review 31 (1910) 79-92.
This address shed interesting light on Crew's epistemology and metaphysical beliefs. He carefully dissociates himself from the metaphysics of a Kantian or Hegelian nature which had brought the term into such disrepute with Helmholtz and Maxwell, for example. Crew identifies the rational basis for the study of physics as a reliance on the observables of nature and a belief in the mechanical, or uniformity, postulate that all phenomena are subject to law, thus ruling out the possibility of genuine "accidents."

1.92 The chemistry profession in early twentieth century America was apparently similarly biased against theorizing, according to Robert E. Kohler, Jr. in his monograph "The Origin of G. N. Lewis's Theory of the Shared Pair Bond" Historical Studies in the Physical Sciences, Vol. 3, 343-376.
Apparently Lewis was discouraged from following up his early attempts at theoretical work and delayed publicizing his ideas for a cubical atom until more than a dozen years had elapsed after his original conception of it. By 1916, under the impetus of the response to the Bohr atomic theory, the time was right in America for the discussion of atomic structure.

1.93 Letter to the author, 3 February 1973.

1.94 E. C. Kemble's graduate studies, about which we shall have more to say later, were unusual in this respect, in that he did not chose his thesis topic until his third year and he took a large portion of his courses in mathematical-theoretical areas in both the mathematics and physics departments. (Private conversation, Fall, 1972 and the transcript of Kemble's graduate record made available, with his permission, by the Registrar, Harvard University,)

1.95 Dissertations in Physics, and Harvard University Department of Physics Doctors of Philosophy and Doctors of Science 1873-1964, pamphlet published by the Harvard Graduate Society for Advanced Study and Research, September, 1965.

1.96 Harvard University Archives

1.97 To W. J. King, deposited at the Center for History of Physics, American Institute of Physics, New York.

1.98 Letter to the author, 7 August 1972.
It came about that, a few years after Webster obtained his degree, permission was granted by the Physics Department at Harvard to E. C. Kemble to write a thesis that was entirely theoretical. As he neared the end of his work, however, Kemble realized that a fellow graduate student had built an apparatus which could be used to test the theory developed by Kemble. Though cooperative effort this was done and the results incorporated into the thesis Kemble submitted in 1917. We shall discuss the specific content of this work in a later section as it related to the development of quantum theory.

1.99 A. G. Webster--In Memoriam, Publication of the Clark University Library, Volume 7, March 1924, contains a list of the twenty-nine students who received Ph.D.'s under Webster's guidance together with the titles of their dissertations.

1.100 "The Laboratory in Modern Education" Johns Hopkins Univeristy Circulars 50 (1886) 103-105; reprinted in The Physical Papers of Henry Augustus Rowland 614-618.

1.101 Pop. Sci. Mon. 84 (1914) 105-117. Another statement by Webster extolling the virtues of the pursuit of science, "Scientific Faith and Works" appeared in Pop. Sci. Mon. 76 (1910) 108-115, 117-123 and is reprinted in Burnham Science in America pp. 260-271.

1.102 See, for example, their autobiographies. Both Millikan and Pupin embody success stories possible only in America at their particular time. They were both keenly aware of the social context in which their scientific activities were carried out. Pupin had risen from a poor immigrant who arrived in New York with only fifteen cents in his pocket to the post of respected professor at Columbia University and financially successful inventor. Millikan's childhood experiences as a son of a preacher who moved his family westward into newly developing country were followed by stimulating academic experiences at Oberlin College, Columbia and Chicago Universities as well as at European centers of learning. By 1920 Millikan was undoubtedly one of America's most influential physicists.

1.103 Specific mention will be made shortly about Millikan's original rejection of the quantum interpretation of his own photoelectric results. The remarks in this section are prompted in part by the writings of Paul Forman concerning the possible connection between the rise of quantum mechanics and the political-philosophical climate of the Weimar Republic (see: "Acausality before Quantum

Mechanics 1919-1925" AAAS address, December 1970 and "Weimar Culture, Causality and Quantum Theory 1918-1927" Historical Studies in Physical Science, Vol. 3, pp. 1-115.) The American outlook was a different one whether or not his had any real relevance to the development of ideas in physics.

1.104 Popular Science Monthly 72 (1908) 193-210.

1.105 "The Predicament of Scholarship in America and One Solution" Science 39 (1914) 587-595.

1.106 The Theory of Electrons and its Applications to the Phenomena of Light and Radiant Heat (Leipzig: B. G. Teubner, 1909) pp. 78-80. In the Preface Lorentz, writing three years after the delivery of the lectures, said that he regarded the scant attention paid to quantum theory as one of the deficiences of his presentation.

1.107 Acht Vorlesungen über Theoretische Physik (Leipzig: S. Hirzel, 1910) Planck's individual lectures were entitled: 1) Einleitung. Reversibilität und Irreversibilität; 2) Thermodyanmische Gleichgewichtzustände in verdünnten Lösungen; 3) Atomische Theorie der Materie; 4) Zustandgleichung eines einatomigen Gases; 5) Wärmestrahlung. Elektrodynamische Theorie; 6) Wärmestrahlung. Statische Theorie; 7) Allgemeine Dynamik. Prinzip der Kleinsten Wirkung; and, 8) Allgemeine Dynamik. Prinzip der Relativität.

1.108 Letter to the author, 3 February 1973.

1.109 The lack of uniformity among the various institutions with regard to the nature of the descriptive material supplied in their course announcements makes it difficult to assess how many others may have also introduced discussion of quantum theory by this time.

1.110 Vorlesungen über die Theorie der Wärmestrahlung (Leipzig: Barth, 1906) This publication was based on a course give by Planck at the University of Berlin in the winter semester of 1905-1906, according to the Vorwort.

1.111 Private conversation, 11 December 1972. Darrow spent the year following his graduation travelling in Europe and informally attending lectures in physics at the Universities of Paris and Berlin. He was sufficiently alerted to quantum theory by his University of Chicago experience to go to hear Planck's own lectures on the subject.

1.112 With George Winchester. "The Influence of Temperature upon Photo-Electric Effects in a Very High Vacuum, and the Order of Photo-Electric Sensitiveness of the Metals." Phil. Mag. 14 (1907) 188-210.

1.113 Physical Review 34 (1912) 65.

1.114 "The Effect of the Character of the Source upon the Velocities of Emission of Electrons Liberated by Ultra-violet Light" Physical Review 35 (1912) 74-76.

1.115 According to Paul Epstein in "Robert Andrews Millikan" N.A.S. Biog. Mem. Vol. 33, p. 257, "spark discharges are liable to falsify the measured potentials by inducing in the apparatus electric oscillations. Indeed, this source of error temporarily led him [Millikan]astray until he corrected for it in 1913." At the April, 1914 APS meeting in Washington Millikan indicated explicitly that he was testing the assertions contained in Einstein's photoelectric equation,{see "Preliminary Report on a Direct Determination of Planck's 'h'" Physical Review 4 (1914) 73-75.}

1.116 Bulletin Nat. Bur. Stds. 7 (1911) 393-406.

1.117 Cochrane Measures for Progress, p. 245.

1.118 The Genesis of Quantum Theory 1899-1913, Claude W. Nash, translator (Cambridge: Massachusetts Institute of Technology Press, 1971).

1.119 Title of conference proceedings (Paris: Gauthier-Villars, 1912) As mentioned earlier there were no Americans present at this meeting.

1.120 Paris: Flammarion, 1913; English translation: Mathematics and Science: Last Essays, John W. Bolduc translator (New York: Dover Publications, 1963).

1.121 During the meeting of the British Association for the Advancement of Science, held in Birmingham in 1913, J. H. Jeans openly credited his conversion to quantum theory to Poincaré's influence. A formal Discussion of Radiation was held as part of the program. Lorentz was another one of the dicussants who favored quatum theory on this occasion. (See Report of the Eighty-third Meeting of the BAAS.)
Additional discussion of the dissemination of quantum theory into French and British scientific circles is contained in "Henri Poincaré and the Quantum Theory" by Russell McCormmack, Isis 58 (Spring 1967) 37-55.

1.122 Physical Review n.s. 1 (1913) 66.

1.123 "Atomic Theories of Radiation" Science 37 (1913) 119-133.

1.124 Autobiography pp. 94-99.

1.125 In the course of his address Millikan took note, with some skepticism, of Poincaré's article, Journal de Physique, se. 5, 2 (1912) 5-34, in which the inevitability of quantum theory was set forth.

1.126 Wien's lectures were published as Neure Probleme der Theoretischen Physik (Leipzig: B. G. Teubner, 1913). Many of the problems centered around quantum theory as evidenced by the titles of the lectures which were: 1) Einleitung; Ableitung der Strahlungsformel; 2) Theorie der spezifischen Wärmen von Debye; 3) Theorie der elektrischen Leitung in Metallen. Zur Elektronentheorie der Metalle; 4) Einsteinsche Schwankungen; 5) Theorie der Röntgenstrahlen. Methoden zur Berechnung der Wellenlänge; 6) Lichtelektrische Wirkung und Lichtemission der Kanalstrahlen.

1.127 Physical Review 3 (1914) 57

1.128 "Physics and Physicists of the Past Fifty Years", Physics Today, (May 1956) 20-28 and History of the Physics Department, Vol.1, Chapter VII, p.11.

1.129 Letter to the author, 3 November 1972. Birge received his Ph.D. from the University of Wisconsin in 1914. Although nominally under the guidance of C.E. Mendenhall, Birge was almost exclusively self-taught in matters of practical spectroscopy and atomic theory.

1.130 Letter to the author, 7 August 1972.

1.131 "The Stark Effect and Atomic Structure", Astrophysics Journal 41 (1915) 359-372.

1.132 Leipzig: S. Hirzel, 1914.

1.133 Physical Review 5 (1915) 72. The speakers and their topics were: H.B.Lemon: The Nicholson Atom; Henry Gale: The Ritz Theory; G. S. Fulcher: The Stark Effect; G. W. Stewart: Energy Relations in Light Excitation by Impact; and Karl Darrow: X-Ray Spectra. The readers attention is called to Appendix IV, Part 5 which contains a list of APS Symposia Topics and Participants.

1.134 Physicist, A. C. Crehore, of the Nela Research Laboratory, Cleveland, Ohio, had proposed an atomic structure involving rings of electrons in orbital motion embedded in a sphere of positive charge, akin to Thomson's model. See Phil. Mag. 26 (1913) 310-332.

1.135 It is worth noting that, by the middle of the second decade of the twentieth century, the reality of atoms was no longer a subject for debate among either physicists or chemists although, only ten years earlier, ln 1904, the St. Louis International Congress was the scene of continuing controversy between the atomists and the energeticists. See Proceedings of the Congress, Appendix II and comments by Millikan, Autobiography, pp. 84,85.

1.136 According to his own statement, in Valence and the Structure of Atoms and Molecules (New York: The Chemical Catalogue Company, 1923; reissued by Dover Publications Inc., 1966) p. 29.

1.137 Jour. Amer. Chem. Soc. 38 (1916) 762-785.

1.138 This point is thoroughly discussed by Kohler in "The Origin of G. N. Lewis's Theory of the Shared Pair Bond," in Historical Studies in the Physical Sciences, Vol. 3, Russell McCormmach, ed. (Philadelphia: University of Pennsylvania Press, 1971) pp.343-376.

1.139 Jour. Amer. Chem. Soc. 41 (1919) 868-934.
I am indebted to Professor J. H. VanVleck for calling to my attention the continued high esteem in which this work of Langmuir is held, as evidenced by the celebration in 1969 of the fiftieth anniversary of its publication during the XXIInd International Congress of Pure and Applied Chemistry, August 20-27. The addresses that comprised the symposium entitled "Fifty Years of Valence Theory" are to be found in Pure and Applied Chemistry 24 (1970) 203-287.

1.140 From Langmuir's summary to paper cited above; also available in Vol. 6 of The Collected Works of Irving Langmuir, C. G. Suits and H. E. Way, eds. (New York: Pergamon Press, 1962) pp. 9-73.

1.141 A comprehensive discussion of Parson's theory and its probable significance for the development of Lewis' ideas is included in Kohler's "The Origin of G. N. Lewis' Theory."

1.142 Smithsonian Misc. Coll. 65 (November 29, 1915) Publication 2371.

1.143 For example, D. L. Webster gave a paper "Parson's Magneton Theory of Atomic Structure" at the Washington, D.C, meeting of the American Physical Society, April 1915; A. P. Wills quoted Parson favorably in his presentation at the symposium to be described in the next paragraph; G. N. Lewis included Parson's ideas in Valence and the Structure of Atoms and Molecules; A. H. Compton mentioned Parson's work in his first paper on the spinning electron, "The Magnetic Electron" Jour. Frank. Inst. 192 (1921) 145-155.

1.144 Physical Review 9 (1917) 172; Millikan's address was published in Physical Review 10 (1917) 194-205; the addresses of Lewis, Humphreys, Wills, Harkins and Jones were printed in Science 46 (1917) 297-302, 273-279, 349-351, 419-427, 443-448 and 493-502, in addition to remarks entitled "Radiation and Matter," Ibid. 347-349, by W. Duane who introduced the program.
Langmuir's presentation was, no doubt, related to a two part paper published by him about that time, "The Constitution and Fundamental Properties of Solids and Liquids" Jour. Amer. Chem. Soc. 38 (1916) 2221-2295, 39 (1917) 1848-1906, both of which are found

also in Vol. 8, The Collected Works of Irving Langmuir. In these papers Langmuir indicated his preference for Lewis' atomic and molecular theory over the large number being advanced by Thomson, Stark, Bohr, Parson, Crehore et al. For some unknown reason Wood's remarks on this occasion were never printed and there is no mention in the Bibliography of his works contained in Seabrook Dr. Wood of a paper that could be correlated with the title listed above.

1.145 Barnett wrote several papers on his work on the inducing of magnetism in rods through rotation. At this very meeting of the American Physical Society he delivered a talk on his latest results and their interpretation "The Magnetization of Iron, Nickel and Cobalt by Rotation and the Nature of the Magnetic Molecule," subsequently published in Physical Review 10 (1971) 7-21.

1.146 "Die universelle Bedeutung des Wirkungsquantums" Tokyo Sugaku Buturigakkawi Kizi 8 (1915) 106-116, cited in Max Jammer The Conceptual Development of Quantum Mechanics (New York: McGraw-Hill Book Company, 1966) p. 92.

1.147 In a supplementary note to his paper "The Bohr Theory and the Approximate Harmonics in the Infra-Red Spectra of Diatomic Gases," Physical Review 15 (1920) 95-109, E. C. Kemble stated that it was only after submitting this paper for publication that he became aware of the generalized formulation of the Bohr quantum condition for steady states of motion which had been proposed independently by W. Wilson in England and A. Sommerfeld in German. In particular, Kemble noted that the 1916 issues of Annalen der Physik, containing Sommerfeld's results, had just arrived at the Jefferson Physical Laboratory of Harvard University.

1.148 Recollections expressed by E. C. Kemble in conversation and informal note, Fall 1972.

1.149 Undated notes in E. C. Kemble's handwriting, found among some of his personal papers which he kindly loaned to the author in July, 1972. The statement quoted was the introductory paragraph to what appears to be a student report on the deriviation of one of the radiation laws, a report made by Kemble while a member of a class studying "Radiation" with G. W. Pierce. More will be said about this course later.

1.150 Leipzig: J. A. Barth, 1913.
The French language publication of the proceedings of the 1911 Solvay conference on quantum theory, cited previously, was also available in America prior to the outbreak of World War I. Another English language publication of this period, relating to quantum theory, was a translation by A. P. Wills of the lectures which Planck had delivered at Columbia University in 1909. (New York: Columbia University Press, 1915).

1.151 Quotation from the preface to Masius' book, The Theory of Heat Radiation by Max Planck (Philadelphia: P. Blakiston's Son and Company 1914; reissued by Dover Publications Inc., 1959).

1.152 London: "The Electricians" Printing and Publishing Company, Ltd., 1914.

1.153 See, for example, catalogues of course offerings of University of Chicago, Harvard University, Johns Hopkins University, Massachusetts Institute of Technology, Princeton University Yale University etc.

1.154 Private communication to author, Fall, 1972.

1.155 Physical Review 3 (1914) 92-102 and 4 (1914) 331-343.

1.156 Ibid. 4 (1914) 145-153.

1.157 Goldberg "The Early Response to Einstein's Special Theory of Relativity."

1.158 Physical Review 6 (1915) 377-389.

1.159 This expression is usually referred to as the "Duane-Hunt Law" although D. L. Webster was also involved in its establishment, according to statements made by Duane himself in the original paper given at the Washington meeting of the American Physical Society, April, 1915, an Abstract of which was published in Physical Review 6 (1915) 166-171, as well as in the paper referred to in Note 1.160.

1.160 Physical Review 7 (1916) 143-146.

1.161 Proc. Nat. Acad. Sci. 2 (1916) 78-83.

1.162 Similar sentiments are found on page 229 of his book The Electron, (Chicago: University of Chicago Press, 1968, a facsimile of the original 1917 edition). There Millikan says: "...we are confronted with the extra-ordinary situation that the semi-corpuscular theory out of which Einstein got his equation seems to be wholly untenable and has in fact been generally abandoned..."

1.163 "Radiation and Atomic Structure" Physical Review 10 (1917) 205.

1.164 "Non-Einsteinian Interpretations of the Photoelectric Effect" Historical and Philosophical Perspectives of Science, Vol. 5 (Minneapolis: University of Minnesota Press, 1970) pp. 246-263.

1.165 Kobenhavn: A. F. Host and Son, 1918-1922, p. 16.

1.166 Braunschweig: Friedr. Vieweg und Sohn, 1919, 1921 et seq.

1.167 Brinsmade was an experimentalist working in the infra-red region of the spectrum, hence it is not surprising that he should have been reading theoretical papers related to that field which, at that time, were neither numerous nor sophisticated.

1.168 Although Bridgman was himself an experimentalist, he was delighted to sponsor Kemble's theoretical investigations and eagerly followed their progress, according to Kemble's recent recollection.

1.169 Nernst Festschrift (Halle: Verlag von Wilhelm Knapp, 1912) 90-98 and Verh. d. D. Phys. Ges. 16 (1914) 640-642.

1.170 Verh. d. D. Phys. Ges. 15 (1913) 710-730, 731-737.

1.171 "The Distribution of angular velocities among diatomic gas molecules" Physical Review 8 (1916) 689-700.
"On the occurrence of harmonics in the infra-red absorption spectra of gasses" Physical Review 8 (1916) 701-714. This was the paper cited by Bohr in 1918.
"The occurrence of harmonics in the infra-red absorption spectrum of diatomic gases" (with J. B. Brinsmade) Proc. N.A.S. 3 (1917) 420-425. This was the paper cited by Sommerfeld.
"A New Formula for the Timperature Variation of the Specific Heat of Hydrogen" Physical Review 11 (1918) 156-158. Abstract of a paper presented at APS meeting held in Rochester, October 26,27, 1917.

It is planned that a copy of Kemble's thesis will be placed on the shelves of the Research Library of the Harvard Physics Department.

1.172 Physical Review 15 (1920) 95-109.

1.173 Harvard University Catalogues, 1919-1920, 1920-1921. For the second year the title of the lecture course was changed to "Quantum Theory with Applications to Series Spectra, Atomic Structure and the Kinetic Theory of Gases."

1.174 The format of Millikan's course, according to John H. VanVleck who was enrolled in it in the summer of 1920, consisted of having individual students make reports on papers related to quantum theory that had appeared in the research literature. VanVleck recalls that his own assignment was to make a report on Sommerfeld's theory of the fine structure of the hydrogen lines, a task whose difficulty was compounded by his own unfamiliarity at the time with scientific German and by the oppressive heat of the Chicago summer.

Professor VanVleck showed me the notebooks that he kept for Millikan's and Kemble's courses. The former was largely descriptive, the latter quite mathematical.

1.175 For an account see Phillip Frank Einstein: His Life and Times (New York: Alfred Knopf, 1967) pp. 140-146.

1.176 Physical Review 15 (1920) 206-216.

1.177 The Development of the Sciences, L. L. Woodruff, ed. (New Haven: Yale University Press, 1923) p. 71.

1.178 American Journal of Science 46 (1918) 303-354.

Addendum:

We note the publication since the preceding pages were prepared of "Physics circa 1900: Personnel, Funding and Productivity of the Academic Establishments" by Paul Forman, John L. Heilbron and Spencer Weart, Hist. Stud. Phys. Sci., VOL 5 (1975). In this work the authors provide a quantitative comparison of certain aspects of professional activity in physics among a dozen nations including the United States. It appears that at that time (when physics had only relatively recently emerged as the discipline we recognize today) American activity was at a respectable level among nations although by no means at the forefront.

Chapter 2

American Physicists and the Old Quantum Theory, 1920 - 1925

Page

Chapter 2

American Physicists and the Old Quantum Theory 1920-1925

The early years of the 1920's marked the beginning of a new era for physics in the United States. By 1920 the older, already established physicists had returned from military service or war-related work to their laboratories and universities. The bonds they had shared in wartime were not dissolved but replaced by other common interests: how was the study and practice of physics to be fostered in this country? and how would the discipline cope with the innovations being forced into it by quantum phenomena?

The role that Americans would play in the evolution of the new physics could not be foreseen in 1920. It turned out, however, that during the next five years sufficient momentum built up within the physics profession in America to ensure its participation in the quantum-mechanical revolution that occurred in the second half-decade of the 1920's. Thus the groundwork was laid during these years for theoretical physics to attain professional status in America and for physics in America to "come of age" by 1930.

Of course, this is a view that emerges in retrospect. The years from 1920 through 1925, as they were experienced, were years of continued growth and development of the institutional aspects of the profession in the United States. Intellectually, throughout the world they were years of tantalizing partial successes and seeming failures, a period of groping for satisfactory answers to the dilemma of the quantum assault on classical physics. On the other hand, they were exciting years with new results and new ideas continually emerging. The generation of young persons who entered the physics profession during this period, unlike its counterpart of twenty-five years earlier, did so in the full realization that physics was not a neat and essentially completed body of knowledge, but rather a set of challenging conundrums. This situation made physics all the more attractive to the newcomers to the profession.

In this chapter we shall examine the development of physics in America during the years when quantum theory was without a correct formal structure, before its current stage of development became "old" by virtue of the "new" quantum mechanics which began to emerge in the Fall of 1925. Here, as in our previous chapter, we begin by surveying broadly the social development of institutions relevant to the full professionalization of physics in America before focusing our attention on the collective and individual responses to quantum physics by members of the American physics profession.

Part I: The Growth of the Physics Profession in America

During the early 1920's the progress of physics in America toward "maturity," in the sense of professionalization, continued in terms of the activity of its professional society, the development of appropriate research journals, the availability of educational opportunities for aspirants to the profession and of facilities for the pursuit of physics by mature researchers and, finally, the relationship of American physicists to the worldwide community of physics.

It will become clear as we examine each of these aspects in the sections to follow that, while physics was indeed prospering in America in the years of rapid growth following the war, internal problems continued to exist and the mainstream of physics remained in Europe.

The American Physical Society

Turning our attention first to the American Physical Society we find the curious situation that while the number of Members grew by about 50 percent in the years 1920-1925 (to about 1260), the number of Fellows increased by only about 7 percent to 491. This would indicate a large increase in the number of Americans becoming sufficiently interested in physics to seek membership in the Society, but a much smaller increase in the number who were adjudged to be contributors to the advancement of the discipline.[1]

During these years the Presidents of the Society were, successively, J. S. Ames, Theodore Lyman, C. E. Mendenhall and D. C. Miller - a group representing the traditional and experimental strength of American physics.

Five or six meetings of the Society were held each year: the "Annual Meeting" continued to be held at the end of December in conjunction with the American Association for the Advancement of Science; in addition, there were regularly scheduled a one-day meeting in New York City in February, a two-day meeting in Washington in April and another two-day meeting in November at some more western location such as Chicago, Cleveland or Ann Arbor. The West Coast Division, formed in 1917, began to hold one-day meetings, usually one in June and another scheduled irregularly during the academic year.[2]

Attendance at some meetings rose to above 250 and the number of papers presented increased to more than 60.[3] It became clear to Harold Webb,[4] who was elected Secretary of the Society in 1923, that parallel simultaneous sessions were in order. The only alternatives would have been to extend the duration of the meetings or to limit severely the time allotted to individual speakers. Webb relates that his suggestion encountered vigorous opposition from physicists who were loathe to see the discipline become compartmentalized. Among these opponents was R. A. Millikan who was finally won over when it appeared that, without such action as parallel sessions, it would be necessary to allow each speaker a mere six minutes. During his tenure as President of the Society, 1921-1922, Lyman introduced the use of an alarm clock to hold speakers to their allotted time.[5] In the interval of just twenty years the problems associated with arranging meetings of the Society had changed from one of so few papers being submitted that speakers had to be recruited[6] to so many that time had to be rationed.

Research Publications

The American physics profession experienced another set of growing pains in the area of research publication which was showing a marked increase in the years 1920-1925. In the Physical Review alone, the number of papers rose from 85 in 1920 to 174 in 1925, requiring that the number of pages to be increased from approximately 1150 to 1750. While the Physical Review was officially sponsored by the American Physical Society, it was not the only medium of publication used by members of the physics profession in America. Papers in physics by American scientists continued to be published in the Astrophysical Journal,

Proceedings of the American Academy of Arts and Sciences, Proceedings of the National Academy of Science, Journal of the Optical Society of America and Journal of the Franklin Institute. In addition, the Massachusetts Institute of Technology began issuing its own Journal of Mathematics and Physics in late 1921. Furthermore, publication in European journals continued to be prized by American authors.

No single journal was available to Americans which satisfied the desire of authors to achieve rapid publication and wide dissemination of their work. These were important considerations, especially in the area of quantum physics which was so active during the 1920's. In particular, there was dissatisfaction[7] with delays frequently associated with publication in the Physical Review. Two elements of editorial policy produced delay. Referees were used to establish the suitability of a paper for publication rather than reliance on the author's own reputation or, in the case of young, unknown authors, the sponsorship of some respected figure in the scientific community. Clearly, circulating a paper to be refereed took time. In addition, other delays were occasioned by requirements, introduced by Gordon Fulcher while Managing Editor from 1923 to 1925, with regard both to the Abstracts which authors were then required to submit with their papers and to the style in which the papers themselves were written.

The fact that delays in Physical Review publication were indeed lengthy and potentially damaging is illustrated by the following example.[8] The paper in which A. H. Compton first fully presented his x-ray scattering results (to be called later the "Compton Effect") and their theoretical interpretation was submitted for publication in

December of 1922, but it was not published until May of 1923. During this five month period of delay P. Debye submitted for publication to *Physikalische Zeitschrift* his theoretical interpretation of Compton's experimental results which had been briefly mentioned by Compton elsewhere.[9] Debye's paper achieved publication within a month after submission, appearing in April of 1923, thus anticipating Compton's paper by one month. This resulted in some references in subsequent literature to the Compton-Debye Effect, although Debye himself gave full credit to Compton.[10]

With regard to reaching the widest audience possible, American physicists were again at a disadvantage, since neither *Physical Review* nor any other American scientific journal was widely read in Europe during the early 1920's.[11] Thus publication in one of the European journals was highly desirable. Some years later one of the American physicists recalled:[12]

"... in 1922 I was greatly pleased that my doctor's thesis was accepted for publication by the *Philosophical Magazine* in England, as we all felt it would have many more readers."

While the availability of the various domestic journals mentioned above, each with its own particular editorial advantages and drawbacks, provided the individual physicist with a good expectation of getting his work published somewhere, it remained a serious disadvantage that the American physics profession did not have a single, strong, rapid and respected research journal during the early 1920's.

The Production of Doctoral Degrees in Physics

An objective measure of American activity in physics is found in the data[13] relating to the production of Doctors of Philosophy and

Doctors of Science. The graph below for the years 1920 - 1926 reveals a relatively steady rate for four years, following the much smaller numbers of the first two years -- readily explained by the dampening effect of the World War on graduate education -- and then a marked increase of about 25 percent in 1926. This marked the beginning of the higher level over production that would characterize the late 1920's.

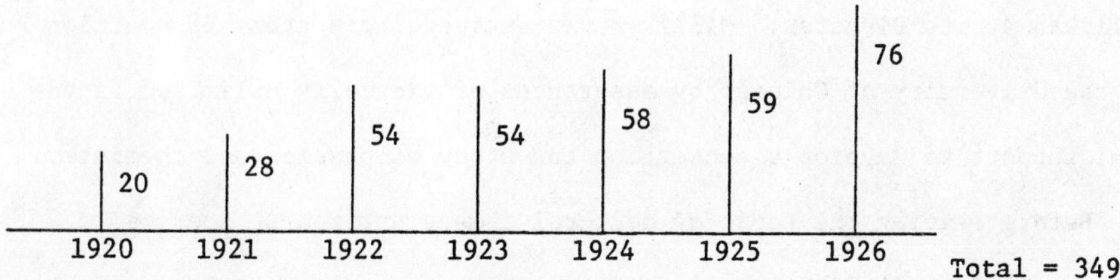

Inquiry into what institutions were responsible for the training of these doctoral candidates discloses that in 1920 and 1921, 17 institutions were involved, of which four universities, Chicago, Johns Hopkins, Cornell and Harvard, accounted for half of the 48 degrees awarded, granting 11, 5, 4 and 4 respectively. By 1926, when 76 degrees were earned in a single academic year, there were 26 institutions involved. In that year the leading institutions were the University of California at Berkeley and the California Institute of Technology, each with 8, followed by Chicago with 7; next were Cornell, Illinois and Michigan, each of which granted 6 degrees. The Johns Hopkins University granted 5 and Harvard 3.

The fact that California suddenly emerged as the leading state for the production of Ph.D.'s in physics is worthy of note and comment. By 1920 R. T. Birge and D. L. Webster had become professors of physics at the University of California in Berkeley and at Stanford University respectively. Birge developed a program centering on molecular spectroscopy and Webster continued to concentrate on the x-ray region of the spectrum. Another, more dramatic change took place accelerating the westward shift of the center of gravity for the study of physics in America when Throop College of Technology was transformed in the early 1920's into the California Institute of Technology with R. A. Millikan as its Director. Millikan was enticed there from his position at the University of Chicago by assurances of virtually unlimited financial support to develop a center for the study of physics and chemistry.[14]

Before leaving the topic of doctoral theses written in America between 1920 and 1926 it should be noted that, for the most part, their content continued in the tradition of strong emphasis on experimental work.[15] Of the total number of 347 theses approved during those years, about a dozen[16] were concerned primarily with the theoretical aspects of quantum physics and some of these contained little mathematical analysis. Apparently the characteristics associated with physics in America that we discussed in our previous chapter (the primacy of experimentation and the weakness of mathematical training) continued to inhibit theoretical activity on the graduate level. There simply were very few American faculty advisors competent and confident enough to supervise dissertations involving theoretical physics.

Postdoctoral Fellowship Programs

Thus far, in considering the institutional changes that took place in the world of physics in America in the early 1920's we have been concerned with the evolution of established entities. Now our attention will focus on an innovation associated with this period: the establishment of sizeable programs of postdoctoral fellowships for the study of physics, and on the effect of these fellowships on the pattern of individual and institutional development. We must, however, first digress in order to explain how these fellowship programs came to be established.

The dominant postdoctoral fellowship program for young American physicists during this period was that sponsored by the National Research Council. We recall from the previous chapter that this agency was inaugurated in 1915, in view of the existing war in Europe, by the United States government and the National Academy of Science for the purpose of making the scientific resources of the nation available for the defense of the nation. Initially the activities of the National Research Council were entirely devoted toward this end, but in 1918, before the end of the war, there was begun discussion of its possible postwar function.[17] It was proposed that the Council should undertake to sponsor such activities as would strengthen and enhance the scientific capabilities of the United States in peace as well as wartime.

Although sanctioned by the Federal Government, the Council was not funded by it; rather, the Council was to rely on private monies. This was subsequently accomplished with grants from the Rockefeller and Carnegie Foundations. Originally, G. E. Vincent, President of the Rockefeller Foundation indicated to R. A. Millikan, then a member of the Executive Board of the National Research Council, that John D.

Rockefeller, Sr. would be inclined to set up and support a central institute of physics and chemistry, comparable with the already existing Rockefeller Institute for Medical Research. Millikan argued against this proposal, urging, instead, that money be given to individual, promising young physicists and chemists who could then choose at which American institution they wished to pursue postdoctoral study. Thus, he reasoned, in a country as large as the United States, no single geographical area would be favored and, furthermore, those institutions which were already started on the road to excellence would be helped along by the presence of these young scholars. Millikan's views prevailed and impetus was given to the concept of postdoctoral funding for individuals which has played so large a role in the scientific development of America.

In retrospect, it appears that this was wise and wholly in keeping with American tradition that there should be a multiplicity and diversity of opportunity for individuals, coupled with a spirit of competition among institutions hoping to attract the young scholars. The impact of the National Research Fellowships thus established, together with those from other sources, on the development of physics in America will become apparent in the discussion which follows.

National Research Fellowships provided stipends of $1800 per year, a generous amount at that time. But more importantly, they allowed the recipients a period of professional growth unencumbered by heavy teaching duties that were normally the lot of young Ph.D.'s. The program was clearly an investment in the future of science in America.

Implementation of a fellowship program in physics, chemistry and mathematics, under the aegis of the National Research Council and funded by the Rockefeller Foundation, came swiftly enough for the first recipients to be named for the academic year 1919-20.[18] Six young physicists received support that first year.

In the years from the inception of the National Research Fellowships in 1919 to the academic year 1925-26 grants were made to 52 young men[19] for one or more years of postdoctoral[20] study in physics at the institution of their own choosing. The initial grant was frequently extended for a second year and there were instances of men being supported for as long as four years.[21] A total of 98 man-years of fellowship aid was awarded at the cost of about $200,000. Some recipients stayed at a single institution while others divided their time among two or three. A few fellows in the early years were allowed to study abroad,[22] but the aim of the program was primarily that of fostering the activity at American institutions.

The chart on the next page gives a picture of how the National Research Fellows in physics distributed themselves among American universities. Two details immediately become obvious: the consistent strength of the established institutions, primarily the University of Chicago, Harvard University and Princeton University, and the blossoming of the new California Institute of Technology.

Institutions did, indeed, feel that the number of National Research Fellows in residence was a measure of the institution's quality. This

2.12

American Institutions chosen by National Research Fellows for Study in Physics

	1919-20	1920-21	1921-22	1922-23	1923-24	1924-25	1925-26	Total
Cal. Inst. Tech.					4	6	8	18
Chicago	1	1	3	3	2	4	5	19
Cornell				1				1
Harvard	2	1	1	2	2	4	4	16
Iowa			1					1
Johns Hopkins					1		2	3
Michigan	1	1	1			1	1	5
Minnesota			1					1
Princeton	1			4	2	1	2	10
Rice			1	1	1			3
Stanford						1		1
Wisconsin				1	1			2
Yale			1	1	1		1	4

Note: the number given in each entry is the number of National Research Fellows in physics in residence in a given year in the physics department of the university in question. The totals represent the number of man-years. The sum of the totals above does not equal the number 98 quoted earlier because of those Fellows who went to Europe and a few others who took their fellowships to the National Bureau of Standards. The information reported here was obtained from _National Research Fellowships 1919-1944_, Neva E. Reynolds, Compiler, Washington, D.C., 1944. See, also, information in Note 2.23

is attested to in the following statement of Theodore Lyman in his Report of the Jefferson Physical Laboratory for the year 1923-24:[23]

"It is interesting to note that while the number of candidates for the degree of Doctor of Philosophy remains comparatively small, the position of this Laboratory is high as measured by the number of National Research Fellows who seek to avail themselves of its facilities. This state of things may be ascribed perhaps to the fact that the candidate for a higher degree is attracted to the institution which offers large scholarships, while the National Research Fellow, who is provided with an adequate income by the foundation which appoints him, chooses the Laboratory which offers the best opportunities for research."

Another example of institutional pride in the presence of National Research Fellows is found in the statement by George E. Hale:[24]

"The privilege of carrying on research at the California Institute [of Technology] is so highly prized that fifteen Research Fellows (the best of the younger physicists and chemists of the country) are sent there this year at the expense of the National Research Council.."

The restriction of National Research Fellows to American institutions, while understandable in terms of the aim of improving and stimulating the study of science in America, was a seriously undesirable feature for the young American who wished to study quantum theory in Europe where the topic was most active. The restriction was not such a drawback for the experimentalists since there were, by that time, very good laboratory facilities available in the United States. As a matter of fact, it was becoming recognized by the 1920's that laboratory facilities for the study of physics in the United States were superior to those at many European institutions, especially those in Germany. When Arnold Sommerfeld visited this country in 1923 he remarked, in an address to graduate students at the University of California at Berkeley:[25]

"In this country you have a special ability for performing the technique of the sciences, you have developed in the last years a very successful sc[h]ool of experimental physics and physical chemistry. In my country, I am afraid, the experimental work in the next years will be almost impossible because of our extreme poverty."

For those Americans wishing to go abroad for postdoctoral study in physics in the early 1920's the financial situation was not easy. Initially, there were available only fellowships given by the American-Scandinavian Foundation or by individual academic institutions and, of course, the private resources of the student and his family to make such study possible.

It was a lucky coincidence that the American-Scandinavian Foundation fellowships became available in the postwar years, just when Bohr's Institute of Theoretical Physics in Copenhagen was becoming so attractive to aspiring theoreticians.[26] It must be noted, however, that the stipends associated with these fellowships were only $1000, from which travel costs had to be met. Consequently the recipients of these fellowships had considerable difficulty in making ends meet.[27] Nevertheless, under these auspices the following Americans were each enabled to spend a year of study at Bohr's Institute: A. D. Udden, 1921-22; R. B. Lindsay, 1922-23; H. C. Urey, 1923-24 and J. A. Clark, 1925-26.[28]

The American-Scandinavian Foundation grants, the first of the postwar subsidies specifically for international education, were soon eclipsed by new programs backed by the Rockefeller and Guggenheim fortunes. We shall defer discussion of the Guggenheim fellowships until our next chapter since, although initiated in 1925, only one grant was made in physics that year.[29] In the subsequent years of the quantum-mechanical revolution and beyond, however, the impact of the Guggenheim awards on the study of physics by Americans came to be of great importance.

The fellowship program sponsored by Rockefeller funds that we shall describe next, was far more comprehensive than those already discussed and was, in itself, only a part of the broad program initiated in 1923 with the founding of the International Education Board. Because this program would come to affect the study of physics at home and abroad, it is appropriate to digress momentarily from our discussion of fellowships for American students of physics to consider the overall scope of this program.[30]

The International Education Board was founded by John D. Rockefeller, Jr. for "the promotion and advancement of education throughout the world" and was, in a sense, an extension of the work that had been carried on in the United States by the General Education Board for more than twenty years. John D. Rockefeller, Sr. had begun that enterprise in 1902 with a charter that restricted its activity to the United States.

During the years 1923 - 1928, when the International Education Board was most active, close to 12 million dollars was committed to the implementation of just the natural science part of a grand "Scheme," conceived and administered by Wycliffe Rose, who served as President of both Boards from 1923 to 1928.

Rose, who had been associated with several earlier health and education programs, believed that education was a prerequisite for all human progress, particularly education in agriculture and the natural sciences. His Scheme placed great emphasis on widespread exchange of personnel and on aid to specific institutions and individuals.

In choosing how aid could be given most effectively, it was decided that exceptional talent should be fostered rather than the improvement of the average. "Make the peaks higher" was Rose's motto.

Individual American scientists[31] were chosen to visit European centers of learning to assess the likelihood of these institutions profiting significantly from sizeable grants at that particular time. As a result of their recommendations some grants were made that had specific relevance to the international development of quantum physics. For example, $45,000 was given to Bohr's Institute of Theoretical Physics to expand its facilities and more than $82,000 was given to the Physics Institute at Göttingen. In addition, a $275,000 grant was made to the Mathematics Institute at Göttingen.[32]

Fellowships awarded by the International Education Board were available to candidates from any nation, and they were free to choose the country and institution to which they would go for study. D. M. Dennison, one of the first American physicists to be so supported, went to Copenhagen to study with Bohr from 1924 - 1926. Now, for the first time, Europeans began to come to study at American institutions. Among the first to come in this capacity were O. Laporte, O. Oldenberg, Herta Sponer and F. Zwicky. Each of these spent one or more years of study in the United States between 1924 and 1926 and eventually each of them became permanent members of the American scientific community.

Laporte originally came to the National Bureau of Standards, Oldenberg and Zwicky to the California Institute of Technology and Herta Sponer to the University of California at Berkeley.[33] While previously most European scientific visitors had largely come to America as missionaries, these young people were the first explorers and potential settlers. This, however, was only the small beginning, not yet seen as a new trend; for the most part American students continued to seek opportunities for study in Europe.

A few Americans obtained support for European travel, not from the large funding organizations discussed above, but from sources within their own universities. For example, John C. Slater and Victor Guillemin held Sheldon Travelling Fellowships from Harvard University.[34] Slater spent a postdoctoral year at Cambridge and Copenhagen in 1923-24, while Guillemin used his grant to go to Munich where he wrote his doctoral dissertation with Sommerfeld in 1926. Ralph Kronig's European travels in 1925, which he described in the "The Turning Point"[35] were made possible by a grant from Columbia University; Jane Dewey was named an International Fellow by Barnard College for two years of study at Bohr's Institute for Theoretical Physics, 1925-27.[36]

The opportunities for study in Europe that were afforded the dozen or so young Americans who went there to study theoretical physics during the early 1920's are noteworthy beyond their specific value to the individuals themselves. The very presence of the young Americans at European theoretical centers was an indication of the growing awareness, on the part of the American physics community, of the limited opportunity for theoretical study at home and of the importance of such study for the coming generation of physicists if America was to move forward into full participation in the new physics that lay ahead.

The fact that sizeable sums of money were now made available, largely through the donations of wealthy Americans, for the support of individuals and institutions engaged in the study of physics shows that, although the plea made by John Tyndall in 1873 for the support of "pure science" in America was not successful at the time,[37] fifty years later the time had arrived for the extensive funding of scientific pursuits which benefited not only Americans but the entire scientific community as well. Several factors contributed to this changed climate: America had survived the World War with far less direct impact than did either its Allies or defeated Germany; the sciences in the United States had become sufficiently "professionalized" to be able to use the munificence profitably; there was a large enough pool of able young scientists to accelerate the general growth of science across the United States; finally, in the case of physics and chemistry, the disciplines themselves were at a crucial stage of development, making them especially attractive to bright young minds. The nature of radiation and the structure of atoms and molecules were the burning issues of the day for physics and chemistry.

<u>International Relations:</u> 1) <u>Visitors From Abroad</u>

Another kind of international contact enjoyed by the American physics profession during the early 1920's came from the visits to America that were made by outstanding European physicists. The chronological list of foreign lecturers in physics coming to the United States, to be found in Appendix II, reveals that in every year from 1921 to 1925 American physicists had an opportunity to hear at least one such speaker. In this section we shall assess the impact of these contacts in a general way, delaying until later a discussion of their significance for the stimulation of American interest in quantum physics.

The resumption, in the postwar years, of the practice of inviting European physicists to lecture at American universities was welcomed by all parties concerned. America was an interesting place to visit, in the opinion of many Europeans, wholly aside from the business of scientific lecturing. Those who had come before, such as H. A. Lorentz and J. J. Thomson,[38] were glad to renew old aquaintances and view the progress that was being made in America on all fronts, but especially in the practice of physics. For those visiting America for the first time, there were other allurements: a new, vast country to be explored and an opportunity to see for themselves what American physicists were doing.

The visits of Einstein and Sommerfeld had political implications as well as scientific significance. Einstein came principally for the purpose of raising money, under the auspices of the Zionist movement, for the establishment of a Hebrew University in Jerusalem.[39] Only incidentally did his schedule allow for scientific contacts, the principal one being his visit to Princeton University where he delivered a series of lectures on relativity[40] and received an honorary doctorate. In the case of Sommerfeld, he came expressly to serve as the Carl Schurz Visiting Professor at the University of Wisconsin during the academic year 1922-23. But Sommerfeld had been deeply concerned over the anti-German sentiment generated in America by the events of the recent war;[41] hence in his letter of acceptance of this position he said that he welcomed the opportunity[42]

"... die wissenschaftlichen Fäden neu zu knüpfen, die durch den Krieg zerrissen sind. Mein höchster Wunsch aber würde es sein, dazu beizutragen, dass sich zwischen meinem und Ihrem Land wieder vertrauensvolle Beziehungen anbahnen mögen, als notwendige Vorbedingung für die Gesundung Deutschlands und für die Kultur der Menschheit."

Whether their prime motivation in coming to America was scientific or otherwise, all the European physicists were warmly received. While it is true that the recent enemy status of the German nation did not go unremarked by some when contemplating the visits of Einstein and Sommerfeld,[43] such episodes were so isolated and so few in number that they contrasted all the more noticeably with the general hearty welcome accorded by their hosts.

An indication of the style in which at least one European scientist was treated is gained from the following:[44]

"I travelled for the rest of my journey in almost regal splendor. My kind friends at the Franklin Institute had placed at my disposal a private train. It had a dining room, a parlour, two bedrooms, a kitchen, a cook, a waiter, and a conductor in charge of the train. I had to travel about a good deal for the rest of my visit, as I had promised to lecture at several universities. All I had to do was to tell the conductor, before I went to bed, where I had to be the next morning and what I wanted for breakfast. When I awoke, I found I was at my destination and the breakfast was ready. In this way I revisited Harvard, Princeton and Yale, lecturing at each of them and assisting at the opening of a magnificent new chemical laboratory at Yale."

Truly the Americans welcomed their European scientific visitors of the early 1920's as honored guests.

Again we find that it was American private wealth that made this activity possible. In addition to the Silliman Lectureship which continued at Yale University, a number of other institutions[45] were becoming able, in the 1920's, to offer sizeable fees to these distinguished visitors. An indication of the order of magnitude can be gained from the following data: Bohr received $3000 from Amherst College[46] and $1250 from Yale University;[47] Sommerfeld received $4000 for the equivalent of one semester as Carl Schurz Professor at the University of Wisconsin.[48] The availability of such fees for lecturing in America accounts for the following remarks in a letter written to Einstein by Hedwig Born:[49]

"My husband feels an inclination to slay the golden calf in America and to earn enough through lecturing to build a small house in Göttingen..."

In many cases it appears that, after receiving adequate funding from one or more institutions to make a trip to America possible, some of these European lecturers were quite willing to give a lecture or two at other institutions which were less well endowed for guest lectures. In this way both the number of American scientists having the opportunity for contact with the guest scientists and the number of diverse institutions that the guests were able to visit and see for themselves was greatly increased. The practically unlimited funds available to Millikan at the California Institute of Technology, already mentioned, enabled him to arrange the visits of Lorentz and Ehrenfest, for example. In those days of transcontinental train travel it was an easy matter for stops to be made along the way for visits to other institutions.

What was the status of physics in America in the eyes of these European visitors? Wherever we have evidence of their assessment it points to the conclusion that America was still regarded as "mission country," especially in the realm of theoretical physics, but that such progress was being made that the outlook was decidedly optimistic.

After Sommerfeld uttered his words of praise for American experimental physics, quoted above on page 2.20, he said:

"On the other hand, in my country we have developed a very successful sc[h]ool of mathematical physicist. The kind reception which I found in your beautiful university and the kind interest which your physical and chemical departments devoted to my lectures seems to indicate that the interest in theoretical work in this country <u>will be in the future</u> as lifely [sic] as the interest in the experimental progresses." (my emphasis)

J. J. Thomson was another who expressed his admiration for the improving scientific environment in America when he visited in 1923, writing:[50]

"It was also most satisfactory to find the great strides that had been made in these universities since my last visit (1903); in each one of them there were many new laboratories - large and well equipped; very many new Professorships had been founded and many more opportunities for advanced study and research made available. Above all, they seemed to be inspired with a genuine enthusiasm for research. The large number of discoveries in Physics and Chemistry of first-rate importance made in the United States during the last ten years proves that full advantage has been taken of these increased opportunities."

Thomson also had high praise for the accomplishments of scientists at industrial laboratories such as General Electric and the Bell Telephone Co., citing as an example the x-ray tubes produced at the former by W. D. Coolidge.

The industrial laboratories of the United States impressed other European visitors as well as Thomson. Otto Oldenberg has recalled:[51]

"Franck, in 1924, was much impressed by the size and high standing of the industrial research labs. They employ outstanding men, Whitney, the founder of the lab, Langmuir, Hull at G.E., and promote research in pure physics and chemistry on the basis of a vague prospect of profits. Thus, later the transistor originated at Bell Lab, nylon at DuPont .... A few years later Born had the opinion that the discovery at Bell Lab by Davisson and Germer required so highly developed techniques in various fields that it could not well be carried out at a university lab, as of 1926, but required the resources of an industrial lab."

Einstein was, of course, the most famous of the European scientists who visited the United States in the early 1920's. Consequently, the opinions of America which he expressed were given widespread attention. Soon after his return to Europe, there was published in the Niewe Rotterdamsche Courant and in the Berliner Tageblatt[52] an unauthorized version of some rather caustic remarks that he was said to have made to a reporter concerning his impressions of America. These were quoted widely in the American press[53] causing some ruffled feelings among the American public and the scientific community.

The reference to Princeton was kindly enough though somewhat condescending:

"... aber ich fand auch Princton (die grosse amerikanishce Universität) schön. Eine noch nicht gerauchte Pfeife. Jung und frisch, von Amerikas Jugend ist noch viel zu erwarten."

What really riled some American scientists was the statement:

"Mein Eindruck vom wissenschaftlichen Leben Amerikas? Nun, ich habe mit grossen Interesse einige ausserordentlich verdienstvolle Professoren kennen gelernt, zum Beispiel Professor Millikan. Professor Michelson in Chicago habe ich leider verfehlt, aber das allgemeine wissenschaftliche Leben Amerikas mit dem in Europe zu vergleichen wäre Unsinn, wie man überhaupt das ganze übrige Leben von Europe mit dem von Amerika nicht vergleichen kann. Es sind nun einmal zwei verschiedene Welten."

Valid though such comments may have been, American feelings were hurt. Some attempts were made to smooth things over by calling attention to how short a time Einstein had spent in the United States and how limited were his scientific contacts.[54] Einstein, embarassed by the publication of this article purporting to express his views, wrote one for the <u>Vössische Zeitung</u>[55] saying that his remarks had been misrepresented. Einstein's article, which bore the headline "Einsteins amerikanische Eindrücke. Was er wirklich sah."[56] included the following comments about the American scientific scene:

"Was mich am meisten erfüllt, wenn ich an Amerika zurückdenke, ist das Gefühl der Dankbarkeit für den warmen und herzlichen Empfang, den ich bei allen Fachgenossen, Behörden und Privatleuten gefunden habe. Ich habe an der Universitäten Princeton, Chicago, an zwei New Yorker Universitäten, an der Columbia Universität, im College of the City in New York, Vorlesungen gehalten und überall grosse Interesse und tiefe Sachkenntnis bei den dortigen Fachgenossen gefunden. Besonders muss ich anerkennen, dass man nichts dabei fand, dass ich deutsch sprach. In der rührendsten Weise bemühten sich alle, die mit mir sprachen, dies auch deutsch zu tun. Alle, die Verbindungen mit Deutschland hatten, sprachen davon und erwähnten die wissenschaftlichen Bonde, die sie an Deutschland knüpften ....

".... Grosse Freude habe ich an den amerikanischen Studenten gehabt. Die Universitäten sind in vielen Fällen Internate. Sie bieten durch die Kameradschaft zwischen Studenten und Professoren ein Bild froher Harmonie. Bei uns findet man vielleicht unter den Studenten mehr individuelles Denken, dafür bei den Amerikaner mehr Urgesundheit und bei allem Interesse für politische Dinge ein massvolles tollerantes Behandeln politischer Fragen.

"Dass die naturwissenschaftliche Forschung in Amerika im Begriff ist, auf manchen Gebieten die Führerrolle zu übernehmen, beruht nicht nur auf den Reichtümern des Landes, sondern auch auf dem Umstande, dass der reiche Amerikaner von der Überzeugung durchdrungen ist, dass er der Allgemeinheit viel schuldig ist. Dieses Gefühl, welches durch die mächtige öffentliche Meinung des Landes wacherhalten wird, bringt es zu wege, dass für öffentliche Zwecke grosse Mittel überall zur Verfügung stehen, ohne dass der Staat sich darum zu kümmern braucht."

In summary, it can be said that the visiting European physicists of the early 1920's came and lectured in sizeable number, enjoyed themsleves, but did not feel that physics had yet "come of age" in America.

2) Honors Accorded American Physicists

Looking now at the aspect of international relations which concerned honors being awarded to American physicists during the period 1920 - 1925, we can point to one Nobel Prize and four Solvay Conference invitations.

The Nobel Prize was awarded to R. A. Millikan in 1923 for his experimental studies of the electronic charge and of the photoelectric effect.[57] This was not the first European honor to be won by Millikan during this period. In 1921 he and Michelson were invited to participate in the third Solvay Conference, held in April of that year.[58] Although neither Millikan nor Michelson were there as principal speakers, Millikan did make extensive commentaries during the discussions of the photoelectric effect and of atomic structure. On the other hand, Michelson remained a quiet spectator.[59]

The fourth Solvay Conference, held in 1924, on the topic of the electrical conductivity of metals, had the American physicists P. W. Bridgman and E. H. Hall as participants and each delivered a principal address.[60]

It certainly was a step forward on the international scene for two American physicists to have been invited to each of these Solvay conferences where the total membership at each was less than two dozen. We note, however, that only American experimentalists were thus honored.

Summary:

There can be little doubt of the healthy growth in the practice of physics in America during the early 1920's. Scientific manpower, research productivity and educational opportunity were all on the increase. Of particular importance were the new programs for post-doctoral study at home and abroad. While it is true that American activity in theoretical areas continued to lag behind the pace set in Europe, experimental physics was flourishing at an increasing number of academic and industrial laboratories. The period was characterized by an external peace and prosperity which was conducive to the free exchange of physicists with the European centers of study. The discipline of physics itself was in a state of excitement over the challenge to classical physics being posed by quantum theory, and the American physics profession, strengthened by postwar cohesiveness and financial backing, was coming closer to full participation on an international scale.

Part II: American Responses to Quantum Theory

The realization that quantum theory was here to stay was greeted with mixed feelings by various members of the American physics community.

"The bizarre and revolutionary character of the quantum theory and quantum experiments gave interest to the field for bright students and simultaneously turned off [many of] the senior professors,"

according to the recollections of E. C. Kemble[61] who was a young instructor in the physics department of Harvard Univeristy during the early 1920's.

In between the two extremes of hostility and enthusiasm was the response of the vast majority of the American physics profession. While many continued to have doubts and misgivings about the direction in which quantum theory seemed to be leading, most American physicists during this period tried to accommodate to the new situation as best they could. No longer, as in the previous decade, was there serious effort to discredit, or find escape routes from, quantum theory.[62] Rather, the emphasis was now placed on encouraging the widest possible study, discussion and testing of the new ideas.

In this section we shall discuss first a few specific instances of continued outright rejection of quantum theory and then focus on the activities within the American physics profession which served to advance the dissemination of the new concepts. We reserve until our next section, Part III, discussion of contributions to the progress of quantum physics that were made by Americans during the early 1920's.

Lingering Doubts

While it was true that quantum theory did provide formal explanations of particular phenomena that had not been successfully explained by classical physics, it did so through the introduction of concepts which seemed completely irreconcilable with classical physics. Whereas classical physics provided a self-consistent frame work that was both powerful and beautiful, quantum theory, prior to the introduction of quantum mechanics, provided, rather, a series of "insights" into disparate situations. The relatively few facts explained by quantum theory seemed to many to be insignificant in comparison with the vast area of observation rationalized by classical physics.[63] Furthermore, the spectroscopic successes of Bohr's atomic theory failed to convince most chemists that planar orbits would ever be able to account for the well-known characteristics of valence and molecular structure. These, then were the bases of the lingering doubts and occasional antagonisms voiced with regard to quantum theory.

In a few instances individuals were seriously alienated by the recent developments in physics. Not surprisingly, some of these were older members of the physics community whose own education had been completed, for the most part, before 1900. Their teaching and research focused on classical areas. They were not willing or able to cope with the changes being forced upon them and longed for the good old days when they understood physics.

A. G. Webster was one of these. In 1921 he wrote an article entitled "Oh Quanta!"[64] which was prompted by Planck's receiving the Nobel Prize in 1918 and contained quite a good account for general audiences of the contributions of such physicists as Maxwell, Wien, Planck, Boltzmann, Nernst and Bohr. But, in addition, we find this remark:

"To understand the theory of quanta requires a knowledge of all the most difficult parts of mathematical physics. I do not half understand it. Do you? But, like the theory of relativity, it is a great thought, worthy of the Nobel Prize."

More emotional was the statement of Carl Barus, who reacted to the suggestion that the conservation of energy may be only a statistical law by writing a letter to <u>Science</u>, saying:[65]

"We who are about to be shelved used to live in this country, peacefully under the constitution and were quite happy in our simplicity. One day a man by the name of Einstein came along and mixed that constitution up. We were told that it had long been an antiquated document anyway. There were difficulties, but eventually we managed to fit in; for they had left us at least with the doctrine of energy. Now, I read that the classical law of the conservation of energy must also go, that at best it is only statistical like the second law of thermodynamics. Truly these young bloods are Balkanizing the whole of physics, and our ancient constitution has gone the way of the mark."

Another American physicist who turned his back on quantum theory was Max Mason, then a professor of mathematical physics at the University of Wisconsin. According to Warren Weaver, a student and later associate of Mason at Wisconsin, Mason was repelled by quantum theory's "messiness" and incompatability with classical physics.[66] Weaver, himself one of the bright young men entering physics in the early 1920's, was strongly influenced by Mason's attitude and remained aloof from the development of quantum theory. To this day Weaver has not accepted it as a permanent part of physics.[67]

Negative reaction to quantum theory as it was applied to atomic structure in the manner of Bohr and Sommerfeld came from chemists and others who tended to prefer the Lewis-Langmuir atomic model.

During the meeting of the American Association for the Advancement of Science that was held in Toronto in December of 1921 there was a symposium on Quantum Theory which brought together representatives of the American Physical Society, the American Chemical Society and the American Mathematical Society. On this occasion R. C. Tolman, speaking for the Chemical Society, described the point of view of the chemists as:[68]

"... largely the negative one, of extreme hostility to the physicists, with their absurd atom, like a pan-cake of rotating electrons, an attitude which is only slightly modified by a pious wish that somehow the vitamine "h" ought to find its way into the vital organs of their own, entirely satisfactory, cubical atom....

"This atom [Bohr's] was constructed by the physicists, like a solar system, with electrons rotating around a central nucleus, partly because the physicists were familiar with the mechanics and mathematics of the solar system, and partly because they were entirely unfamiliar with the actual facts concerning the behavior of atoms in chemical combination. No chemist would be willing to think of a carbon atom as a positive nucleus with a ring of electrons rotating around it in a single plane. The carbon atom must have tetrahedral properties..."

In the course of his talk Tolman also criticized quantum theory in general for the arbitrariness with which "h" was introduced into the study of various phenomena; for its contradiction of the undulatory theory of light; and for its "cavalier" treatment of the classical dynamics characterized by Hamilton's principle and the Lagrange function.

There were also physicists who continued to reject or remain skeptical of Bohr's theory. D. L. Webster was one who "was not ever converted to the Bohr-Sommerfeld type model."[69] K. T. Compton's course at Princeton University, entitled "Atomic Structure, including Lewis, Langmuir and Bohr," begun in 1919-20, continued to be given through the year 1925-26, according to the course catalogues of the period.

At the University of California at Berkeley R. T. Birge successfully sought converts to Bohr's theory from among members of the physics and chemistry departments.[70] Apparently, even G. N. Lewis himself came to the conclusion that "every essential element of conflict between the views of the physicist and the chemist [with regard to the arrangement of electrons within atoms]" was removed when Bohr introduced the notion of shell structure for multi-electron atoms.[71]

R. A. Millikan, on the other hand, a consistent adherent of Bohr's atomic theory,[72] retained his misgivings about the validity of quantum emission and absorption. In December of 1919 he stated flatly "the absorption of energy cannot take place quantum-wise at all" (emphasis Millikan's).[73] In his Nobel Lecture, delivered in May, 1924, after describing how the recently discovered Compton Effect would seem to support the Einstein conception of localized quanta, Millikan concluded:[74]

"But until it can account for the facts of interference and the other effects which have seemed thus far to be irreconcilable with it, we must withhold our full assent."

In Millikan's case his qualms about quantum theory did not cause him to turn his back in rejection. Rather, his approach was to continue to seek experimental evidence relating to quantum phenomena and to encourage American intercourse with European physicists who were up-to-date on the latest developments of quantum theory.[75]

The Dissemination of Quantum Theory

The American physics profession, in general, shared the conviction that familiarity with quantum developments was essential to the continuing healthy growth of physics in America. Steps were taken toward this end in terms of symposia, publications, foreign lecturers and curriculum innovation. All of these measures contributed to the dissemination of quantum theory among the older, established members of the physics profession as well as among the new, postwar generation of young physicists who were eager to become involved with the new concepts. Since these activities comprised the intellectual background against which individual Americans made contributions to quantum developments it is appropriate at this time to consider some of the details of these programs.

1) Symposia

The custom of incorporating panel discussions of topics with wide current interest into the programs of the American Physical Society meetings was given new impetus after the end of the war in order to counteract the negative effects of the recent interruptions of research activity and of free flow of scientific journals from Europe. Furthermore, the early 1920's saw the emergence of so many new ideas that it is not surprising to note (see Appendix IV) that these were occasions of lively discussion in almost every year 1919-1925, many of which related specifically to quantum theory developments. It is difficult to assess how much real impact these programs had on the dissemination of quantum physics in America, but the fact that they were held and the identities of the panelists provide insight into the perspectives of the American physics profession with regard to what topics were considered to be of prime importance and which individuals were recognized as authorities at the time.

2) Publications

Clearly there was a great need for published, coordinated presentations that could be used to bring established members of the physics profession up to date on recent developments and would serve as texts, or related reading, for students, especially those at the graduate level.

The first such book to come on the scene was Atombau und Spektrallinien by Arnold Sommerfeld,[76] professor of theoretical physics at the University of Munich. Its first edition appeared in September of 1919, the second a year later, the third in January of 1922 and the fourth in October 1924. It became almost a year book of quantum theory. Each new edition was eagerly awaited. Its value for the American audience was increased by the appearance in June 1923 of an English translation of the third edition by Henry L. Brose of the University of Nottingham, England under the title Atomic Structure and Spectral Lines.[77] Virtually all members of the physics profession who were active during the 1920's recall this work as having been most valuable for their study of quantum physics.[78]

Sommerfeld's presentation was very well suited to well prepared graduate students in physics. Something else was needed for less advanced students of physics, chemists and others who needed a more gradual introduction to the material. Publications responsive to this need came from the National Research Council and, in at least one case, from the American Chemical Society.[79] Among the earliest of these was an 81 page pamphlet by E. P. Adams "The Quantum Theory," based on a series of talks on quantum theory which Adams had delivered at Princeton University.[80]

The content of a number of other National Research Council Bulletins that were issued during this period of the early 1920's[81] included discussions of topics which had bearing on quantum theory. Most of these laid heavy emphasis on experimental findings and tended to present theoretical discussion in a tentative tone. Their very existence, however, clearly indicate that the American physics community during this period was alert to quantum theory development and sought to insure its dissemination among American readers.

As the "old" quantum theory period was drawing to a close there appeared two National Research Council Bulletins which departed from the pattern of the previous ones in two respects: they were far more comprehensive, being actually books rather than pamphlets, and they were predominantly theoretical. They were written by authors well grounded in quantum theory. The first of these was "Quantum Principles and Line Spectra," by John H. VanVleck;[82] the second, "Molecular Spectra in Gases," by E. C. Kemble, R. T. Birge, W. F. Colby, F. W. Loomis and L. Page.[83]

Both these works reflect the difficulties of trying to write such tomes. VanVleck wrote in his Preface, dated August 7, 1925:

"The writer has struggled to keep the material of this Bulletin up-to-date, but the quantum theory is so alive that it develops and changes almost overnight."

Little did VanVleck realize as he wrote these words that within a matter of weeks matrix mechanics would appear and signal the beginning of the "revolution." In the case of "Molecular Spectra in Gases," the writing of it not only suffered the difficulties associated with a committee authorship[84] but, after years of preparation, the manuscript was completed just as the new quantum theory broke upon the scene, necessitating a last-minute up-dating of several sections.[85]

Although it must have been a frustrating experience for all the authors involved in these two Bulletins to have completed their arduous tasks of summarizing the achievements of the old quantum theory just at the time when it was being replaced, these works were of considerable value in the years immediately ahead, serving as transitional presentations until the time when the new quantum mechanics was sufficiently developed to warrant the writing of books entirely from the new perspective. Furthermore, these two Bulletins were not merely surveys of the existing literature but were subjective, creative enterprises on the part of the authors involved giving their perspective on topics that were very much in a state of flux. The fact that American authors were now writing such theoretical presentations represented a marked departure from the preceedings decades.

Another book written by American physicists during this period that was instrumental in the dissemination of quantum physics was The Origin of Spectra by P. D. Foote and F. L. Mohler of the National Bureau of Standards in Washington.[86] In the Preface the authors stated their intention to present the material from an experimental, rather than a mathematical, point of view. This was especially fitting since they themselves had done considerable work in the field of ionization and excitation potentials and had thoroughly familiarized themselves with the work of others. Given the American tradition of reliance on experimental evidence rather than mathematical argument, it seems reasonable to claim that this book carried considerable weight in favor of the acceptance of the Bohr theory of atomic structure, since this was the point of view adopted by the authors as providing, through the energy level concept, a practical way of explaining the observed spectra of very many elements.

One more type of publication serving the physics community in its attempt to keep abreast of rapidly occurring developments needs to be mentioned - Review Articles published in established journals. This kind of publication became increasingly important in later years, and we shall have occasion to discuss specific instances in subsequent chapters.

The earliest works of this genre, at least in American journals, were those of K. K. Darrow, whose series entitled "Contemporary Advances in Physics" appeared in the Bell System Technical Journal from 1922 until 1939.[87] This journal was a quarterly which began publication in July 1922. Almost every issue contained an article written in Darrow's lucid style about some topic of wide interest in the physics community. Of course, not all related articles to quantum theory, but there were four on atomic structure and one on waves and quanta in the period of the "old quantum theory."[88]

In 1926 Darrow published a book Introduction to Contemporary Physics[89] which covered much of the same ground as his articles. This was a book that could be read with profit by students relatively early in their study of physics, thus serving as an introduction and inspiration before they moved on to more mathematically sophisticated and detailed presentations.

The considerable number of publications relating to quantum physics that were written by American authors during the early 1920's for readership at various levels stands as testimony to a high degree of interest in the topic among the scientific community. Although these works later became outdated and retain primarily historical interest now, they were exceedingly important in preparing the American physics community to appreciate and cope with the new quantum mechanics when it came. Our discussion of the dissemination of quantum physics in America through publications has naturally stressed the role of American publications, but it must also be recognized that those American physicists who wished to be really up-to-date on quantum theory as it was developing during this period relied heavily on a new monthly journal of the Deutsche Physikalische Gesellschaft, Zeitschrift für Physik, which began publication in 1920 and was considered the most lively of the European journals in the area of quantum theory.[90]

3) Foreign Lecturers

Each one of the stream of foreign lecturers in physics who came to the United States during the early 1920's stimulated a general awareness of the current liveliness of the discipline but some were especially effective in assisting and encouraging Americans trying to cope with the development of quantum theory. Born,[91] Ehrenfest and Sommerfeld, in particular, stand out in that capacity.

Einstein's visit was primarily non-scientific and his few scientific lectures were confined to relativity theory. Lorentz included only meager discussion of quantum theory within the broad context of all of theoretical physics. J. J. Thomson never mentioned quantum theory at all in five lectures discussing the role of the electron in chemistry.

Bohr's visit in the Fall of 1923 was of relatively short duration and his major presentations at Amherst College and Yale University were aimed at a general rather than scientific audience. Bohr's appearances before scientific audiences were few in number. He gave two lectures at Harvard University on "The Theory of Spectra and Atomic Constitution,"[92] and he addressed a meeting of the American Physical Society in Chicago on "The Quantum Theory of Atoms with Several Electrons" drawing an audience of 350.[93] Bohr seems to have made pleasant contact with a number of individual American scientists, including Langmuir[94] and Lyman,[95] during his visit, but Bohr was generally acknowledged to be a difficult lecturer to listen to. Although his command of the English language was excellent, his accent, coupled with poor enunciation and low volume, made him difficult to understand.[96]

Bohr's visit did include one episode that was to have its effect on the development of quantum study by Americans. During a stop in Ann Arbor Bohr met young David Dennison, then a doctoral candidate writing his thesis under Oskar Klein, a former assistant to Bohr. Bohr was sufficiently impressed with Dennison's work-in-progress that arrangements were undertaken for Dennison to go to Copenhagen for postdoctoral work the following year.[97]

Paul Ehrenfest came to the California Institute of Technology with the title of Research Associate in the Spring of 1924.[98] He travelled widely, however, lecturing at other institutions such as the University of California at Berkeley,[99] the University of Minnesota,[100] Columbia[101] and Harvard[102] Universities, on the topic of Quantum Statistics. His warm personality and words of encouragement are remembered by several American physicists with whom he came in contact.[103] An unusual situation arose during Ehrenfest's visit when he, in collaboration with Paul Epstein, undertook to follow up a theoretical idea for an additional quantum condition that had been proposed by an American experimentalist, William Duane.[104] For European visitors to be impressed on encountering innovative theoretical ideas on this side of the Atlantic certainly represented a break with the recent tradition. Another such instance occurred during Sommerfeld's visit.

Of all the European lecturers in physics who came to the United States in the early 1920's, Arnold Sommerfeld stayed the longest and lectured at the largest number of institutions. In view of the widespread popularity of his Atombau und Spektrallinien, already noted, he was an especially prominent figure in promoting the acceptance of quantum theory before American audiences.

It is important, however, to note, from the perspective of the advent of contributions to quantum theory by Americans and the recognition thereof by Europeans, that one of the most significant aspects of Sommerfeld's visit was his response to the recent work of A. H. Compton.

Sommerfeld met Compton personally during the Christmas holidays of 1922 in Urbana where Sommerfeld had gone to visit his former pupil, Jakob Kunz.[105] A month later, Compton wrote to Sommerfeld saying that he was sending him a copy of his paper "Quantum Theory of X-ray Scattering" which he had recently submitted to the Physical Review.[106] Compton added:[107]

"I should be very glad indeed if you care to publish a discussion of the problem of the scattering of X-rays. I feel that we are here comparatively well informed concerning the experimental status of the problem, but I myself am not sufficiently familiar with the developments of the quantum theory to feel that I can apply it to the best advantage. I should, therefore, be only too glad if someone like yourself would contribute to the discussion of this important subject."

Apparently Sommerfeld was very much interested in the results Compton was obtaining. He recognized their importance for quantum theory and was soon[108] convinced of their validity.

Sommerfeld began lecturing on Compton's work while still in America. R. T. Birge has related[109] that Sommerfeld delivered eight lectures at the University of California at Berkeley in February of 1923 on the general topic "Atomic Structure and Radiation;"

"When he arrived Compton had just discovered the "Compton Effect" ... Hence there was, at the time, an animated discussion as to whether light ... was a particle or a wave. As soon as Sommerfeld arrived we therefore asked him that question, to which he replied, 'you will learn that this afternoon.' Then, at his first lecture, that afternoon, he derived the now well-known formula for the Compton Effect, treating light (radiation) as a particle."

Upon returning to Europe, Sommerfeld presented his enthusiastic view of the "Compton Effect" though he related:[110]

"In unserem Münchner Colloquium stiess ich bei W. Wien und Zenneck mit meiner Auffassung des Compton-Effektes auf stärksten Widerspruch."

Further evidence of Sommerfeld's role in publicizing Compton's work in Europe is contained in the following recollections of Otto Oldenberg:[111]

"The ... conversation took place at Göttingen, where I was assistant at Franck's laboratory, at 1923 or 24. Sommerfeld, coming back from USA, visited Franck and Born. Reporting on American physics, he emphasized the great importance of Compton's discovery which assigned momenta as well as energies to the discrete quanta."

When the next edition of <u>Atombau</u> came out in October of 1924 Compton's work rated ten pages of discussion. It is curious to note that this was at the time when Compton's results were not universally accepted in America, owing to the opposition of William Duane who had not been able to reproduce Compton's findings in his own laboratory.[112]

A number of facets of Sommerfeld's visit indicate the importance of his trip for the dissemination of quantum physics in America. The total time he spent in the United States was more than six months. The number and geographical diversity of the institutions at which he lectured was significant - more than half a dozen from coast to coast. And the level of his presentation was aimed at advanced graduate students and faculty. Incidentally, his visit was also responsible for a young American's subsequently going to Europe and writing his doctoral thesis under Sommerfeld's direction. Victor Guilemin, an undergraduate at the University of Wisconsin when Sommerfeld came there, has recently recalled:[113]

"I was fortunate in making his (Sommerfeld's) personal acquaintance during that year (1923) and I became aware of his outstanding capabilities as both a great scientist and a great teacher. The Institut für Theoretische Physik, under Sommerfeld's guidance, attracted young physicists from all over the world. In 1925-26, Heisenberg was a student at the Institut, and many of the men who contributed actively to the development of quantum mechanics were frequent visitors. Consequently quantum mechanics has been, to me, not something I read about; I was 'there' when it was born."

But, from the point of view of international recognition of American achievement in quantum physics, Sommerfeld's publicizing of Compton's work stands out as a high point.

4) Curriculum Innovation

Although the research in physics carried on at American universities continued to be predominantly experimental during the early 1920's, there is considerable evidence that theoretical training, especially in quantum theory, was becoming recognized as an essential part of the graduate curriculum. On the next page is a chart summarizing the offerings relevant to quantum theory that were available at a number of institutions and the names of the faculty members responsible for them.

COURSES RELATING TO QUANTUM PHYSICS AT CERTAIN AMERICAN UNIVERSITIES ca. 1923 - 1925

Institution	Lecturer	Topics Listed, or Course Titles
Univ. of Calif. Berkeley	R.T. Birge W.H.Williams	Radiation and Atomic Structure. Quantum Theory
Cal. Inst. Tech.	P. Epstein R.C. Tolman	Heat Radiation and Quantum Theory; Physical Optics and the Quantum Theory of Spectral Lines.
Chicago	A.H. Compton F.C. Hoyt (summer 1925)	X-Rays and Quantum Theory; Dynamics and Quantum Theory.
Cornell	E.H. Kennard	Kinetic Theory of Matter and Quantum Theory.
Harvard	E.C. Kemble G.W. Pierce J.C. Slater	Quantum Theory and Radiation, Series Spectra, Atomic Structure, Kinetic Theory of Gases and Infra-red Spectra.
Mass. Inst. Tech.	H.B. Phillips*	Quantum Theory.
Michigan	W.F. Colby O. Klein E.F. Barker	Quantum Mechanics**; Atomic Structure.
Minnesota	G. Breit J.H. Van Vleck	Radiation, Atomic Structure.
Princeton	E.P. Adams K.T. Compton	Statistical Mechanics, Kinetic Theory of Gases and Quantum Theory; Atomic Structure.
Stanford	D.L. Webster	Modern Physics.
Yale	L. Page	Radiation and Atomic Structure.

COMMENT: The information tabulated in this chart was obtained from the annual catalogues of the individual institutions.

* Phillips was a member of the mathematics department who undertook to teach this course in the absence of a suitable member of the physics department.

** Note the pre-1926 use of the term quantum <u>mechanics</u>.

While it is not claimed that this listing is all-inclusive (nor that the quality of the courses was uniform among the institutions listed), the absence of some institutions such as Columbia University and the Johns Hopkins University is noteworthy. Both of these institutions had faculty members capable of giving theoretical courses of a classical nature, but turned to outside help to cope with quantum physics.[114]

At Columbia University, some attempt was made to alleviate the lack of instruction in the new physics by means of visitors to the summer session. For example, in the years 1922-26 Columbia had the following summer visiting lecturers:[115]

1922 W.F.G. Swann, of the University of Minnesota, who lectured on "Radiation and Atomic Structure."

1923 P.D. Foote, of the National Bureau of Standards, who lectured on "Spectroscopy and Radiation" and "Atomic Structure in Modern Physics."

1924 A.L. Hughes, of Washington University, St. Louis, who lectured on "The Bohr Theory" and "Phenomena Relating to Ions, Electrons and Radiation."

1925 G.N. Lewis, of the University of California at Berkeley, who lectured on "Molecular and Atomic Theory" and "Thermodynamics."

1926 L. Page, of Yale University, who lectured on "Quantum Theory."

The idea of using the summer session as an opportunity to attract superior lecturers who would enrich the regular offerings of the physics department was popular with many institutions. The University of Chicago and the University of California at Berkeley had well established summer sessions during this period. The University of Michigan Summer School, which later became of great importance to American theoretical physics (see Chapters 3 and 4), began having guest lecturers in physics in 1923. K. T. Compton from Princeton University and F. A. Saunders from Harvard University were the first such visitors.[116]

While the presence of illustrious guest lecturers can be stimulating to both students and regular faculty, an institution which hopes to develop a strong program in theoretical physics must have permanent faculty members who can develop a suitable sequence of courses and who can supervise students desiring to do theoretical thesis research.

An outstanding example of a physics department which took strong steps to improve its theoretical competence was the University of Minnesota: In 1923 Minnesota succeeded in attracting two young theoreticians to its physics faculty as assistant professors, rather than at the more lowly rank of instructor, and most importantly, with responsibility for graduate instruction only. One was John H. Van Vleck who while a student of E. C. Kemble at Harvard University, had, just the year before, completed the first American Ph.D. thesis that was entirely theoretical and dealt with quantum theory. The second was Gregory Breit, who had taken his training through the doctorate at the Johns Hopkins University in 1921.[117] He was then awarded a National Research Fellowship for two years which enabled him to go to Leiden (1921-22), where he worked with Ehrenfest, and to Harvard University (1922-23). During his year at Harvard, Breit published four papers dealing with quantum theory.[118]

With the presence of these two "young bloods" Minnesota was able to expand its offerings in theoretical physics greatly beyond their previous intermediate level "Theoretical Physics" given by John Tate. Graduate level courses then became available in advanced analytical mechanics, the theory of elasticity, hydrodynamics and kinetic theory, thermodynamics and kinetic theory, the theory of atomic structure, the mathematical theory of electricity and magnetism, electron theory and the theory of relativity.[119]

Unfortunately, Breit stayed only one year at Minnesota before returning east as a mathematical physicist at the Carnegie Institution in Washington. Van Vleck remained on the faculty at Minnesota for five years lecturing and supervising the theses of several graduate students during the difficult years of transition from the old quantum theory to the new quantum mechanics.

5) <u>Faculty Seminars</u>

Although the number of American physics faculty members prepared to teach courses in quantum physics was small during this period -- rarely more than one per institution -- many other faculty members were sufficiently interested in trying to keep abreast of the new developments to become participants in locally organized faculty seminars.

One such group met in Cambridge with E. C. Kemble as the guiding figure. The group, which held bi-weekly evening meetings in the years 1923 through 1925, came principally from Harvard and the Massachusetts Institute of Technology. In addition to physicists of a practical as well as theoretical bent, the group contained several mathematicians and at least one astronomer and one biophysicist.[120]

Two such groups met at Princeton University. The activity of one centered around a study of the current issues of the Zeitschrift für Physik and was called by some the "Princeton'sche Physikalische Gesellschaft." The second was organized by H. N. Russell who at the time was working on the details of the Russell-Saunders coupling.[121]

This kind of activity was especially important during this period for American physicists who were laboring under the dual handicap of geographical isolation from the mainstream of physics in Europe and from the lack of an established theoretical tradition in America.

Summary:

By 1925 there were very few members of the physics profession in America who entertained any serious doubts about quantum physics becoming a permanent and important part of the discipline, despite the persistent difficulties associated with quantum theory at that time in terms of its own internal problems and its continued lack of compatibility with classical physics. Furthermore, in contrast with the situation some ten years earlier when America's newly developing, though largely skeptical, interest in quantum theory was abruptly interrupted by the World War, the social context for the study of physics in general in America was enjoying a half decade of remarkable stability. There was time available and channels open for becoming informed about the new physics. In addition, there were individuals who were not content merely to absorb the new developments being made by others, but themselves actively sought to participate in solving the problems associated with quantum physics at that time. Our attention will now focus on these individuals, their efforts and their contributions.

Part III: American Participation in Quantum Theory Development.

During the early 1920's there was, for the first time in the history of the development of physics in America, a number of physicists seriously engaged in attacking problems of a theoretical nature. Although the circumstances in which they had to labor were far less desirable than those enjoyed by their European counterparts, these American physicists did produce a respectable amount of what were, at the time at least, significant results relevant to quantum theory. The fact that so much of what was produced in theoretical areas by all physicists during this half decade was so soon replaced should not be allowed to diminish an appreciation of the significance of this emerging theoretical activity in America for the eventual "coming of age" of physics in America. Without the background provided by this group of pioneers the next phase of American activity in physics, that associated with quantum mechanics, would not have been possible. Furthermore, from the perspective of the international development of quantum physics there were some American contributions which were influential in charting the course of the transition into quantum mechanics and, in a few exceptional cases, some attained a viability which survived "the revolution."

Problems associated with the old quantum theory attracted not only the young aspiring theoreticians in America, but also some of the older and experimentally oriented members of the profession. A broad survey of the journals commonly used by the American physics profession at the time[122] reveals that there were about four dozen American physicists who published one or more papers relating to quantum theory. About 25 per cent of these authors can be considered as having produced substantial results. In the discussion which follows we shall bring out not only the identities of these authors and the nature of their work, but also the circumstances under which their work was undertaken.

Some Circumstantial Difficulties

Wholly aside form the internal problems associated with quantum theory, the theoretical activity that was carried on by American physicists in the early 1920's was burdened by a number of difficulties peculiar to the American scene.

First of all, there was the isolation from the European maninstream of activity, mitigated only occasionally by the possibility of European travel or by visits from European theoreticians touring in this country. Not only was there the obvious lack of personal communication with active leaders in the field, but also the distance between Europe and America, especially the West Coast, meant an often significant delay of about six weeks in the arrival of the latest Zeitschrift für Physik, for example. This fact was commented upon by Gerhard Dieke in a letter to S. A. Goudsmit, written from Berkeley, as follows:[123]

"All of you are still much better off in Europe because all the new work appears there a few weeks sooner."

In addition, the wide geographical dispersal within the American physics community itself meant that the density of theoreticians was very small. Only in rare instances did American theorists, trying to cope with quantum theory, consistently enjoy the stimulation of informally exchanging ideas with kindred spirits.[124] The low density of Americans who were significantly productive in quantum theory made it virtually impossible to attain a state corresponding to a "critical mass" which would constitute a "school of theoretical physics" in America.

America had no real theoretical center comparable to those in Europe in those days, but the Harvard University Physics Department was the closest approach to one. In the eyes of one European physicist, it was a "Schauplatz der Quantentheorie," the only American institution to be listed as such among such European centers as Berlin, Cambridge (England), Copenhagen, Göttingen, Leiden and Utrecht, Munich and Zurich. According to Friedrich Hund[125]:

"Zur Harvard-Universität gehörte in den zwanziger Jahren eine bedeutende spektroskopische Forschungsstätte, deren Ruf vor allem Th. Lyman begründet hatte. H. N. Russell und F. A. Saunders fanden dort das Vektormodell des Atoms mit zwei äusseren Elektronen. Etwa gleichzeitig kam W. Duane der Wellenmechanik recht nahe. Als angehende Physiker lernten R. S. Mulliken, J. C. Slater und J. H. van Vleck die Quantentheorie bei E. C. Kemble; Mulliken wandte sie bald auf die Molekeln, Slater auf den festen Körper an, und van Vleck behandelte mit ihr grundsätzliche Fragen."

Possible Routes into Active Participation: 1) European Study

The situation that prevailed in America in the early 1920's with regard to theoretical physics made it difficult for a young person attracted to that field to make a start in it. A few did it simply by going off to one of the European centers. Gregory Breit and Frank Hoyt began publishing papers in quantum theory during their stays in Leiden and Copenhagen respectively on National Research Council fellowships before those awards were limited to tenure at American institutions.[126]

2) Individual Self-instruction

At least one young American physicist entered the world of theoretical physics through self study which proved fruitful in a remarkably short time. F. Wheeler Loomis had received his Ph.D. from Harvard University in 1917 for an experimental thesis on the specific heat of mercury written under the guidance of H. N. Davis. Loomis recently recalled that his shift from thermodynamics to quantum theory came about in the following way:[127]

"I did not like or respect the kind of work that went on there [at Westinghouse] so I found another job as professor of physics at uptown New York University. I spent the summer of 1920 alone in Cambridge which, because of summer vacation, was practically empty of physicists. So I had no one to talk to about my ambition to catch up with the current physics. I remember that I wasted a lot of time on a very bad book on band spectra by an Englishman whose name I have forgotten. While trying to understand band spectra I recalled hearing a paper at a Physical Society meeting in which weak lines were reported which did not fit into the then current theory. This was just the time when the isotopes of chlorine had been discovered. It occurred to me that the weak lines might be due to the molecule HCl^{37}. Since none of my friends who had worked with band spectra were in town I promptly published my conclusion in Nature and later in the Astrophysical Journal. When term reopened in September the only person I found to discuss it with was Kemble."

Loomis' initial venture into quantum theory was sufficiently noteworthy to warrant citation by Sommerfeld in Atombau und Spektrallinien.[128] The fact that Lommis' publication in this instance predated a similar one by A. Kratzer,[129] a European physicist well established in the study of band spectra, is also interesting to note. More often than not it was the American who came in second in such publication priority situations.

3) Doctoral Dissertation

Despite the scarcity of established figures in the field of quantum theory at American institutions under whom they might study, more than a dozen young men, between 1922 and 1926, wrote doctoral dissertations that related to contemporary quantum theory.

On the next two pages is a chart giving information on fourteen theses dealing with the old quantum theory that were written by Americans during this period.[130]

These theses represented a very small proportion of the more than three hundred physics theses written in America during those years, but the fact that the physics departments of American universities were willing to award doctoral degrees for theoretical work represents a noteworthy departure from the attitude of the previous decades.

The theses themselves involved varying levels of sophistication and some of the young authors listed did not continue to be active in quantum theory in their subsequent careers. The most notable examples of those who did go on into quantum mechanics were E.U. Condon, D.M. Dennison, G.H. Dieke and J.H. Van Vleck.

Another interesting observation that can be made concerns the variety of backgrounds of the men who supervised these dissertations. Birge and Kemble, we recall from Chapter 1, were early advocates of the study of quantum theory. Much of their students' work at this time related to their own study of band spectra. Van Vleck, Kemble's first doctoral student, went on to supervise his own first doctoral student, Hutchisson four years later. The two Comptons, although primarily

American Doctoral Dissertations in Physics

Related to the old Quantum Theory 1922-1926

Name Year	Institution Supervisor	Subject Related Publication
J. H. Van Vleck 1922	Harvard E. C. Kemble	Crossed Orbit Helium Model Phil. Mag. 44 (1922) 842-869
L. A. Turner 1923	Princeton K. T. Compton	Sizes of Alkali Metal Atoms J. Astrophys. 58 (1923) 176-194
D. M. Dennison 1924	Michigan O. Klein and W. Colby	Molecular Structure of Methane J. Astrophys. 62 (1925) 84-103
L. B. Ham 1924	Illinois J. Kunz	Atomic Mass Variations Phys. Rev. 25 (1925) 762-767
R. B. Lindsay 1924	Mass. Inst. Tech. H. B. Phillips	Penetrating Atomic Orbits Jour. Math. Phys. M.I.T. 3 (1924) 191-236
M. S. Vallarta 1924	Mass. Inst. Tech. H. B. Phillips, J. Lipka and N. Wiener	General Relativity and Calculus of Perturbations Applied to Bohr's Atomic Model Jour. Math. Phys. M.I.T. 4 (1925) 65-83

Name / Year	Institution / Adviser	Thesis / Publication
C. M. Blackburn 1925	Chicago H. G. Lemon	Band Spectra Analysis Proc. Nat. Acad. Sci. 11 (1925) 28-34 J. Astrophys. 62 (1925) 61-64
F. W. Bubb 1925	Chicago A. H. Compton	Vector Quantum Theory Phys. Rev. 24 (1924) 177-189 Phil. Mag. 49 (1925) 824-838
D. G. Bourgin 1926	Harvard E. C. Kemble	HCl Band Line Intensities Phys. Rev. 29 (1927) 794-816
E. U. Condon 1926	California-Berkeley R. T. Birge	Intensity Distribution-Band Spectra Phys. Rev. 28 (1926) 1182-1201
G. H. Dieke 1926	California-Berkeley R. T. Birge	Ultra-violet Spectrum of H_2 Zeit. f. Phys. 40 (1927) 299-308
E. Hutchisson 1926	Minnesota J. H. Van Vleck	Specific Heat and Crossed Orbit Model of Hydrogen Phys. Rev. 28 (1926) 1022-1029; 29 (1927) 270-284
W. C. Pomeroy 1926	California-Berkeley R. T. Birge	Band Spectrum of Al_2O_3 Phys. Rev. 29 (1927) 59-78
J. D. Shea 1926	California-Berkeley R. T. Birge	Structure of Swan Bands Phys. Rev. 27 (1926) 245A

experimentalists themselves, were very much alert to the development of modern physics. Phillips, himself a mathematician, was eager to assist physics students at the Massachusetts Institute of Technology who wanted to work in quantum theory. In a sense, these few individuals laid the groundwork for the rise of theoretical physics that was to take place during the next decade.

Aside from the question of the availability of suitable thesis advisors, there were other conditions associated with the American educational scene which made it less conducive to the development of young theorists than was the case at European universities.

One practical disadvantage for a young potential American theorist, as contrasted with his European counterpart, lay in the American educational system which puts four years of college level study between high school and university graduate level training. A young person going from a gymnasium to a university in Europe came in contact with the frontiers of theoretical physics significantly sooner than would have been possible in America. For example, in 1919 when S. A. Goudsmit was 17 years old, he began studying with Ehrenfest and published his first research paper the following year.[131] At the age of 19 Werner Heisenberg was accepted as a student by Sommerfeld.[132]

Furthermore, the physics curriculum at most American institutions tended to reinforce the delay by an apparent insistence that the older, established classical physics be properly mastered in order to qualify for a bachelor's degree and to pass comprehensive examinations preliminary to serious doctoral work. This was not the case in Europe, and some physicists educated there ended their training with significant gaps in their own range of knowledge in physics.[133] Of course, European university education proceeded on the assumption that the individual could, on his own, master particular topics for which he might later feel the need. It is somewhat surprising, nevertheless, to read[134] that Gerhard Dieke who had come from the University of Leiden to be an assistant in the physics department of the University of California feared that he did not know the material on which he was supposed to be assisting students.

Finally, the weakness in the mathematical training of most American students of physics, which we discussed in our previous chapter, had not been overcome in the early 1920's. J. J. Thomson commented upon this situation after his visit in 1923 and added a further interesting criticism of American universitites as follows:[135]

"The university courses are governed by the theory that there is something undemocratic in distinguishing between the training of an honours and a pass man and an athlete. It has been tried but it is not popular. It is contrary, they say, to democratic principles to make these distinctions. Democratic principles are mysterious!! I found that, though it is not democratic to distinguish between the training of a good mathematician and a poor one, yet no one thinks it undemocratic to distinguish at the university between the training of a good football player and a bad one."

European Experiences of Urey, Lindsay, Dennison and Slater

Some of the young Americans who went to Europe to study quantum theory have expressed the opinion that they were not adequately prepared for the sophisticated theoretical environment in which they found themselves. One of these was Harold C. Urey who was primarily a chemist by previous training but who had written a thesis at Berkeley on the distribution of electrons in various orbits of the hydrogen atom.[136] While at Copenhagen on an American-Scandinavian Fellowship Urey worked successfully on the problem of the effect on the hydrogen spectrum of crossed electric and magnetic fields,[137] but he has said recently:[138]

"As a matter of fact one of the things that I really learned at Copenhagen was that I did not have the mathematical equipment and the mathematical ability to be an effective theoretical person. This is the reason I stopped."

R. B. Lindsay was another American-Scandinavian Fellow who felt that, despite his previous study of the latest edition of Sommerfeld's Atombau, he was "certainly pretty ignorant upon arrival in Copenhagen." Lindsay than continues:[139]

"When he found the depth of my ignorance, Professor Bohr suggested what he thought was the kind of problem that I might successfully work on. That was an attempt to do something with polyelectronic atoms in a semi-quantitative way."

During his year in Copenhagen, in addition to collaborating with his wife in translating a book by Kramers and Holst into English[140] and attending lectures on theoretical physics by Kramers, Lindsay was able to make very good progress in calculating the details of the elliptic orbits of the outer electrons of alkali atoms. In particular, he arrived at

"the concept of self-consistency, which, so far as I was aware, had not been suggested or used by any theoretical atomic investigator at that time"

and

"ultimately D. R. Hartree applied the scheme to the quantum-mechanical evaluation of atomic fields under the title 'self-consistent field' method."[141]

When Lindsay's year as an American-Scandinavian Fellow came to an end and a teaching position awaited him at Yale University, his work had not yet reached a decisive stage. It was possible, however, for him to complete sufficient additional material after his return to the United States to have his work qualify, with Bohr's approval, as a doctoral thesis at the Massachusetts Institute of Technology in 1924.[142] For a while after this Lindsay continued to work on the problem of penetrating orbits but, after the advent of quantum mechanics, his interests turned increasingly to acoustics and considerations of the history and philosophy of physics.

One-year stays at European theoretical centers, such as Urey and Lindsay had as American-Scandinavian Fellows, could scarcely have been expected to provide a satisfactory route into quantum theory on a permanent professional level for young Americans, especially those who had not already benefited from the best theoretical training available in America at that time. David Dennison's theoretical career developed much more satisfactorily when, after having completed his doctoral dissertation at the University of Michigan, he was able to spend three years studying at European theoretical centers as an International Education Board and University of Michigan Fellow.

John C. Slater's European experience with quantum theory differed from all of those we have thus far discussed, and contained its own particular disappointment and frustration. Slater had written an experimental thesis under P. W. Bridgman at Harvard University,[143] but, since he was already very much interested in quantum theory when he began his graduate studies in 1920, he took the quantum theory course that E. C. Kemble was giving at that time.[144] Upon completion of his degree Slater was given a Sheldon Travelling Fellowship by Harvard University which enabled him to spend the academic year 1923-24 at Cambridge and Copenhagen.

By the time he arrived in Copenhagen Slater had partially worked out a theory of radiation of his own which involved the concept of the atom behaving as a virtual oscillator which sets up a radiation field in advance of the discontinuous radiation of energy quanta. The fact that a young American arrived in Copenhagen with a provocative idea was itself unusual. Both Bohr and Kramers looked favorably on Slater's idea of virtual oscillators but not on the idea of discrete quanta. In the published version of the now well-known joint paper "The Quantum Theory of Radiation" by N. Bohr, H. A. Kramers and J. C. Slater,[145] quanta were omitted and the conservation of energy and momentum were regarded as only statistically valid.

Within the following year, however, the reality of quanta and the strict validity of energy and momentum conservation were experimentally confirmed through the work of A. H. Compton, A. W. Simon, W. Bothe and H. Geiger. At that time Slater was able to demonstrate that his original conception was still valid. In fact, it has remained so, surviving the quantum-mechanical revolution as few other ideas have, in addition to having played a role in the conceptual development of the new quantum mechanics.[147]

Instances of Lack of Recognition and Duplication of Ideas

1) The Helium Atom

During these years of the early 1920's when American physicists were beginning to publish papers on various aspects of quantum theory it was inevitable that, in such an active field, there would arise instances where the same idea would occur independently on both sides of the Atlantic. For example, E. C. Kemble[148] and Niels Bohr[149] both proposed a crossed orbit model of the helium atom which was sometimes referred to as the Bohr-Kemble" model.[150]

Much more disappointing were the instances when American productivity was entirely overlooked. In some cases, such as Lindsay's analysis of penetrating orbits, this might be ascribed to the fact, mentioned earlier, that American journals of physics were not widely read in Europe at that time.[151] On the other hand, that could not be given as the excuse for the oversights of Van Vleck's analysis of the crossed orbit helium atom model that was published in the _Philosophical Magazine_ and yet ignored by Bohr in the reference cited in Note 2.150 where Bohr states "Kemble has not, however, investigated this notion further." Kemble did not, but his graduate student Van Vleck did.

Max Born was another European author who failed to give proper attention to Kemble's and Van Vleck's work on helium in his book _Vorlesungen über Atommechanik_.[152] Van Vleck called this to Born's attention in an exchange of correspondence that took place during Born's visit to America in 1926. Born responded cordially to the young Van Vleck, who was than at the University of Minnesota, apologizing for his oversight and saying that he looked forward to meeting Van Vleck when he travelled west.[153]

The search for a satisfactory model for the helium atom was a task which occupied a number of scientists during the early 1920's and it turned out to be a most troublesome one. Indeed, the failure of the old quantum theory to provide a solution to this problem (i.e. a model which gave a theoretically calculated value for the ionization potential which agreed with the experimentally known value) can be considered one of the most compelling reasons to seek a new quantum theory.[154]

Until that time arrived, however, new models continued to be proposed by scientists in Europe and America. It is interesting to note that Sommerfeld chose to publish his version in the Journal of the Optical Society of America.[155]

Irving Langmuir was another American scientist who tackled the problem of helium's atomic structure.[156] Langmuir's proposals appeared in 1921 and, of course, failed. His attempts hold interest, however, in that he did impose quantum conditions on the electrons - one of the very few times[157] Langmuir ever used quantum theory in any way - and in that he clearly was considering a dynamic, Bohr-type atom, rather than a static one.[158]

Now, half a century later, one might wonder a bit at the large number of publications setting forth models of the helium atom that were obviously unsatisfactory, even at that time. From the contemporary point of view, however, it is clear that it was important to make known which ideas had been tried and the extent to which they failed.

At the time these publications appeared the scientific world was committed to the concept of a clearly definable physical model, involving electrons occupying specific positions which varied with time along well defined orbits.

2) Electron Spin

Another topic which arose within the context of the old quantum theory and about which some Americans arrived at their own conclusions, quite independently of European acitivty on the topic, was electron spin. The development of this concept and its integration into the body of scientific understanding provides an excellent illustration of some of the criteria that must be met before an idea can become a fruitful one. Among these criteria are: the idea must be adequately publicized in appreciative circles and its timing must be such that its usefulness becomes apparent within a short time.[159]

From the perspective of these criteria there is no doubt that the introduction of electron spin by Goudsmit and Uhlenbeck in 1925 was the fruitful one. From our perspective, however, of examining the American scientific environment and productivity of ideas during the early 1920's it will be worthwhile now to describe some of the instances when American physicists suggested that electrons spin. It will become clear in each case why their efforts failed to stimulate significant response within the international scientific community.

In 1921 A. H. Compton published a paper entitled "The Magnetic Electron"[160] in which he suggested that "the electron itself, spinning like a tiny gyroscope, is probably the ultimate magnetic particle."

Compton's conclusion was based on crystal studies conducted by himself and O. W. Rognley[161] and others. In addition, he believed his idea was applicable to the case of free electrons (beta rays) on the basis of their observed helical tracks.[162]

Compton's suggestion seems to have elicited little response at home or abroad[163] and his own attention at that time was becoming increasingly focussed on his X-Ray studies. In the meantime, E. H. Kennard of Cornell University, made the suggestion that a spinning electron could be invoked to account for ferromagnetism, in a paper delivered at the meeting of the American Physical Society held in Toronto in December of 1921. An abstract of this paper was published[164] but Kennard's idea was pursued no further, at least in print, by himself or others.

The idea of associating spin with electron motion as a possible interpretation of the fourth quantum number in the Pauli Exclusion Principle occurred, in early 1925, to Ralph Kronig soon after he arrived in Europe, following the completion of his doctoral studies at Columbia University. When Pauli was told of this he failed to regard it as anything more than "ein ganz witziger Einfall" having no connection with reality.[165] Furthermore, Kronig received no encouragement to pursue the idea when he discussed it with Bohr, Kramers and Heisenberg. Admittedly there were serious difficulties associated with implementing the idea at that time, and Kronig was not only discouraged from publishing anything on this topic but also became sufficiently convinced that the idea must be wrong that he wrote a Letter to the Editor of Nature[166] in which he opposed the electron spin concept after it was advanced by Goudsmit and Uhlenbeck.

In December of 1925 F. R. Bichowsky, a physical chemist at the Johns Hopkins University, and H.C. Urey, responding to the original Uhlenbeck and Goudsmit note,[167] submitted for publication a paper entitled "A Possible Explanation of the Relativity Doublets and Anomalous Zeeman Effect by Means of a Magnetic Electron" which contianed the following remarks:[168]

"This idea [electron spin] had also occurred to us quite independently for largely the same reasons as those given by these authors. We have, however, carried the idea somewhat further than the authors have reported in their brief note."

It is obvious that the topic of electron spin was a very lively one in late 1925. J. C. Slater was another American physicist who, at that time, was involved in considerations quite similar to, but independent of, those of Goudsmit and Uhlenbeck. Slater's work in this area was published and soon after cited by Goudsmit and Uhlenbeck.[169]

The recounting of the American activity relative to spinning electrons set forth here must not be interpreted as detracting in any way from the value and importance of the work of Goudsmit and Uhlenbeck relative to this topic. Rather, it is our intention to use this discussion to point up the characteristics of American activity in quantum physics at that time. Since there happened to be so many Americans involved, it serves that purpose admirably. In short, the discussion of this topic delineates both the strengths and weaknesses of the American scientific enterprise as it was evolving within the total setting of quantum theory.

In particular, we may note as strong assets the facts that growing numbers of young Americans were not only quite up-to-date on the current topics of interest in the world of physics, but also were thinking creatively about them. Furthermore, European physicists were beginning to take notice of this activity across the Atlantic.

Meanwhile, the geographic isolation, domestic and international, in which American physicists operated probably had a dampening effect. In the cases of Compton and Kennard, there was no exchange or follow-up of their ideas among themselves or others. Perhaps, if they had been closer to the mainstream of activity in theoretical physics, their ideas might have had greater impact. But this is by no means certain, for in the case of Kronig we have another bright young American arriving in Europe with a potentially good idea, only to find the leading European physicists at that time not ready to appreciate its significance.

On the other hand, by 1925, Americans were beginning to come closer to normal participation in an active scientific field in that they were experiencing "near misses" in priority of publication. This situation also prevailed during the years following the advent of quantum mechanics and we shall have more to say about it in our next chapter.

<u>American Contributions Gaining International Recognition</u>

Now, having thus far concentrated principally on the difficulties Americans experienced during the early 1920's in attaining a fully recognized, participating status within the worldwide physics community, we turn to an examination of clearly recognized American successes in the area of quantum physics. These success fall into three categories: contributions in the realm of theory made by members of the small, but growing, number of American theoreticians; contributions in the realm of theory made by American physicists who were primarily experimentalists; and, finally, contributions in the realm of experiment, the traditional stronghold of American physics, which had particular relevance to the development of quantum theory.

The cases which will be discussed below have been chosen to provide examples which illustrate our theme of the progress that took place during the early 1920's toward intellectual coming of age for physics in America. The fact that, for each of the three categories listed above, we provide only two or three examples does not mean that there were no others; rather, it signifies that, in this presentation, we make no attempt to be exhaustive.

1) Band Spectra Studies: Mulliken and Condon

Of all the possible aspects of quantum physics the one that occupied the attention of the largest number of American physicists during the early 1920's was that of molecular structure of gases as evidenced by their band spectra. This is not surprising since there were several American laboratories established in previous decades where spectroscopic studies of very high caliber were carried out. At these sites established techniques were adapted to the study of the absorption spectrum of numerous gases with particular emphasis on the characteristic separation of lines and relative intensity distribution within the observed bands.

We have already discussed the early work of E. C. Kemble in providing theoretical explanation for such data by means of quantum theory. During the early 1920's Kemble continued his own work on interpreting the details of the HCl spectrum as far as was possible within the idiom of the old quantum theory. In addition, and perhaps more importantly, Kemble exerted considerable influence on several members of the new generation of American theoreticians. We have already mentioned some of them, but, in connection with the topic of molecular theory, the name of Robert S. Mulliken, who came to the Harvard Physics Department as a National Research Fellow in 1923, stands out.

Mulliken has recently recalled his early studies at Harvard as follows:[170]

"These people [the members of the Harvard Physics Department], especially Saunders and Kemble, were exceedingly kind and helpful to the ignorant novice in spectroscopy and quantum theory. (I had first heard about quantum theory about 1920 while I was a graduate student at Chicago, in lectures by a great enthusiast, Professor R. A. Millikan; but at the time the subject seemed to me a disorganized chaos.)"

Mulliken then goes on to describe his studies of CN and NO molecules in collaboration with other National Research Fellows, S. K. Allison, F. A. Jenkins and L. A. Turner, all of whom were skilled experimentalists.

During this period, Mulliken's growing interest in the effect of isotopes on the structure of band spectra led him to recognize an error in the British physicist W. Jevons' interpretation of the BN spectrum. An exchange of letters ensued in Nature[171] and the controversy was eventually settled in Mulliken's favor during a visit that he made to England in the summer of 1925.[172] In a later chapter we shall discuss details of further successful studies by Mulliken that were the outgrowth of his early work on molecular structure and for which he was awarded the Nobel Prize in Chemistry in 1966.

Geographically far away from Cambridge, Massachusetts, but closely allied in interests, was a group at the University of California at Berkeley under the leadership of R. T. Birge. We have already mentioned the National Research Council Bulletin, "Molecular Spectra of Gases," on which Kemble and Birge worked so dilligently.[173] Birge was, and still is, a most gracious letter writer; this author can vouch for his kind and helpful response to written inquiries in recent years.[174] R. S. Mulliken, in his article cited above, specifically recalls extensive correspondence that he had with Birge in 1924 which included advice on how to handle Mulliken's controversy with Jevons.

Birge was recognized as one of the few American physicists during the early 1920's who was enthusiastically committed to, and well informed about, quantum theory. After Gerhard Dieke arrived at Berkeley from Leiden in 1925, he wrote back such comments about Birge as:[175]

"Birge is the man who imported the quantum theory here and who is about the only one who has some knowledge of it."

and[176]

"The only one who does know something about modern quantum theory is Birge and he has for the past few years tried to popularize it in which to some extent he has been successful ... Birge is really the only one with whom one can discuss them [topics in modern physics]."

Birge gave courses dealing with quantum theory, and supervised four doctoral dissertations using it during the period before the advent of quantum mechanics. In 1925 Hertha Sponer came to Birge's group as one of the first International Education Board Fellows to come to America. It was in this setting, that E. U. Condon entered the world of theoretical physics as a graduate student at the University of California in Berkeley during the final years of the old quantum theory. The details associated with the writing of Condon's thesis have been set forth by Condon and by Birge[177] and we shall not recount them here. The importance of Condon's thesis from our perspective lies in the fact that a young American graduate student of physics in 1925 was quickly able to grasp and extend a new idea recently suggested by the European physicist, James Franck, so successfully that he has come to share in the honor associated with the discovery of the "Franck-Condon Principle,"[178] a principle which emerged in the context of the old quantum theory but which was later assimilated permanently into quantum mechanics.[179]

2) Van Vleck's Extension of the Correspondence Principle

Finally, we can cite an instance where the work of a young American theoretician not only gained recognition and appreciation by the members of the European scientific community, but also has been credited with playing a role in the evolution of quantum mechanics.[180] One of the guidelines widely used by all quantum theorists in the early 1920's was Bohr's Correspondence Principle. In 1924 J. H. Van Vleck published two papers in which he considered the relationship of this principle to the phenomena of absorption. In particular, Van Vleck showed that[181]

"If we want to estimate the absorption by means of the Principle of Correspondence, we have to compare the absorption, computed classically, with the difference between absorption and induced emission, computed from Einstein's formulae. In the limit of high quantum numbers, this difference must become equal to the classical absorption."

In recent years P. Jordan has described the contribution of this work by Van Vleck to the background from which emerged the quantum mechanical papers that he wrote in collaboration with Born and Heisenberg as follows:[182]

"... Andererseits nahm Van Vleck folgendes Problem in Angriff: nach Einstein erforderte ja das Plancksche Gesetz bestimmte Proportionalitäten zwischen den Prozessen positiver Absorption, negativer Absorption und spontaner Emission. Bohr hatte diese Einsteinschen Ergebnisse lange Zeit recht skeptisch beurteilt, aber nun zeigte Van Vleck dass man auch diese Einsteinschen Gesetze durch verschärfende Anwendung des Korrespondenzprinzips rechtfertigen konnte.

"Heisenberg unternahm nun den Versuch, die Intensitätsgesetze der Balmer-Serie in ihrer exakten Form zu erraten auf Grund von Korrespondenz-Betrachtungen: ... [leading to Heisenberg's first paper in quantum mechanics][183]

"Born und ich waren dafür sehr aufgeschlossen, weil wir zusammen uns gerade damit beschäftigt hatten, die erwähnte amerikanische Untersuchung weiterzuführen;..."

At the time when Van Vleck did this work he was in his first year at the University of Minnesota, where he and Gregory Breit comprised the new graduate faculty of theoretical physics. They must have discussed this work while it was in progress, for Van Vleck included in the first paper the words "As suggested in part to the author by Dr. Breit..."[184]

Wholly aside from the intrinsic significance of these papers by Van Vleck, it is interesting to note that they were published in American journals of physics and yet European physicists did become aware of them.[185] Apparently, by the middle of the 1920's, distinguished members of the European scientific community were coming to realize that enough good theoretical quantum physics was being done in America to warrant reading American journals.

When we come to cite contributions in the realm of quantum theory made by American physicists who were primarily experimentalists, the names of A. H. Compton, William Duane, H. N. Russell and F. A. Saunders stand out. As might be expected, the theoretical papers which they published contained little mathematical sophistication but did provide important conceptual insights. In all cases they sought to give straightforward, physical explanations for clearly recognized experimental results.

3) The Compton Effect

While A. H. Compton's earlier, theoretical postulate of a spinning electron gained only slight, belated attention, quite different were the circumstances associated with his discovery of the "Compton Effect." Response, at home and abroad, came rapidly and, when Compton's results were finally completely validated, their significance for the reality of localized quanta was seen to be profound. We have already mentioned Compton's work in connection with the fate of the Bohr-Kramers-Slater paper (page 2.58).

Without any doubt the discovery of the Compton Effect was of tremendous importance in the conceptual development of quantum theory. When viewed within the context of other developments taking place in physics in those years, it is seen that the discovery of the Compton Effect fits in nicely with the work of L. de Broglie which associated waves with matter, such as electrons. Some authors, in recent years, however, have gone further and ascribed a definite causal role to Compton's work which is not supported by the testimony of de Broglie himself. For example, in a volume entitled The Questioners: Physicists and the Quantum Theory, Barbara L. Cline makes the statement[186]:

"And Louis de Broglie, pondering Compton's findings, was the one who asked the right question: 'Can it be that matter also has a double - a dual - character?'"

When I asked M. de Broglie to comment on this statement, he wrote (in part)[187]:

"La découverte par A. H. Compton de l'effet qui porte son nom m'a été connue très rapidement puisque je travaillais alors au Laboratoire de mon frère qui était spécialiste des rayons X. Mais cette découverte, qui m'a naturellement intéressé parce qu'elle apportait une preuve nouvelle de l'existance des photons, ne m'a aucunement servi dans mes premières recherches sur la Mécanique ondulatoire." (my emphasis)

De Broglie stressed instead, the importance of Einstein's influence in his own conceptual development. (It is curious to note that Compton did not seem to have been aware of Einstein's paper in which he ascribed momentum to photons in 1917.[188])

Before leaving the topic of the discovery of the Compton Effect we should point out to how great an extent Compton, at that time, worked in isolation from the rest of the world of physics. Compton did spend the year 1919-20 at the Cavendish Laboratory as a National Research Fellow, but upon his return to America he specifically chose to accept a position at Washington University in St. Louis where he would be very much on his own.[189] He did, however, serve as chairman of the National Research Council Committee on X-Rays and Radioactivity, and wrote the report "Secondary Radiations Produced by X-Rays" which was issued as National Research Council Bulletin No. 20. It is not clear to which foreign journals Compton had access during his Washington University period; but, as mentioned above, when assigning momentum = h/c to quanta, Compton did not refer to Einstein's 1917 paper in which the same formula appeared. In all likelihood Compton arrived at his formulation entirely on his own from physical consideration of the collision process.

4) Duane's Quantization of Momentum Transfer

In our previous chapter we discussed the experimental work of William Duane which led to the "Duane-Hunt Law," and Duane's seeming reluctance to embrace an explanation for it in terms of quantum theory. By the 1920's Duane was still engaged in X-Ray studies and, as already noted, was opposed to accepting the validity of the Compton Effect. That Duane's reluctance in this case was not based on a continued rejection of quanta will be seen when we now examine a theoretical proposal he made in 1923. By a curious coincidence the names of Duane and Compton became linked not only in the context of Duane's opposition to the Compton Effect but also because Compton was one of the several scientists who took up and extended the idea proposed by Duane which we now describe.

Duane's excursion into theoretical physics was primarily in the form of a paper entitled "The Transfer in Quanta of Radiation Momentum to Matter."[190] He shared the concern, which was widespread in the physics community, for seeking a solution to the apparently contradictory wave and particle aspects of radiation. Duane's proposal involved the drastic one of replacing the whole wave terminology with a new quantum condition. This would in no way conflict with those already in use which quantized changes in energy and orbital angular momentum. In his own words:[190]

"The fundamental hypothesis of the theory now presented is that the momentum of radiation is transferred to and from matter in quanta, and further, that the laws of the conservation of energy and of momentum apply to these transfers."

Using this assumption Duane was able to arrive at Bragg's law of the reflection of x-rays from a crystal and also to explain the action of a diffraction grating.

The quantitative relation between the momentum exchange and the diffracting material, which Duane proposed, was exceedingly simple: The amount by which the momentum changed = h/a, where "h" is Planck's constant and "a" is the grating space. This idea of quantizing exchanges of linear momentum between an x-ray quantum and a crystal, or grating, was certainly an innovative one which did not go unnoticed, although it was far afield from the mainstream of physical thought at the time and from the subsequent development of quantum mechanics.

Duane's idea was taken up and expanded upon by others such as G. Breit,[191] A. H. Compton,[192] as well as by Ehrenfest and Epstein.[193] In his Nobel Prize Address[194] Millikan took note of these papers saying

"Possibly the recent steps taken by Duane, Compton, Epstein and Ehrenfest may ultimately bear fruit in bringing even interference under the control of localized light quanta. But as yet the path is dark."

With the advent of the new quantum mechanics Duane's approach slipped into a kind of limbo from which it was occasionally recalled. Heisenberg mentioned it, in passing, in the lectures he delivered at the University of Chicago in 1929.[195] Later,[196] Heisenberg recalled that Gregor Wentzel was especially fascinated by Duane's proposal and discussed it with himself and Pauli. In Heisenberg's words:

"It was such a funny idea that the periodicity of a grating could be used like a real mechanical periodicity, so that you could apply the integral of pdq. Very nice idea, a very brilliant idea, because it's so against normal possibilities."

It also has recently come to light[197] that in 1925 Born commented to Einstein:

"I am also speculating a little about de Broglie's waves. It seems to me that a connection of a completely formal kind exists between these and that other mystical explanation of reflection, diffraction and interference using 'spatial' quantization which Compton and Duane proposed and which has been more closely studied by Epstein and Ehrenfest."

In the commentary which Born prepared to accompany the publication of his correspondence with Einstein, he remarks:[198]

"The connection I suggested with Duane's and Compton's 'spatial quantization' does indeed exist: de Broglie's spin quantum condition is exactly the same thing, but differently and more intuitively expressed. While Duane speaks of conceptual decomposition of a radiation process into harmonic componenets, de Broglie regards these as real, material waves which are supposed to replace the particles."

In Compton's Nobel Prize Lecture[199] of 1927, he also called attention to the similarity of the proposals made by de Broglie in his thesis at the University of Paris in 1924 and by Duane in the second paper cited in Note 2.190.

Duane's most ardent champion over the years has been Alfred Landé, the more so as Landé has become increasingly disenchanted with what he calls the "Copenhagen language" of a dualistic character of nature. Typical of Landé's recent statements is the following:[200]

"Duane's great achievement met with little response because the photon theory, which does not account for the electromagnetic qualities of radiation, could not even at that time (1923) be taken seriously. Thus, it was soon forgotten, after having been mentioned only once, in Heisenberg's Chicago Lectures of 1931. It was for a long time ignored in the literature in spite of its containing the key to a mechanical particle explanation of matter diffraction. It was only 32 years later that the writer in 1955 transferred Duane's theory from light, where it is out of place, to matter where it yields the clue to a unitary mechanical particle explanation of electron diffraction without appealing to wave interference."

The entire episode of Duane's proposal and its impact on other physicists is a curious one in the history of science: an older,[201] experimental American physicist, using little mathematical sophistication, introduced an idea that stimulated widespread response at the time and has maintained a viability for fifty years; it has neither been absorbed into current theory nor has it been buried and forgotten. Perhaps the fact that Duane was not a member of any theoretical group made possible his unusual approach and the fact that America did not have a strong theoretical tradition made it possible that an experimentalist such as Duane did not feel hesitant to publish a bold idea. It will be remembered from our discussion of Compton's work and attitude that, at times, the American isolation could be turned to advantage. It is interesting to note, further, that both Duane and Compton were willing to accept localized quanta before Bohr, and other Europeans, were ready to do so. And, as noted earlier, Slater was temporarily dissuaded from using localized quanta when he took his ideas to Copenhagen in 1924.

5) Russell - Saunders Coupling

As our third example of a theoretical contribution to quantum physics by Americans who were not theoretical physicists we have the concept of electron "coupling" that was introduced by H. N. Russell and F. A. Saunders.[202] By 1925 most physicists had accepted the conceptual scheme of Bohr and Sommerfeld for an atomic model that would account for the observed spectral lines of various elements. It had been generally assumed that each line corresponded to radiation given off by a single electron in dropping from an excited energy state to a lower state which might be either its normal, unexcited state or another excited state of lower energy than the first. Energy level schemes were devised and selection rules for "allowed" changes in quantum numbers had been discovered.

Saunders was a laboratory spectroscopist at Harvard and Russell was an astrophysicist at Princeton when their joint paper was published. In their analysis of the spectra of calcium, strontium and barium, all of which have two valence electrons, they realized that some lines, hitherto unexplained, could be accounted for on the basis that "both valency electrons may jump at once from outer to inner orbits while the net energy lost is radiated as a single quantum." In the formalism of a vector model for the atom this "Russell-Saunders Coupling" could be interpreted in terms of an atom having "total quantum numbers" that corresponded to various possible combinations of the individual quantum numbers of the two electrons involved.

When Russell and Saunders published their paper in the January 1925 issue of the Astrophysical Journal Heisenberg was preparing a paper which was a generalized discussion of the structure of spectral multiplets and the anomolous Zeeman effect.[203] When submitting it for publication on April 10, 1925 Heisenberg took note of the paper by Russell and Saunders, calling it "eine eben erschiene reichhaltige Arbeit."

It happened that the proposal made by Russell and Saunders was similar to one of three that Heisenberg had developed. We have here not only an example of a very prompt citation of American work in a European journal, but also further evidence that American journals were being read by some European physicists. By the time that Sommerfeld's Atombau reached its fifth edition in 1931, it contained a fourteen page section entitled "Russell-Saunders Coupling."

6) Millikan and Bowen's Study of "Stripped" Atoms

Coming now to our third and last category of contributions to quantum physics that were made by Americans in the early 1920's, we find that two more groups of experiments should be recognized in addition to the experimental areas of X-Ray, line and band spectroscopy, where we have already discussed some examples of American activity.

One particular group of American experiments which attracted European attention were those involving a "hot spark" technique developed by R. A. Millikan and I. S. Bowen.[204] This technique made possible spectral studies of atoms that had been singly or multiply ionized. In some cases relatively heavy atoms were stripped of all but one electron, thus providing interesting tests of the Bohr-Sommerfeld atomic theory. The results obtained by Millikan and Bowen were instrumental in leading to the abandonment of the belief that the fourth quantum number, needed for spectral classification, belonged to the atomic core, thus paving the way for the introduction of electron spin, according to the view of Paul Epstein.[205]

7) Critical Potential Studies: Compton and Mohler

Another experimental field where American studies held considerable interest for the testing of quantum theory was the investigation of electron impact phenomena. Groups at Princeton University and at the National Bureau of Standards were especially active and noted for the reliability of the results they reported. The National Research Council Bulletin No. 48, "Critical Potentials" by K. T. Compton and F. L. Mohler, provided a comprehensive summary of the state of knowledge in that field in 1924. This Bulletin held the rare distinction of being translated into the German language.[206] In addition, ample evidence that American work in this field was widely recognized may be found in the numerous references to American publications in the section "Kritische Potentiale und Spektralterme von Atomen" in the 1926 Handbuch der Physik, Band XXIII, Quanten.[207]

From the discussion of American contributions to the theoretical and experimetal aspects of quantum physics it thus becomes clear that the years 1920-1925 marked a new era for physics in America in terms of the scope of activity and degree of international recognition. In contrast with the activity of previous decades, many American physicists were now actively concerned with problems associated with quantum physics, in its theoretical as well as its experimental aspect, and in several instances their contributions received international recognition.

Summary

By 1925 physics in America was well on its way toward full maturity. Among the assets which can be recognized were:

a well established and continuously growing professional society;

scientific journals carrying the research results obtained by American physicists to an ever widening international audience;

the availability of sizeable funds from private American wealth for the support of promising American post-doctoral students at domestic and foreign institutions, and for the support of foreign physicists at all levels to come to American institutions;

improved channels for the dissemination of latest scientific developments, especially in the area of quantum physics;

the continued excellence of experimental work by American physicists, coupled with the appearance of a group of theoreticians who were beginning to attain professional status at home and to achieve results that were worthy of international recognition;

a larger (even if not yet fully adequate) number of faculty and of students engaged in theoretical pursuits;

the emergence of some "proto-centers" for the study of theoretical physics.

Among the difficulties which still tended to inhibit Americans from functioning on a par with European physicists were:

the inadequate mathematical preparation of many American physicists;

the lack of centers devoted primarily to theoretical physics and under the guidance of established authorities in the field;

the geographic isolation from each other and from the mainstream of physics in Europe;

the small number of well trained theoreticians.

By the end of 1925 there were many intimations of the potential that existed in America for the elimination of the drawbacks and for the continued expansion of the assets. In our next chapter, which deals with the years immediately following the advent of the quantum mechanics, 1926 - 1929, we shall see that developments at home and abroad reinforced each other to such an extent that not only did physics in America reach maturity at home but also America began its rise toward eventual top rank in the world of physics.

Chapter 2: Notes and References

2.1 The term "Fellow" appears for the first time in the Minutes of the Chicago Meeting, Nov. 25 and 26, 1921 in Phys. Rev. 19 (1922) 241.
I have been unable to determine precisely when the action was taken to adopt this term in place of the former designation of "Regular Member," as contrasted with "Associate Member," a distinction established in 1904.
In that year, during the APS Meeting held in New York on February 27, the By-Laws of the Society were amended to permit the election of Associate Members, according to the Minutes of that meeting in Phys. Rev. 18 (1904) 295-296.
It was stipulated that election to either level of membership would be by favorable vote of a majority of the Council of the Society, following an initiative petition for the candidate's entry signed by two members of the society.
According to Articles III of the present Constitution of the Society "Membership" is available to young graduate students, teachers of physics and other interested persons. "Fellowship" may be attained either at initial entry into the Society or through transfer by Council action from the ranks of Membership. In either case, conferring the title "Fellow" implies that the recipient has established a record of contributions to the profession, usually in the form of research publications.
Article IV provides for an election mechanism, similar to that described above, involving Council action on nominations submitted by two Fellows or Members.
Presently the ratio of Fellows to Members is about 1:10. Until about 1917 Regular Members outnumbered Associate Members.
See Bulletin of the American Physical Society 18 (1973) 930, 942.
Appendix IV, Part 3 of this thesis contains a graph of the APS Membership 1900 to 1935.

2.2 The programs of the meetings indicate that the West Coast activity was closely tied to the increasing importance of the University of California at Berkeley, the California Institute of Technology and Stanford University.

2.3 According to Proceedings of the American Physical Society in Phys. Rev. 15-28 (1920-1926)

2.4 See Interview with Harold Webb conducted by Charles Weiner, 24 April 1970 and Autobiographical Notes prepared by Webb in 1962; both on deposit at the Center for the History of Physics, American Institute of Physics.

2.5 Recollection of J.H. Van Vleck, shared with the author in private conversation, January 1973.

2.6 Ernest Merritt "Early Days of the Physical Society," p. 146.

2.7 In this connection, it is of interest to note that the Proceedings of the National Academy of Science were inaugurated specifically as a channel for prompt, brief publication of research results achieved either by members of the Academy or by other Americans whose work some member might deem to be sufficiently important. In addition, the Proceedings were to be actively circulated internationally for the stated purpose of calling attention to American scientific achievement. (E.B. Wilson History of the Proceedings of the National Academy of Science, Washington, D.C.: National Academy of Science, 1966)
With regard to publication difficulties with the Physical Review, in the AHQP interview conducted with Paul Epstein, June 2, 1962, p.3 he voiced an additional criticism of refereeing as follows:
"And they did not judge the paper in the office, but sent it around the country to referees, which is a very doubtful procedure from the point of view of priority, because before they begin to set it, they have already sent it all over the country, and a number of people have already read it. Now the Americans seem to be exceptionally honest, but in Germany it wouldn't have done, it would have appeared somewhere else without the author's name."
Epstein was also outspokenly critical of the Managing Editor's own stylistic requirements, which will be described below.

2.8 Full discussion of this episode will be found in R.H. Stuewer The Compton Effect: Turning Point in Physics (New York: Science History Publications, 1975).
I am indebted to Professor Stuewer for making available some of the chapters of this work while still in manuscript form.

2.9 National Research Council Bulletin No. 20, October 1922.
I have been unable to determine how Debye happened to be aware of this publication. It seems likely, however, that it was through Sommerfeld who was in the United States at the time and met with A.H. Compton late in 1922. (Information supplied by Ulrich Benz who is writing a biography of Sommerfeld as his doctoral thesis at the University of Stuttgart. Mr. Benz has most graciously made available to me some pages from the preliminary version of his thesis which relate to Sommerfeld's

visit to the United States in 1922-23.) Also, Sommerfeld, himself, in his "Autobiographische Skizze" (Gesammelte Schriften, Vol. IV, Braunschweig: Friedr. Vieweg u. Sohn, 1968, p. 678), when discussing the Compton Effect, refers to an earlier (1911) intimation of the duality of light which he and Debye had shared.

2.10 AHQP Interview with Debye, 3 May 1962, cited by Stuewer.

2.11 See, for example, the following AHQP Interviews: Heisenberg, 11 February 1963, p. 5; Uhlenbeck, 9 December 1963, p. 20; and Rabi, 8 December 1963, p. 29.

2.12 J.H. Van Vleck, "American Physics Comes of Age," Physics Today, June 1964, p. 22.

2.13 The data presented here were drawn from the National Research Council Reprint and Circular Series as follows:
No. 12: "Doctorates Conferred in the Sciences in 1920 by American Universities" compiled by the Research Information Service of the National Research Council, Callie Hull, Technical Assistant. (Also published in Science 52 (1920) 478-483, 514-517).
No. 26: Similarly for 1921, compiled by Clarence J. West and Callie Hull. (Also published in Science 55 (1922) 271-279).
No. 42: Similarly for 1921-22.
No. 75: Similarly for 1925-26.
According to a note in No. 75 the data for the years 1922-25 are available only on file in the office of the Research Information Service of the National Research Council. An inquiry directed to that office, September 10, 1971, elicited the response, October 13, 1971, that detailed information for those years was not now available to anyone outside that office. The information had been collected but never organized for publication. The 1925-26 report did, however, give total numbers of doctorates in individual sciences for the missing years, thereby making possible the statement regarding the total for the years 1920-1926.
Note: A second, unsuccessful attempt was made to secure the information for the missing years 1922-25 from the National Research Council in May, 1974. The response at that time stated "We do not have the titles of doctoral dissertations for the early period, 1920 to 1957. Even if we had collected the titles, we would be unable to furnish you with titles and authors because of our confidentiality requirements."

2.14 According to Judith R. Goodstein, Institute Archivist at the California Institute of Technology, in a letter to the author, 29 February 1972, there are no "serious histories" of "Caltech" or of its physics department. In a subsequent letter, 8 May 1974, Dr. Goodstein noted that she and Robert Kargon were submitting a proposal to write a history of Caltech in the 1920's, but indicated that the work would not be completed for some time.
At the present time it is possible, however, to present some details of the emergence of the California Institute of Technology during the 1920's as one of the strongest American institutions for the study of physics, based on the following sources:
Millikan, Autobiography, especially pp. 207-237;
Warren Weaver, Scene of Change: A Lifetime in American Science, (New York: Chas. Scribner's Sons, 1970) pp. 36-48; and
Imra W. Buwalda "The Roots of the California Institute of Technology" a three part article in the C.I.T. publication Engineering and Science, (October 1966) 8-12, (November 1966) 20-26, (December 1966) 18-23. Note: copies of the Buwalda articles were kindly made available to this author by Dr. Goodstein.

"Throop University" was founded in 1891 in Pasadena, California by Amos G. Throop, a retired wealthy businessman from Chicago. It was soon after renamed "Throop Polytechnic Institute" and given a clear mandate by its trustees to put its money and its energy into a practical manual training program.
By 1906 it was decided to feature a "College of Science and Engineering" with hopes of becoming "The MIT of the West." George E. Hale became the moving spirit for this development. In 1913 the institution's name was changed to "Throop College of Technology." A year later, Arthur A. Noyes, distinguished chemist and vice-president of the Massachusetts Institute of Technology, was drawn to the Pasadena institution, first on a part-time, but later (1919) full-time, basis.
Hale and Noyes became closely associated with R.A. Millikan during World War I in the course of the activity of the National Research Council, with the result that Millikan began spending a quarter of his time at Throop in 1920.
In addition to the funds initially provided by Throop himself, over the years wealthy Californians contributed generously to the endowment of the growing institution. A particularly generous gift was made by Arthur Fleming in 1921 of his entire fortune, valued at about $4 million, contingent (according to Millikan) on Millikan's coming to Pasadena permanently. Millikan was assured of receiving close to $100,000 annually to develop "the strongest possible department of physics."
Millikan agreed to come and immediately began to assemble a physics faculty that included European and American members on visiting or permanent status.

Millikan, a fine experimentalist himself, built an excellent laboratory facility. He was, however, also sensitive to the need for a strong theoretical competence within the Institute. Thus Richard Tolman and Paul Epstein became regular faculty members of the California Institute of Technology, as it was called after February, 1920, and throughout the ensuing years there was a steady stream of visiting lecturers from Europe including such figures as Lorentz, Einstein, Sommerfeld, Ehrenfest and many others.
In addition, "Caltech" became in the years ahead, an attractive place of study for young physicists with postdoctoral fellowships. In the course of this thesis we shall have numerous occasions to mention specifically the accomplishments of physicists at the California Institute of Technology that related to quantum physics.

2.15 Due to the lack of information on individual theses for the years 1922-25 in the National Research Council Reprint and Circular Series, as commented on in Note 2.13, other sources had to be consulted to determine the titles, authors and degree granting institutions for the individual theses written during those years. Two were used:

Marckworth, Dissertations in Physics; and

A List of American Doctoral Dissertations Printed in 1922, 1923, 1924, 1925 (Washington, D.C.: Library of Congress, individual years).

In each of these the data were less readily accessible than in the N.R.C. publications, cited above. The first is arranged alphabetically by author; the second has no subdivisions within the broad category of "Science." In most cases the titles of the theses were sufficient to determine whether they were theoretical or not. Wherever doubt occurred, the content was ascertained by direct inspection of those deposited at Harvard University and the Massachusetts Institute of Technology; for others, the literature was searched for published versions with the aid of Science Abstracts.

2.16 Detail on these theses will be found on pages 52 and 53 of this chapter in the section dealing with American participation in quantum physics.

2.17 The transformation of the National Research Council from wartime to peacetime activity is described by several authors, such as Millikan and Pupin in their autobiographies, and in the Minutes of a number of N.R.C. Committees published in Proc. Nat. Acad. Sci. 5 (1919).

The subject is also discussed by Robert Kargon in his editorial Introduction to The Maturing of American Science (Washington, D.C.: The American Association for the Advancement of Science 1974) and by A.H. Dupree in Science in the Federal Government (New York: Belknap Press, 1964) pp. 326-330.

Among the peacetime activities sponsored by the National Research Council, two had particular relevance to the study of quantum physics in America and they will be frequently mentioned in the course of this presentation. They were the formation of committees of scientists to prepare reports on particular topics for publication as issues of the N.R.C. Bulletin and the establishment of the National Research Fellowship Program.

In the case of the latter some confusion exists with regard to whether recipients should take their fellowships to American institutions. Our discussion of the matter has relied heavily on the account provided by R.A. Millikan in his Autobiography, pp. 180-184, of the setting up of the program and his own role in setting its policy. (Millikan was a member of the N.R.C. Fellowship Board 1919-1937 according to Reynolds publication cited below.) According to Millikan they were clearly expected to do so. However, as stated below in Note 2.22, three of the early recipients of physics fellowships did go to European institutions. Also, I am indebted to Nathan Reingold for informing me (Letter, 28 March 1975) that his investigations at the Archives of NAS-NRC show that no such limitation was imposed, and Reingold notes that 19 percent of the Fellows went to foreign institutions.

On the other hand, David Dennison has stated in his AHQP Interview, Session 2, 28 January 1964, and in the "Transcript of an informal talk given by D.M. Dennison, November 15, 1968, for Professor Tomozawa's preseminar class for first-year graduate students" (a copy of which Professor Dennison kindly provided to the author) that he turned down the N.R.C. fellowship awarded to him in 1924 when he learned that he would have to take it at an American institution.

According to the Foreword of National Research Fellowships 1919-1944, Neva E. Reynolds, compiler (Washington, D.C.: National Research Council, 1944) "They [the Fellowships] were established after long consideration of various steps that might be taken to strengthen the foundations of science in the United States" and "In general, the purposes of the National Research fellowships are:

1. To promote the training in research of young men and women of promise;
2. To increase knowledge of the fundamental sciences upon which scientific progress of all kinds depends;
3. To encourage research in the educational institutions of this country."

It is interesting to note in connection with our discussion in Chapter 1, page 41, of the suggestion that America needed a center for the study of physics that was separate from industries and from academic institutions, that when one was essentially offered to the physics community, it was turned down, according to Millikan's account.

2.18 Information presented on fellowship awards was obtained from Reynolds National Research Fellowships 1919-1944, and "The National Research Fellowships" by Myron J. Rand The Scientific Monthly 73 (1951) 71-80.

2.19 The ratio of applicants to appointments for these fellowships was about seven to one, according to data in the Rand article cited above.

2.20 F.W. Bubb was an exception; he received his Ph.D. at the end of his year at Chicago.

2.21 L.F. Curtiss, E.H. Kurth, L.B. Loeb and G.P. Paine each were 4-year Fellows.

2.22 G. Breit (Leiden), A.H. Compton (Cambridge) and F.C. Hoyt (Copenhagen)

2.23 From Reports of the President and the Treasurer of Harvard College 1923-24, Harvard University Archives.
It must be noted that two Fellows (G.L. Clark who came to work with W. Duane and R.S. Mulliken who came to work with E.C. Kemble) who had received their awards as chemists chose to come to the Harvard Physics Department in 1923-24, thus making the total 4, rather than the 2 entered on the chart; in 1924-25 there were 5 N.R.C. Fellows in the Physics Department of Harvard University and in 1925-26 there were 6.

2.24 From the Minutes of the September 18, 1926 meeting of the Board of Trustees of the Huntington Library, San Marino, California. Text provided by that Library's Archivist in letter of 8 June 1973 to the author. (Millikan, in his Autobiography, p. 221, erroneously associated this statement with 1923.)

2.25 AHQP Microfilm 23, Section 6.

2.26 Erik J. Friis The American Scandanavian Foundation 1910-1960: A Brief History (New York: American Scandanavian Foundation, 1961) and American Scandanavian Foundation Annual Reports 1912-1936.

2.27 See, for example, AHQP Interview with H.C. Urey, March 24, 1964, Sessions 1, page 5. Two of the other American-Scandanavian Fellows did some translation work during their year abroad. Probably this was financially indispensible to these young men, but it must have distracted their attention from the principal purpose of going to Copenhagen to study theoretical physics. Udden prepared the translation of three essays by Bohr, based on talks Bohr had delivered to the Physical Societies of Copenhagen and Berlin (The Theory of Spectra and Atomic Constitution, Cambridge: The University Press, 1st edition 1922, 2nd edition, 1924). Lindsay, together with his wife Rachel, who had accompanied him to Copenhagen, translated an elementary presentation of Bohr's theory which had been prepared by H.A. Kramers and Helge Holst. (The Atom and the Bohr Theory of its Structure, London: Glydendal, 1923). Lindsay has stated (letter to the author, 15 May 1976) "We did the translation largely in order to get better acquainted with Kramers and Holst, to improve our knowledge of Danish, and to make Bohr's work better known to the general English speaking world. Naturally we were glad to accept payment on the principle that 'the laborer is worthy of his hire'. But the money was not indispensable. Moreover, the work on the translation did not distract me at all from the work on my research topic."

2.28 J.C. Slater's name is included in the list of Fellows in Appendix II of Friis American Scandanavian Foundation. However, Professor Slater has informed me, in a letter of 21 May 1973, that the American Scandanavian Foundation gave him an honorary appointment without stipend but did assist in making the arrangements for his visit to Bohr's Institute; his financial support came from Harvard University and private funds.

2.29 Directory of Fellows of John Simon Guggenheim Memorial Foundation 1925-1967 (New York: John Simon Guggenheim Memorial Foundation, 1968).
Annual Reports of the John Simon Guggenheim Memorial Foundation 1925-1936.
Gerhard Rollefson, who had received his doctorate in chemistry at the University of California at Berkeley in 1923 (Amer. Men of Sci. 7th edition, 1944) went to Göttingen to do experimental work with J. Franck in 1925.

2.30 The information presented in this discussion of the International Education Board was obtained from

George W. Gray Education on an International Scale: A History of the International Education Board (New York: Harcourt, Brace and Co., 1941) and

Raymond B. Fosdick Adventure in Giving: The Story of the General Education Board (New York: Harper and Row, 1962).

2.31 G.D. Birkhoff of Harvard University surveyed the mathematical capabilities, while C.E. Mendenhall of the University of Wisconsin did the same for physics.

2.32 These Göttingen grants, plus a chemistry grant to Munich, would seem to contradict the thesis propounded by Joseph Haberer ("Politicalization in Science" Science 178 (1972) 713-724) that German science was blacklisted in the postwar years by the Allied Nations, including the United States. Such a policy may have been in effect within the International Union of Pure and Applied Physics, but not among those responsible for the International Education Board Funds.

2.33 Information on individual recipients given in The Rockefeller Foundation Directory of Fellowship Awards 1917-1950 (New York: The Rockefeller Foundation, 1951).

2.34 Records of the Harvard University Physics Department; letters to the author from J.C. Slater, 21 May 1973; V. Guillemin, 31 May 1972.

2.35 Theoretical Physics in the Twentieth Century: A Memorial Volume to Wolfgang Pauli, M. Fierz and V.F. Weiskopf, Eds. (New York: Interscience Publishers Inc. 1960) 5-39.

2.36 American Men and Women of Science, 12th Edition, 1972.

2.37 Sopka, "An Apostle of Science Visits America," p. 373.

2.38 Details to be found in "Reminiscences" by Lorentz' daughter, G.L. deHaas-Lorentz, in H.A. Lorentz: Impressions of his Life and Work, pp. 96,97; and in Thomson, Recollections and Reflections, pp. 243-266.

2.39 Einstein's American visit is well described in Einstein: His Life and Times by Phillip Frank (New York: Alfred A. Knopf, Inc., 1947) pp. 176-186.

2.40 These lectures were subsequently translated and published as The Meaning of Relativity: Four Lectures Delivered at Princeton University May, 1921, E.P. Adams, Trans. (Princeton: University Press, 1923).

2.41 See for example, references to "Lusitania Medaille" pp. 81-86 in Albert Einstein/Arnold Sommerfeld Briefwechsel herausgegeben und kommentiert von Armin Hermann (Basel: Schwabe and Co. Verlag, 1968) and Amerikanische Eindrücke (AHQP Microfilm 23, section 6). I am indebted to Roger Stuewer for calling the latter to my attention.

2.42 Sommerfeld letter to E.A. Birge, President of the University of Wisconsin, 17 July 1922; quoted by Ulrich Benz in the thesis he is writing on the life of Sommerfeld.

2.43 See pages 183-184 of the biography by Frank; also, Benz describes "der einzige feindselige Akt" experienced by Sommerfeld as the retraction, by Michelson, of an invitation to Sommerfeld for a visit to the University of Chicago, causing Sommerfeld to decide against attending the November meeting of the American Physical Society that was scheduled there.

2.44 Thomson, Recollections and Reflections 246-247. (Probably it was a private car rather than a whole train that was put at his disposal.) Thomson seems to have erred in his recollection of revisiting and lecturing at Harvard. Correspondence in the Lyman Papers indicate that his visit to the Boston area involved a reception at the American Academy of Arts and Sciences where he met with scientists from a number of institutions in the Boston-Cambridge area. Further, the Harvard University Gazette has no mention of a lecture by Thomson, as it surely would have carried, if there was one.

2.45 For example, Amherst College inaugurated its John W. Simpson Foundation Lectureship with Bohr's visit in 1923 (Amherst Student September 24, 1923, page 1) and Lorentz' visit to Cornell University in 1926 was made possible by the Jacob H. Schiff Foundation (The Cornell Daily Sun, September 30, 1926, page 1).

2.46 Letter of Alexander Meiklejohn, President of Amherst College to James Angell, President of Yale University, 31 May 1923; located in the Yale University Presidential Records.

2.47 Letter of James Angell to Niels Bohr, 21 April 1923, also in the Yale University Presidential Records; the full fee of $2500 for delivering the Silliman Lectures was never collected by Bohr because he failed to write up the lectures for publication, as stipulated in the letter cited.

2.48 Letter of E.A. Birge, President of the University of Wisconsin, to A. Sommerfeld, 13 August 1922, quoted by U. Benz in his thesis.

2.49 _The Born-Einstein Letters_, Irene Born trans. (New York: Walker and Co. 1971) letter no. 22, p. 36.

2.50 _Recollections and Reflections_, pp. 247-249; it should be noted that Thomson wrote these words in 1936, hence his reference to "the last ten years" would mean 1926-36, not 1913-23.

2.51 Recollections of Otto Oldenberg, letter to the author 16 January 1972.

2.52 July 7, 1921, p. 2.

2.53 _New York Times_: July 8, 9:1; July 9, 7:6; July 12, 12:15, July 31, sect. 2, 4:3.

2.54 See, for example, _New York Times_ July 9, 7:6.

2.55 July 10, 1921, Supplement 1; copy obtained from microfilm of _Vössische Zeitung_ at the United States Library of Congress.

2.56 This whole episode is surrounded with confusion that lasted well after 1921. In particular, there is an article on page 3 of A. Einstein _Ideas and Opinions_, Sonja Bargmann, trans. (New York: Crown Publishers Inc., 1954) which is entitled "My First Impression of the U.S.A." and carries the notation "An interview for _Nieuwe Rotterdamsche Courant_, 1921. Appeared in _Berliner_

Tageblatt, July 7, 1921." This article is, in fact, neither that interview nor the subsequent rebuttal that was published in Vössische Zeitung. I have been unable to determine the true source of this text which appeared in the 1933 German version of Mein Weltbild without identification. L. Loeb (AHQP Interview, 7 August 1962, pp. 16,17) sheds some light on the background of the original account by revealing that the Nieuwe Rotterdamsche Courant reporter was a niece of H.A. Lorentz. Her meeting with Einstein had been arranged by her uncle on an informal basis. Einstein apparently spoke freely, not anticipating that his words would be published. (Loeb also remarks on the supercilious attitude of many European visitors toward science in America, and similar sentiments were expressed by R.T. Birge in a letter of 5 September 1925 to E.C. Kemble, AHQP Microfilm 50, Section 5).

2.57 It is curious to note that Millikan gives scarcely any mention of this award in his autobiography, merely using a quotation from his Nobel Lecture in explicating his work for the reader. Apparently he did not go to Stockholm to receive the award but delegated the American ambassador in Sweden to receive it for him. According to J.W.M. Dumond, Millikan apologized to the Nobel Prize Committee for his absence saying, he could not afford the loss of time (the trip would have taken a month in those days) from his research and from his students and responsibilities at the California Institute of Technology (see p. xliv of Editor's Introduction to R.A. Millikan The Electron (Chicago: The University of Chicago Press, 1968). Millikan did deliver his Nobel Lecture the following May 23, according to Nobel Lectures: Physics, 1922-1941 (Amsterdam: Elseview Publishing Co., 1965). It was appropriately titled "The electron and the light-quant from the experimental point of view."

2.58 Atomes et Electrons (Paris: Gauthier-Villars et Cie., 1923).

2.59 According to Michelson's daughter (D.M. Livingston, The Master of Light, 291-2) he felt somewhat lost during the Conference discussions of electron theory and atomic structure that were so far removed from his own professional work. In addition to attending the Solvay Conference while in Europe at that time Michelson delivered lectures at the Sorbonne as an exchange professor.

2.60 Conductibilite Electrique des Metaux (Paris: Gauthier-Villars et Cie, 1927).
Bridgman's address was entitled "Rapport sur les phenomenes de conductibilite dans les metaux et leur explication theorique," pp. 67-114; it was followed by general discussion, pp. 115-134. Hall's address was entitled "La conduction metallique et les effets transversaux du champ magnetique," pp. 303-349, with discussion, pp. 350-362. Hall's presentation was primarily concerned with his own discovery known as the "Hall Effect."

2.61 Private communication to the author, December, 1972.

2.62 A possible exception to this statement was Leigh Page. Although well informed on quantum physics, as evidenced by his contributions to the National Research Council Bulletins on "Atomic Structure" and "The Molecular Spectra of Gases," Page continued to try to find a way to by-pass Bohr's radiationless orbits with a system that would not conflict with classical electrodynamics. See, for example, L. Page "Radiation from a Group of Electrons" Phys. Rev. 20 (1922) 18-25. Page never used quantum theory in the publication of his original work.

2.63 Remarks in this paragraph were paraphrased from the Kemble communication cited in Note 2.61.

2.64 The Weekly Review 4 (1921) 537-538. Webster prided himself on his classical knowledge. He was an expert linguist, both modern and classical. The rather fanciful title "Oh Quanta!," came from the writings of Abelard. Two years later Webster, who had long held a leading position in the American physics profession, committed suicide, apparently in despair over his present status in the world of physics. (See suicide note published in the Worcester Evening Gazette, May 15, 1923; p.1).

2.65 Science 55 (1922) 19. Barus had just read the National Research Council Bulletins No. 5 on "Quantum Theory" by E.P. Adams and No. 14 on "Atomic Structure" by L. Page and D.L. Webster. It was Webster who had suggested abandoning strict conservation of energy. Incidentally, Webster was led to this idea completely by his own considerations and independently of any European influence. (Letter to the author, 27 February 1973).

2.66 "Max Mason" N.A.S. Biog. Mem. Vol. 37 p. 219 and Warren Weaver Scene of Change: A Lifetime in American Science, p. 28.

2.67 Ibid. p. 57.

2.68 Jour. Opt. Soc. Amer. 6 (1922) 211-228. Despite the views expressed by Tolman on this occasion, he was responsible for introducing about 1923 a course based on Sommerfeld's Atombau und Spektrallinien for Chemistry students at the California Institute of Technology. So far as I have been able to determine, no other American chemistry departments at that time had a comparable course. Linus Pauling, a student in Tolman's course, has generously shared with the author his recollections of learning about the old quantum theory at the California Institute of Technology. (Letters to the author, 8 May 1974 and 23 May 1974).

2.69 Letter to the author, 27 February 1973.

2.70 Birge History of the Physics Department Vol. II, pp. VII,11-13.

2.71 Valence and the Structure of Atoms and Molecules (New York: The Chemical Catalog Co. 1923, reprinted by Dover Publications 1966) p. 56.
H.C. Urey (AHQP Interview, March 24, 1964, p. 3) recalls G.N. Lewis announcing in a seminar at Berkeley that he now believed "Bohr was right" in about 1922.

2.72 See, for example, the following addresses by Millikan during this period: "Atomism in Modern Physics" Faraday Lecture delivered before the Fellows of the Chemical Society, London, June 12, 1924, published in Jour. of Chem. Soc. 125 (1924) 1405-1417, and "The Physicist's Conception of an Atom" delivered in April 1924 before the American Chemical Society and published in Science 59 (1924) 473-476.

2.73 Science 51 (1920) 505.

2.74 Nobel Lectures in Physics 1922-1941 (Amsterdam: Elsevier Publishing Co., 1965) p. 65.

2.75 We shall mention later Millikan's investigation of "stripped atoms" that were carried out during this period.
We have already discussed, in Note 2.14, his bringing European theoreticians to the California Institute of Technology.

2.76 Braunschweig: Friedr. Vieweg und Sohn. 1919, 1920, 1922, 1924.

2.77 London Methuen and Co. Ltd., 1923.

2.78 See, for example, interviews in AHQP and AIP-CHP with R. Kronig, R.B. Lindsay, L. Pauling and H.C. Urey; R.S. Mulliken "Molecular Scientists and Molecular Science: Some Reminiscences" Jour. of Chem. Phys. 43 (1965) S2-S11. G.R. Harrison, E.C. Kemble, J.C. Slater, and J.H. Van Vleck, in private conversations with the author, have also attested to the importance of Sommerfeld's book for their own study of quantum theory in the early 1920's.

2.79 The Division of Physical Sciences of the National Research Council appointed committees of physicists responsible for surveying the current state of development of specific topics and writing reports of their findings. The final reports were then published as individual issues of the National Research Council Bulletin and put on sale at a nominal price. (Information from individual Bulletins and from the Minutes of the meeting of the Executive Board of the Nation Research Council, June 10, 1919, Proc. Nat. Acad. Sci. 5 (1919) 464.)

The American Chemical Society, working in collaboration with the Interallied Conference of Pure and Applied Chemistry and the National Research Council, undertook, beginning in 1919, to sponsor a Series of Scientific and Technologic Monographs on chemical topics selected for current interest and written by authors recognized as authorities on those topics. These Monographs were published by the Chemical Catalog Co. of New York City. (Information from P.D. Foote and F.L. Mohler The Origin of Spectra, New York: The Chemical Catalog Co. Inc., 1922, General Introduction).

2.80 N.R.C. Bulletin No. 5, October, 1920. A second edition was issued as N.R.C. Bulletin No. 39, November, 1923. With regard to Adams' presentations of quantum theory, Carl Eckart (AHQP Interview, 31 May 1962, p. 5) recalled that they were used in a course Adams gave during those years at Princeton University. In addition, Eckart made the following remark:
"(it) was in many ways highly original; this report and his course both stressed the relation between Poincaré's integral invariants and the Bohr-Sommerfeld quantum integral."
Apparently there was another early American publication on quantum theory but I have not been able to locate a copy of it. Foote and Mohler (The Origin of Spectra, p. 50) cite L. Silberstein, Report on the Quantum Theory of Spectra, 42 pp., 1920 and call it "a splendid synopsis of the theoretical development to this date."

2.81 See, for example, the following issues:
No. 6, November 1920, "Data Relating to X-Ray Spectra" by Wm. Duane;
No. 7, December 1920, "Intensity of Emission of X-Rays and their Reflection from Crystals" by Bergen Davis;
No.14, July 1921, "A General Survey of the Present State of the Atomic Structure Problem" by D.L. Webster and L. Page;
No.20, October 1922, "Secondary Radiations Produced by X-Rays" by A.H. Compton;
No.48, September 1924, "Critical Potentials" by K.T. Compton and F.L. Mohler.

2.82 N.R.C. Bulletin No. 54, March, 1926.

2.83 N.R.C. Bulletin No. 57, December, 1926.

2.84 See Birge-Kemble Correspondence, AHQP Microfilm 50, Sections 1-6.

2.85 According to the Preface, written by E.C. Kemble, who was chairman of the Committee on Radiation in Gases.

2.86 Foote and Mohler's book was one of the series of monographs commissioned by the American Chemical Society. (See Note 2.79.) R.S. Mulliken has called it "delightfully written and illustrated ... a splendid source from which at that time to learn the current state of knowledge on quantum phenomena and their explanation." (Jour. Chem. Phys. 43 (1965) S5.) Linus Pauling has stated (letter to the author, 8 May 1974)
"As I recall, it was during my first year [1922] that Tolman gave a course on atomic structure and quantum theory, using the recently published book by Foote and Mohler, ..., The Origin of Spectra."
The role of the National Bureau of Standards in atomic research is worthy of comment at this time: In 1922, Foote and Mohler, who had been conducting excitation and ionization studies there since 1918, were named sole members of an atomic physics section within the optics division. Apparently there was need to justify such pure research activity within the practical mandate of the Bureau. S.W. Stratton, head of the Bureau at the time, did this by arguing that an understanding of the nature of collisions between atoms and electrons might lead to more efficient illuminants and better "radio bulb" design. (From the National Bureau of Standards Annual Report 1922, quoted by Cochrane in Measures for Progress, p. 249.) Though small in membership, the atomic physics section carried out some excellent work for a number of years.

2.87 I am indebted to E.M. Purcell for calling to my attention the importance of Darrow's articles during private conversation in the Spring of 1972. Professor Purcell recalled the influence which Darrow's articles had in stimulating his early interest in physics. Other established members of the American physics profession today have also spoken of their personal experiences associated with discovering Darrow's presentations in the Bell System Technical Journal. For example, see quotation of W.H. Brattain on p. 27 of Charles Weiner "How the Transistor Emerged" IEEE Spectrum 10 (1973) 24-33; and paraphrased comments of C.H. Townes on p. 25 of J.H. Van Vleck "Karl Kechner Darrow: Writer, Councilor and Secretary," Physics Today (April 1967) 23-26.

2.88 Volumes 3, 4, and 5, 1924-26.

2.89 New York: D. Van Nostrand. A second edition was published in 1939.

2.90 See, for example, comment by J.H. Van Vleck in "American Physics Comes of Age," p. 26 to the effect that "... forty years ago I eagerly awaited the next issue of the Zeitschrift für Physik, which would reveal almost anything of signigicance in theoretical physics..." (this was written in 1964); and Carl Eckart's description of the sharing at Princeton University of copies of the Zeitschrift für Physik that belonged to H.D. Smyth in Eckart's AHQP Interview, 31 May 1962, p. 5.

2.91 Since Born's visit, which occurred in the Fall of 1925, came just at the time that the new matrix mechanics was breaking, we shall defer discussion of his visit until our next chapter.

2.92 Harvard University Gazette, October 20, 1923.

2.93 Phys. Rev. 23 (1924) 104.

2.94 Albert Rosenfeld, "The Quintessence of Irving Langmuir," in The Collected Works of Irving Langmuir, Vol 12, p. 125.

2.95 Correspondence between Bohr and Lyman in the Lyman Papers, located in the Harvard University Physics Department.

2.96 F.C. Hoyt, Letter to the author, 18 January 1973. Hoyt, a National Research Fellow to Copenhagen at the time, served as Bohr's travelling companion during this visit to America.

2.97 D.M. Dennison "Recollections of physics and of physicists during the 1920's" Amer. Jour. of Phys. 42 (1974) 1051-1056.

2.98 Bulletin of the California Institute of Technology, Annual Catalogue, December, 1923.

2.99 Birge, History of the Physics Department, Vol. II, p. VIII (15).

2.100 Recollection of J.H. Van Vleck shared with the author in private conversation.

2.101 R. Kronig, "Thr Turning Point," p. 17.

2.102 Harvard University Gazette, April 18, 1924.

2.103 Personal recollections of E.C. Kemble and J.H. Van Vleck shared with the author in private conversations; comment by R. Kronig in "The Turning Point" p. 17: "... a meeting with Ehrenfest, lecturing in America in 1924, had made a strong impression on me, and I took to heart his suggestion to go abroad and in particular to visit his institute."

2.104 Proc. Nat. Acad. Sci. 10 (1924) 133-139; 13 (1927) 400. For discussion of Duane's idea see pages 2.72 - 2.75

2.105 Information concerning the relations between Sommerfeld and Compton described in this paragraph was kindly supplied to me by Ulrich Benz.

2.106 21 (1923) 483-502. This was the paper referred to earlier as having suffered a five month delay in publication.

2.107 This letter, quoted by Benz and dated 17 January 1923, provides an interesting example of an American physicist expressing his feeling of inferiority in theoretical matters when corresponding with a European theorist such as Sommerfeld.

2.108 Roger Stuewer, in his book <u>The Compton Effect: Turing Point in Physics</u>, p. 241 quotes a letter from Sommerfeld to Bohr, dated 21 January 1923, in which Sommerfeld expressed some reservations about Compton's findings.

2.109 <u>History of the Physics Department</u> Vol. II, p. VII (28).

2.110 <u>Gesammelte Schriften</u> Vol. IV, p. 678.

2.111 Letter to the author, 2 November 1971.

2.112 Later, when Duane had observed the shifted wavelength associated with the Compton Effect, he publicly withdrew his opposition during a meeting of the American Physical Society in December of 1924.
"Once convinced of the true state of affairs, Duane was most generous in his admission of error, and the unreservedness of his announcement of his change of position at a meeting of the Physical Society must remain a pleasant memory to all who heard it."
P.W. Bridgman "William Duane" <u>N.A.S. Biog. Mem</u>. vol. 18, pp. 23-41.

2.113 Letter to the author, 31 May 1972. Guillemin's thesis, for which he was awarded a doctor's degree in 1926, was a theoretical treatment of the methane molecule. His results were published in <u>Ann. der Phys</u>. <u>81</u> (1926) 173-204.

2.114 At Columbia University, George Pegram, Michael Pupin and A.P. Wills gave theoretical courses of a classical nature. J.S. Ames was responsible for the theoretical offerings at the Johns Hopkins University. Contact was maintained between Johns Hopkins and some of the more modern physicists at the National Bureau of Standards and the Carnegie Institution in Washington during this period, according to the recollections of Merle Tuve, shared with the author in private conversation, 24 January 1975. Gregory Breit began travelling to Baltimore once a week to give a course "Advanced Atomic Theories" by 1926, (<u>The Johns Hopkins University Circular</u> 368, March 1926). Otto Laporte (AHQP Interview, 31 January 1964, Session 2, p. 23) described some of the relationship between Johns Hopkins and the National Bureau of Standards while he was at the latter as an International Education Board Fellow.

2.115 Information obtained from individual Columbia University Catalogues for those years.

2.116 S.A. Goudsmit, "The Michigan Symposium in Theoretical Physics," Michigan Alumnus Quarterly Review, May 20, 1961; copy kindly supplied by the author.

2.117 Breit's thesis, dealing with the mathematical and theoretical aspects of the distributed capacity of inductance coils, was published in Phys. Rev. 17 (1921) 649-677.

2.118 A complete Bibliography of Breit's publications, which are many and varied, may be found in Facets of Physics, D.A. Bromley and V.W. Hughes, eds., (New York: Academic Press, 1970), a Festschrift containing the proceedings of a symposium held May 3, 1968 on the occasion of Breit's retirement from the faculty of Yale University.

2.119 Henry A. Erikson, History of the Minnesota Physics Department, an unpublished manuscript deposited at the Center for the History of Physics, American Institute of Physics.

2.120 Information from materials in the files of E.C. Kemble, kindly made available by Professor Kemble; also on AHQP Microfilm 55, sections 3 and 4.

2.121 Information from AHQP Interview with Carl Eckart, 31 May 1962, page 5.

2.122 Astrophysical Journal 51-62 (1920-25); Physical Review Subject Index 17-80 (1920-50) 353-354, "Quantum Theory Preceding Quantum Mechanics;" Proceedings of the National Academy of Science 6-11 (1920-25); Journal of the Optical Society of America 4-11 (1920-25).

2.123 20 January 1927. Original letter in Dutch and English translation (used in quotation) are on deposit at the Center for the History of Physics, American Institute of Physics.

2.124 That was to be one of the main attractions of the University of Minnesota plan to have both Breit and Van Vleck join the faculty as theoreticians.

2.125 Friedrich Hund, Geschichte der Quantentheorie (Mannheim: Bibliographisches Institut AG, 1967).

2.126 Breit spent the second year of his fellowship at Harvard University while Hoyt spent two years, 1922-24, in Copenhagen before continuing as a National Research Fellow at the University of Chicago.

2.127 Letter to the author, 20 February 1973; Nature 106 (1920) 179 and Jour. Astrophys. 52 (1920) 248.
Since the first edition of Sommerfeld's Atombau was published in 1919 one might wonder why Loomis did not use that work. It appears that very few copies of the 1919 edition came to America. The Harvard University Library does not have a copy. A.H. Compton brought one back from Cambridge in 1920 which he believed to be the first in America (AHQP Interview with C. Eckart, 31 May 1962). Langmuir also seems to have had access to a copy as he cites it in a footnote to his paper "The Structure of the Helium Atom," Phys. Rev. 17 (1921) 339-353.

2.128 Third Edition, 1922, p. 519

2.129 Zeit. f. Phys. 3 (1920) 460.

2.130 This chart includes all of the relevant theses written by American physics students of which I am presently aware. The sources of my information were numerous and included all of those listed in Notes 2.13 and 2.15. In addition, I have obtained pertinent information directly from E. Hutchisson, M.S. Vallarta and J.H. Van Vleck. Reminiscence articles, as well as AHQP and CHP Interviews have been helpful in the cases of E.U. Condon, D.M. Dennison and R.B. Lindsay. R.T. Brige, in his History of the Physics Department provided some information on the dissertations written under his direction. More will be said later about Birge's activities during this period that related to quantum theory.

2.131 Daniel Lang "A Farewell to String and Sealing Wax II" The New Yorker, November 14, 1953, pp. 46-67.

2.132 W. Heisenberg Physics and Beyond: Encounters and Conversations, A.J. Pomerans, translator, (New York: Harper and Row Publishers, 1971).

2.133 In private conversation, 14 November 1973, S.A. Goudsmit stated that he did not learn Maxwell's electromagnetic theory during his student years.

2.134 Letter from G.Dieke to S.A. Goudsmit, 15 October 1925, on deposit at the Center for History of Physics, American Institute of Physics.

2.135 From an address delivered to the Institute of Physics by Thomson after his return to England; quoted in Lord Rayleigh, The Life of Sir J.J. Thomson (Cambridge: The University Press, 1942) p. 226.

2.136 Astrophys. Jour. 59 (1924) 1-10. A second part of Urey's thesis dealing with the heat capacities and entropies of diatomic and polyatomic gases was published in Jour. Amer. Chem. Soc. 45 (1923) 1445-1455.
So far as I have been able to ascertain, Urey was the first American student of chemistry to use quantum theory in his doctoral dissertation.
Professor Urey has kindly provided the following comments on his thesis in a letter to the author, 10 July 1974:
"My major professor at Berkeley involved a rather curious situation. G.N. Lewis was my official advisor, but I talked with him very little. The professor I talked to mostly was Professor Olsen. I did my work rather independently of any direction. Tolman finally visited the laboratory, read my dissertation and made criticisms. I got more criticism from him than from anyone else.
I have always thought my dissertation was a rather weak one, but it was referred to favorably by Professor Sir Ralph Fowler of England in his book. It was in connection with this dissertation that I learned enough statistical mechanics to be able to calculate the vapor pressure of H_2 and HD, and thus be able to discover heavy hydrogen."
Notes: R.H. Fowler referrred to Urey's work on pp. 62, 186, and 348 of Statistical Mechanics: The Theory of the Properties of Matter in Equilibrium (Cambridge, Eng.: University Press, 1929).
Urey received the Nobel Prize in chemistry in 1934 for his discovery of heavy hydrogen.

2.137 "On the Effect of Perturbing Electric Fields on the Zeeman Effect of the Hydrogen Spectrum" Det Kgl. Danske Videnskabernes Selskab., Mathematisk-fysiske Meddelser VI, 2 (1924) 1-19. Urey published an extension of his studies to ionized helium which appeared in Zeitschrift für Physik 29 (1924) 86-90 as "Über den störenden Einfluss eines elektrischen Feldes auf den Zeeman-effekt von Spektrallinien" translated from Urey's English manuscript by Werner Kuhn.

2.138 AHQP Interview, 24 March 1964, Session 1, page 6. Actually Urey did maintain close ties with quantum theory as it developed and published some further papers after his return to the United States. Also, in 1930, he coauthored with A.E. Ruark a textbook Atoms, Molecules and Quanta (New York: McGraw-Hill Publishers, 1930). His principal work, however, turned increasingly toward experimental studies, among which was the discovery of heavy hydrogen.

2.139 Quotations from CHP Interview, 9 July 1964, page 14.

2.140 See Note 2.27.

2.141 CHP Autobiography, page 65.

2.142 Bohr's approval was given during his visit to Yale as Silliman Lecturer in the Fall of 1923 (CHP Interview, 9 July 1964, page 18) and Lindsay received his degree the following June. (See chart on page 2.52)
H.B. Phillips of the mathematics department of the Massachusetts Institute of Technology, one of the very few American mathematicians interested in quantum theory at that time, served as Lindsay's formal thesis supervisor. H.M. Goodwin of the physics department joined Phillips in signing the certificate of acceptance. (Thesis deposited in the Archives at the Massachusetts Institute of Technology.)

2.143 Slater's thesis bore the title "The Compressibility of the Alkali Halides." He was sufficiently well versed in the crystal theories of Born, Landé and others to come to the conclusion that his results were "beyond the power of present theories." (From Thesis deposited in the Archives of Harvard University.)

2.144 In his "Introduction of Prof. J.H. Van Vleck, Sanibel Symposium, Monday, January 18, 1971" Internat. Jour. Qu. Chem. 5 (1971) 1-2, Slater recalled that "E.C. Kemble ... was giving one of the earliest, and best, courses in quantum [theory] which were then being given in the country, and was turning the department into quite a center of theory."
In private conversation, 11 August 1971, J.C. Slater shared with me information about his early interest in theoretical physics as follows: while an undergraduate at the University of Rochester, after hearing about the Bohr atom, Slater looked up the original papers on it. This led to his writing a senior prize-winning paper that was based on his own experimental study of hydrogen spectral lines at various potentials and pressures and included an attempt at a simple theory to explain the widths and intensities of the lines.

2.145 Phil. Mag. 47 (1924) 785-802; Zeit. f. Phys. 24 (1924) 69-87.

2.146 Phys. Rev. 26 (1925) 289-299; Zeit. f. Phys. 32 (1925) 639-663.

2.147 Further discussion of this episode is found in Sources of Quantum Mechanics B.L. van der Waerden ed. (New York: Dover Publications Inc. 1967) pp. 11-14, and in the following contemporary and recent publications by Slater:
"Radiation and Atoms" Nature 113 (1924) 307-308
"The Nature of Radiation" Ibid. 116 (1925) 278
"Quantum Physics in America Between the Wars" Internat. Jour. Qu. Chem. 1S (1967) 1-23
"The Development of Quantum Mechanics in the Period 1924-1926" in the de Broglie Festschrift Wave Mechanics: The First Fifty Years W.C. Price, S.S. Chissick and T. Ravensdale, editors (New York: Halsted, 1973) pp. 19-25
Solid State and Molecular Theory: A Scientific Biography (New York: John Wiley and Sons, 1975) pp. 8-12.
See also AHQP Interview with Slater, 3 October 1963, p. 30ff.

2.148 Phil. Mag. 42 (1921) 123.

2.149 Zeit. f. Phys. 9 (1922) 1-67.

2.150 Sommerfeld Atombau, 4th edition, 1924, p. 198. Bohr also took note of the similarity of their ideas in The Theory of Spectra and Atomic Constitution (Cambridge: University Press, 2nd eition 1924) p. 88.

2.151 J.H. Van Vleck commented on the neglect of Lindsay's work in "Quantum Principles and Line Spectra" p. 83 and in a review of A. Lande Die Neure Entwicklung der Quantentheorie (Dresden and Leipzig: Verlag von Theodor Steinkopff 1926) published in Phys. Rev. 27 (1926) 635.

2.152 Berlin: Verlag von Julius Springer, 1925. In the text, p. 332, Born had given all the credit for the idea to Bohr and the working out of the details to Kramers. In the footnote citing Kramers paper, Zeit. f. Phys. 13 (1923) 312-341, Born made reference to a publication by Van Vleck in Phys. Rev. 21 (1923) 372 that was only an abstract of a paper by Van Vleck delivered at the American Physical Society meeting of December, 1922, which had no close relevance to the crossed orbit model of helium. Born made no mention of the Phil. Mag. 44 (1922) 842-869 publication by Van Vleck which carried through the detailed calculations on the crossed orbit helium model as proposed by Kemble in Phil. Mag. 42 (1921) 123-133.

2.153 AHQP Microfilm 49.

2.154 Henry Small "The Helium Atom in the Old Quantum Theory," an unpublished Ph.D. Thesis, University of Wisconsin, 1971, contains a complete discussion of the various attempts that were made, and of the impact of their failure on the old quantum theory.

2.155 "The Model of the Neutral Helium Atom," J. Opt. Soc. Amer. 7 (1923) 509-516. Sommerfeld's model had the two electrons travelling along coplanar, elliptical paths with one electron at aphelion when the other was at perihelion.

2.156 Phys. Rev. 17 (1921) 339-353.
Langmuir suggested two possible arrangements: both electrons oscillating in semi-circles lying in the same plane and with the positive nucleus at the center or the electrons executing full circular paths in planes that were parallel to each other and with the positive nucleus located midway between the planes.

2.157 P.W. Bridgman "Some of the Physical Aspects of the Work of Langmuir" Coll. Works of Irving Langmuir, Vol. 12, p. 484 describes another brief foray into quantum theory that Langmuir made, briefly but enthusiastically, in 1920.

2.158 D.M. Dennison, in AHQP Interview, Session 1, 27 January 1964, pp. 8,9, relates his experiences while working summers in Langmuir's laboratory at the time when Langmuir was attempting to develop his ideas about the helium atom. Dennison, at the time an undergraduate at Swarthmore College, assisted Langmuir in carrying out the p dq integrations.

2.159 The idea of evaluating scientific contributions on the basis of these criteria, rather than priority considerations, was propounded by S.A. Goudsmit in a lecture delivered in Boston on November 14, 1973, during a Conference on Magnetism and Magnetic Materials held at the Hotel Statler.

2.160 Jour. Frank. Inst. 192 (1921) 145-155. This paper, published in August 1921, was based on a talk delivered by Compton in December 1920 at the annual AAAS meeting, according to a note in the published version.
We omit from our discussion consideration of Parson's spinning ring of electric charge (although Compton referred to it in his article) because of its shape.
It should be noted that Compton's conception of the electron involved a rather large size for it - about 10^{-10} centimeters in radius.

2.161 Phys. Rev. 16 (1920) 464-476.

2.162 Compton had submitted a paper dealing with this point in August 1920, while he was still at the Cavendish Laboratory. It was published in February 1921 as "Possible Magnetic Polarity of Free Electrons" Phil. Mag. 41 (1921) 279-281.
Compton argued that the helical tracks of beta rays indicated the existance of a local magnetic field of uniform strength. He inferred that it was the constant spin of the electron itself which induced such a field.

2.163 Compton's suggestion was not mentioned either in the N.R.C. Bulletin No. 18 "Theories of Magnetism" by A.P. Wills, S.J. Barnett, L.R. Ingersoll, J. Kunz, S.L. Quimby, E.M. Terry and S.R. Williams, or in the N.R.C. Bulletin No. 54, "Quantum Principles and Line Spectra" by J.H. Van Vleck.
A British publication, E.C. Stoner Magnetism and Atomic Structure (London: Methuen and Co. Ltd. 1926), while recognizing the experimental work of Compton and Rognley, made no mention of Compton's interpretation of their results.

When Goudsmit and Uhlenbeck published their second paper dealing with electron spin, "Spinning Electrons and the Structure of Spectra," Nature 117 (1926) 264, they stated at the outset: "So far as we know, the idea of a quantized spinning of the electron was put forth for the first time by A.K. [sic] Compton (Jour. Frank. Inst. Aug. 1921, p. 145) who pointed out the possible bearing of this idea on the origin of the natural unit of magnetism. Without being aware of Compton's suggestion, we have directed attention in a recent note (Naturwissenschaften Nov. 20, 1925) to the possibility of applying the spinning electron to interpret a number of features of the quantum theory of the Zeeman effect ... and also of the analysis of complex spectra in general."

Thus we see that the same idea arose in two quite different contexts. Since there was much greater interest in spectral and quantum considerations in Europe in 1925 than there was in magnetic considerations in America (or Europe) in 1921, it was relatively easy for Compton's proposal to be overlooked. Also, the Journal of the Franklin Institute was even less likely to come to European attention, perhaps, than the Physical Review or the Proceedings of the National Academy of Science might have been.

2.164 "Moment of Momentum of Magnetic Electrons" Phys. Rev. 19 (1922) 420.

2.165 Kronig "The Turning Point" contains his recollections of the circumstances surrounding this episode. The quotation from Pauli remark is taken from page 21.
B.L. van der Waerden article "Exclusion Principle and Spin" in the same volume, pp. 199-244 also discussed the response to Kronig's suggestion by European physicists.

2.166 "Spinning Electrons and the Structure of Spectra" Nature 117 (1926) 550.

2.167 Naturwissenschaften 47 (1925) 953.

2.168 Proc. Nat. Acad. Sci. 12 (1926) 80-85.

2.169 Slater's paper was entitled "Interpretation of the Hydrogen and Helium Spectra" and published in Proc. Nat. Acad. Sci. 11 (1925) 732-738 in December, 1925. It was cited by Goudsmit and Uhlenbeck in "Spinning Electrons and the Structure of Spectra" Nature 117 (1926) 264-265, the issue of February 20, 1926, and also in "Over Het Roteerende Electrons en de Structuur der Spectra" Physica 6 (1926) 273-290. The latter paper also makes note of the suggestion of electron spin by Parson and by Compton.

2.170 "Molecular Scientists and Molecular Science: Some Reminiscences" Jour. of Chem. Phys. 43 (1965) S2-S11.

2.171 Nature 114 (1924) 349.
Nature 113 (1924) 744,785.

2.172 Details related to this episode are contained in the reference cited in Note 2.170.

2.173 AHQO Microfilm 50 includes the correspondence that took place between Birge and Kemble relating to that undertaking.

2.174 Letters from R.T. Birge to the author, 5 July 1972; 25 July 1972; 3 November 1972; 30 November 1972 and 13 December 1972.

2.175 Letter in Dutch language from G. Dieke to S.A. Goudsmit, 15 August 1925, on deposit at the Center for History of Physics, American Institute of Physics; quotation taken from the English translation provided by the Center.

2.176 G. Dieke to S.A. Goudsmit Letter, 26 October 1925, deposited and translated as described in preceding Note.

2.177 E.U. Condon "The Franck-Condon Principle and Related Topics" Amer. Jour. Phys. 15 (1947) 365-408;
"Reminiscences of a Life in and out of Quantum Mechanics" Internat. Jour. Qu. Chem. Symp. No. 7 (1973) 7-22;
R.T. Birge History of the Physics Department Vol. II, pp. VIII, 41,42, (revised).

2.178 Franck's original idea provided a mechanism for explaining the photochemical dissociation of a diatomic iodine molecule, initially in a non-vibrating state, in terms of a process involving the excitation of one of its electrons by the absorption of energy in the form of light. Condon extended this notion to cases of vibrating initial states and to cases of emission as well as absorption. Condon's analysis was remarkably successful in explaining the observed intensity distribution within band spectra patterns in terms of the probabilities of transitions between various allowed states.

2.179 E.U. Condon, "Nuclear Motion Associated with Electron Transitions in Diatomic Molecules" Phys. Rev. 32 (1928) 858-872.

2.180 Sources of Quantum Mechanics, B.L. van der Waerden, ed. (Amsterdam: North Holland Publishing Co., 1967) pp. 16-18.

2.181 Ibid. p. 17.
Van Vleck's two papers on this topic were
"A Correspondence Principle for Absorption" Jour. Opt. Soc. Amer. 9 (1924) 27-30
"The Absorption of Radiation by Multiple Periodic Orbits and its Relation to the Correspondence Principle and the Rayleigh-Jeans Law Parts I and II" Phys. Rev. 24 (1924) 330-365
Part I of this article is reprinted in van der Waerden Sources pp. 203-222.

2.182 van der Waerden Sources, p. 17.

2.183 Zeit. f. Phys. 33 (1925) 879-893. An English translation of this article is found in van der Waerden Sources pp. 261-276.

2.184 Professor Van Vleck has kindly provided me with the following information about the writing of this paper:

Van Vleck's paper on the correspondence principle for absorption has a rather curious origin. In a conversation he had with Breit at Minnesota in the winter of 1923-24 he remarked to Breit that there didn't seem to be anything in classical theory that corresponded to Einstein's negative absorption (now usually called induced emission) in quantum theory. Breit replied that classically there could be fluctuations up and fluctuations down. He had in mind the fact that with certain phase relations a mechanical system can do work on an electric field, although on averaging over all phases the net work done is by the field.

(Conceivably Breit was aware of this fact because he had been in the Netherlands two years earlier and so became acquainted with Fokker's calculation of absorption by a rotator, where a fluctuation theory is used which is quite similar to that in Einstein's theory of Brownian motion.) The fluctuations up and down, Van Vleck took to mean the quantum transitions up and down from a given stationary state, and so he examined the difference between the transitions up and down from a given stationary state, comparing the transitions in which the changes in quantum numbers were the same except for sign. Van Vleck cites Breit in a preliminary note published in the Journal of the Optical Society of America, but does not in his longer article. As he remembers it, he wanted to thank Breit in the latter, but Breit objected on the ground that the phase fluctuations he had in mind were quite different from the difference effect employed by Van Vleck and so, overmodestly, felt no acknowledgement was in order.
In footnote 43 of his paper, in the Physical Review, Van Vleck proposed relations equivalent to the well known sum rules of Thomas and Kuhn before they did (cf. Van Vleck Quantum Principles, p. 152) but rejected them on two grounds (a) disagreement with some spurious experimental data on x-ray absorption and (b) because they seemed to yield finite transition probabilities to non-existent states. The difficulty (b) arose because he implicitly evaluated an additive constant by incorrectly demanding that the quantum theoretical transition probability agree with some average of a classical formula involving Fourier components (his Eq. (10)). (In the particular case of the harmonic oscillator, this constant is essentially the same as the "const" in Eq. (20) of Heisenberg's celebrated paper in vol. 33 of Zeits f. Physik.) This constant cancels out of and so cannot be determined just by use of the Thomas-Kuhn relation. Instead it should be determined by the boundary condition that the transition probability to non-existent states should vanish, as was properly done by Heisenberg and by Thomas and Reiche for the harmonic oscillator. On the other hand their treatments of the two dimensional rotator are open to objection.

2.185 It is not clear precisely how soon Van Vleck's idea received the attention of European physicists. We do know, however, that the papers by Van Vleck, discussed by van der Waerden and commented upon by Jordan, were published in April and October of 1924. Jordan's papers with Born and with Born and Heisenberg were submitted for publication in September and November, respectively, 1925.

2.186 New York: Thomas Y. Crowell Co., 1965, p. 182.

2.187 Letter to the author, 30 April 1971.

M. de Broglie's entire letter is of such great interest that we now present the rest of its contents.

"Madame,

J'ai reçu votre lettre avec un très grand interêt parce que je crois que l'histoire des origines de la Mécanique ondulatoire est un des chapitres les plus intéressants de l'histoire contemporaine des Sciences. La façon dont les choses se sont passées à cette époque et les idées qui m'ont guidé avant, pendant et tout de suite après ma thèse de Doctorat sont généralement aujourd'hui ignorées, même des spécialistes de la Physique théorique quantique.

[Passage quoted on p. 2.71]. Les travaux qui m'ont le plus guidé dans ces recherches sont naturellement ceux de Planck sur les quanta, ceux d'Einstein sur les photons et le double aspect ondulatoire et corpusculaire de la lumière et ceux de Bohr sur la théorie quantique de l'atome. J'avais beaucoup d'admiration pour Einstein et j'avais beaucoup étudié la théorie de la Relativité, surtout sous sa forme restreinte, notamment en suivant les beaux cours de Paul Langevin au Collège de France. Je suis convaincu que je n'aurais pas découvert les principes de la Mécanique ondulatoire si je n'avais pas bien connu les résultats de la théorie de la Relativité. C'est, en effet, en remarquant que la formule de transformation relativiste de la fréquence d'une onde est différente de la formule de transformation relativiste de la fréquence d'une horloge, que j'ai eu l'idée que la particule se déplaçait dans son onde de telle facon que la vibration interne dont je la dotais reste constamment en phase avec celle de l'onde. C'est cette idée qui m'a conduit à la formule $p = \frac{h}{\lambda}$ qui porte aujourd'hui mon nom. Et cela est fort peu connu.

Dans l'espoir que cet envoi pourra vous intéresser, je vous envois le texte d'un article que j'ai publié assez récemment dans la Revue française de l'Electricité. Vous y apprendrez, si vous ne le saviez pas déjà, qu'à partir de 1928 j'ai abandonné mes idées primitives pour enseigner celles de l'Ecole de Copenhague qui sont aujourd'hui considérées comme "orthodoxes" par beaucoup de théoriciens de la physique. Mais vous y verrez aussi que j'ai repris depuis environ vingt ans les idées qui m'avaient guidé dans ma jeunesse et que j'ai pu les développer beaucoup, surtout dans les dernières années. Et, bien que mes derniers travaux ne soient encore très connus, je suis aujourd'hui persuadé que c'est dans cette voie que l'on parviendra à bien comprendre la véritable nature de la coexistence des ondes et des particules découverte par Einstein dans le cas de la lumière et étendue par moi à toutes les particules.

J'ai fait récemment une conférence où j'ai assez longuement développé les idées contenues dans l'article que je vous envoie. Je ne manquerai pas de vous envoyer un tirage à part de cet article dès que j'en aurai à ma disposition. Et, naturellement, je serai très heureux de vous fournir tous les renseignements qui pourraient vous intéresser.

Veuillez agréer, Madame, l'expression de mes sentiments très dévoués.

Louis de Broglie "

Note: The articles which M. de Broglie was so kind as to send to me were:

"Vue d'ensemble sur l'histoire et l'interprétation de la mécanique ondulatoire" (referred to in the above letter as having been published in la Revue française de l'Electricité.)

"Les ondes de la mécanique ondulatoire", Revue Science Progrès Découverte N° 3432, (Avril 1971) 39-47.

2.188 "Zur Quantentheorie der Strahlung" Phys. Zeit. 18 (1918) 121-128.

2.189 Compton, himself, has provided interesting insight into the circumstances surrounding his going to Washington University and his own approach to doing physics. In the section entitled "Personal Reminiscences" of the book The Cosmos of Arthur Holly Compton, Marjorie Johnston ed., (New York: Alfred A. Knopf, 1967) he has this to say on page 31:
"In 1920 Washington University, at least so far as its college of liberal arts was concerned, was a small kind of place. I went there, frankly, because I did not want to become confused in being in a center where there was so much going on along the line in which I was involved that I would be led away by the thinking of the time. At Washington University there were small but reasonably adequate facilities to do the kind of work that I wanted to do, and I was far enough removed from such places as Chicago, or Princeton, or Harvard, where I felt I would have been thrown into the pattern of thought of the period and would not have been able to develop what I had conceived of as my own contribution."
By 1923, however, he seems to have been ready to transfer to the University of Chicago, where he remained until 1945.

Additional information specifically relevant to the discovery of the Compton Effect may be found in the following publications:
A.H. Compton, "The Scattering of X-Rays as Particles" Amer. Jour. Phys. 29 (1961) 817-820
A.A. Bartlett, "Compton Effect: Historical Background" Ibid. 32 (1964) 120-127
as well as the previously cited book by Stuewer The Compton Effect.

2.190 Proc. Nat. Acad. Sci. 9 (1923) 158-164; quotation from p. 158. Duane wrote one other paper using this hypothesis, "The Calculation of the X-Ray Diffracting Power at Points in a Crystal" Ibid. 11 (1925) 489-493.

2.191 "The Interference of Light and the Quantum Theory" Proc. Nat. Acad. Sci. 9 (1923) 238-243.

2.192 "The Quantum Integral and Diffraction by a Crystal" Ibid. 359-362; and "A Quantum Theory of Uniform Rectilinear Motion" Phys. Rev. 23 (1924) 118, the abstract of a talk given by Compton at the American Physical Society meeting in Chicago, November, 1923.

2.193 "The Quantum Theory of Fraunhofer Diffraction" Proc. Nat. Acad. Sci. 10 (1924) 133-139 and "Remarks on the Quantum Theory of Diffraction," Ibid. 13 (1927) 400-408.

2.194 Nobel Lectures in Physics 1922-1941, p. 65.

2.195 The Physical Principles of the Quantum Theory, C. Eckart and F.C. Hoyt, trans. (New York: Dover Publications, 196) p. 77.

2.196 AHQP Interviews of 11 February 1963, p. 21 and 22 February 1963 p. 30; quotation is from the former.

2.197 Born-Einstein Letters, No. 49, 15 July 1925, pp. 83-86.

2.198 Ibid., p. 86.

2.199 Nobel Lectures in Physics 1922-1941, p. 189

2.200 "Unity in Quantum Theory" Foundations of Physics 1 (1971) 191-202; this was preceeded by a series "Quantum Fact and Fiction, I-III" in Amer. Jour. Phys. 33 (1965) 123-127; 34 (1966) 1160-1163; 37 (1969) 541-548.
Landé's views are of interest in the context of the various criticisms in the past twenty years, by de Broglie and others, of the philosophical aspects of quantum mechanics. Landé has disclosed, in a letter to the author, 18 February 1972, that Part IV of his series "Quantum Fact and Fiction" was rejected for publication although he had received more than 500 requests for reprints of I-III from all around the world. Instead, he published the material in a different form in "Unity in Quantum Theory," cited above and the "The Decline and Fall of Quantum Dualism" in Philosophy of Science 38 (1971) 221-223, according to a second letter to the author 16 March 1972.

2.201 Duane was 51 years old in 1923.

2.202 "New Regularities in the Spectra of the Alkaline Earths" Astrophys. Jour. 61 (1925) 38-69; quotation taken from Abstract on page 38.

2.203 Zeits. f. Phys. 32 (1925) 841-860.

2.204 Phys. Rev. 23 (1924) 1-34; 24 (1924) 209-222, 223-228; 25(1925) 295-305, 591-599, 600-605; 26 (1925) 150-164, 310-318. Phil. Mag. 48 (1924) 259-264; 49 (1925) 923-935. Proc. Nat. Acad. Sci. 11 (1925) 329-334. These were all referenced by Pauli in his article "Quantentheorie" in Handbuch der Physik, H. Geiger and K. Scheel, eds. Vol. 23 (Berlin: Springer Verlag, 1926) 1-278; reprinted in Collected Scientific of Wolfgang Pauli R. Kronig and V. Weisskopf, eds. (New York: John Wiley and Sons Inc. 1964) 269-548, Volume I.

2.205 Paul Epstein, "Robert Andrews Millikan" N.A.S. Biog. Mem. Vol. 33.

2.206 Anregungs- und Ionisierungsspannungen (Berlin, 1926).

2.207 Berlin: Verlag von Julius Springer, 1926; pp. 696-721. This section was written by J. Franck and P. Jordan.

Chapter 3

The Impact of Quantum Mechanics on Physics in America, 1926-1929

Chapter 3

The Impact of Quantum Mechanics on Physics in America 1926-1929

Introduction

The new quantum mechanics which dominated the world of physics in the latter part of the 1920's, began in late 1925 with the introduction, by Heisenberg, Born and Jordan, of matrix mechanics into the domain of atomic physics. The early successes achieved by these new and unfamiliar concepts and techniques aroused great interest and some bewilderment among physicists everywhere. The coming of Schrödinger wave mechanics within the next year provided a welcome alternative to physicists who felt more at home with differential equations than with matrix algebra. Nevertheless, it was clear that a new era had dawned in physics. There could no longer be any hope of salvaging a strictly classical physics based on the ideas of Newton and Maxwell nor of continuing to develop a quantum physics in the style of Bohr. Along with the abandonment of a theoretical structure based on a physical model that could be easily visualized came the necessity of using the fairly sophisticated mathematical techniques implicit in the new formulations.

From our previous discussion of American physicists' preference for physical model formulation and of the frequent weakness in the mathematical training of American physicists, it is obvious that the advent of quantum mechanics was bound to have a profound influence on the course of the development of physics in the United States. One the one hand, the new modes of thought required by quantum mechanics could have had a dampening and discouraging effect on American progress toward maturity in physics. On the other hand, the radical innovations associated with quantum mechanics might serve to accelerate the development of physics in the United States if two conditions were met: 1) that the physics profession as a whole was receptive to the new trend toward abstraction, and 2) that there was a sufficient number of Americans skilled in the new methods to assure its assimilation into the enterprise of physics in this country.

In this chapter we shall see that the latter alternative did prevail. Superimposed on the general growth pattern that physics in America had been experiencing for more than a quarter of a century there was, during the period 1926-1929, a marked increase in the level of interest and productivity in physics in America that was specifically generated by the recent and continuing development of quantum mechanics. Whereas confusion and disappointment had often characterized the earlier experience with the old quantum theory, the impressive successes of the new quantum mechanics produced astonishment and excitement. To be sure, the transition into the new modes of thought was not achieved without some difficulty either for the majority of American physicists whose private outlook, but not their research endeavor, had to be altered, or for the still small, but rapidly increasing, number of eager young theorists who sought personally to utilize and extend the domain of quantum mechanics. Amid the intellectual ferment that attended the introduction of quantum mechanics, theoretical physics attained professional status in America during the late 1920's.

While it is not possible to state unequivocally that physics in America "came of age" during the final years of the 1920's, there were unmistakable signs visible by 1930 that the study and practice of all aspects of physics in America then had sufficient breadth and depth to assure a coequal status for America within the international community of physics in the years ahead. By 1930 the maturity of physics in America was as definitely established as had been the birth of physics in America by 1900.

The timing of these developments within physics in America was especially fortuitous in the light of two completely external circumstances. Physics in America had become strong enough not only to survive, but to continue to grow despite the period of severe economic depression that began in 1929. In addition, America had come to be regarded by Europeans as an attractive place to pursue the study of physics even before the political situation in Germany and Italy deteriorated so badly that many physicists felt forced to flee their native lands in the 1930's. The period 1926-1929 saw the beginning of the sizable influx of European physicists into the United States as permanent residents, a phenomenon that would later influence not only the academic study of physics in America, but also the course of world history in the days of World War II.

In this chapter we shall again follow the pattern of examining the characteristics of the institutional growth of the American physics profession before focussing, first, broadly on the response of American physicists to the new quantum mechanics and, then, more specifically on the involvement of individual Americans in developing its usefulness.

Part I: Institutional Developments within the American Physics Profession

All of the institutions associated with the enterprise of physics in America which had been established prior to 1926 experienced continued growth during the late 1920's. In addition, the drawbacks that we noted in our previous chapter attended research publication and study abroad were practically eliminated by 1929. American disadvantages due to the lack of strong mathematical training for physics students and of centers for the study of theoretical physics were made acutely obvious by the highly theoretical and mathematical nature of quantum mechanics and resulted in steps being taken in the direction of correcting these deficiencies. The circumstances under which Americans pursued all phases of physics were markedly improved in the late 1920's through the growth of domestic institutions and increased intercourse with foreign physicists.

The American Physical Society

Between 1926 and 1929 Membership in the American Physical Society grew from 1358 to 1844.[1] During the same period Fellowship in the Society rose from 487 to 522 with a singularly large increase taking place in 1928 when 63 physicists were so designated.[2] Two of this number were the European physicists, Erwin Schrödinger and Hermann Weyl, who were in this country at that time. Many of the Americans honored were young men who had entered the physics profession in the decade following the World War and who were working on either the experimental or theoretical aspects of quantum physics. Notable among them were several whose names we have already met in our previous chapter or will do so in this, for example, E. U. Condon, L. H. Germer, F. C. Hoyt, G. R. Harrison, F. A. Jenkins, F. W. Loomis, A. E. Ruark, J. C. Slater, and H. C. Urey.[3]

During the period 1926-1929 the Presidents of the Society were D. C. Miller who had assumed office in 1925, K. T. Compton who served during 1927 and 1928, and H. G. Gale who began his tenure in 1929. While all three were experimentalists, Compton's research was clearly tied to the newer, quantum physics. Furthermore, Compton was elected at the much younger age of 39 than was Gale and Miller who were 55 and 59 respectively at the time they took office. Harold Webb continued as Secretary until 1929 when he was succeeded by W. L. Severinghaus.

The pattern of meetings already established was continued during this period, namely, a February meeting in New York City, an April meeting in Washington, D.C., a November meeting in the midwest and "the annual meeting" in December held wherever the American Association for the Advancement of Science had chosen to meet. Also, the West Coast Division continued to hold two meetings a year. In addition to the joint meetings with Section B of AAAS, other joint meetings were held during this period with the Optical Society of America and with the American Mathematical Society.[4] Attendance at the larger meetings numbered well over 300 and as many as 135 papers were delivered during a single meeting.[5]

A change in format of APS meetings was provided for in December, 1927 at the annual meeting when a resolution was passed to the effect "that one session of each meeting be devoted to invited papers summarizing various fields of research."[6] It was felt that meetings made up exclusively of ten-minute papers had led to a loss of interest on the part of the membership at large. By this time it was clear that physics had developed to such a complex stage that, unaided, an individual could not longer hope to keep abreast of the latest work on all fronts. While it had been true for many years that invited speakers had taken part in those Physical Society meetings held jointly with the American Association for the Advancement of Science and other scientific societies, the passing of this resolution was a move to make presentations with broad interest a part of every American Physical Society meeting. Eventually this was accomplished. Early examples of Physical Society meetings featuring such programs were those held in Minneapolis at the end of November, 1928 and at Washington, D. C. in April of 1929. At the former, six speakers presented a discussion of quantum mechanics and electron impact phenomena while, at the latter, seven speakers took part in a symposium on dielectrics.[7]

By the end of the third decade of its existence the American Physical Society had become a thriving organization. Its total membership was well above 2,000. It enjoyed cordial relations with other American scientific professional societies. Its meetings were well attended and frequently had European physicists as guest participants. There was a conscious effort on the part of those responsible for the programs to make the meetings attractive to all members of the profession whether or not they themselves were delivering papers. An additional social dimension was added on at least one occasion when, not only were Werner Heisenberg, Friedrich Hund, L. S. Ornstein and Arnold Sommerfeld present as guests of honor, but, after the formal dinner that was part of the program, Heisenberg and Sommerfeld joined APS Vice-president W. F. G. Swann in providing a musical program.[8]

Serial Publications

The years from 1926-1929 saw the beginning of a new era in the field of American scientific publications sponsored by the physics profession. Innovations were accomplished during that period which brought about the beginning of the rapid ascent toward worldwide prestige that was soon to be enjoyed by American journals of physics. Much of this rise can be attributed to the efforts of one particularly dedicated member of the profession, John T. Tate who, in 1926, became Managing Editor of the American Physical Society with responsibility for its journal Physical Review.

Assisting Tate there was a Board of Editors which consisted of about eight members at any one time, each of whom usually served for a two-year period on an overlapping schedule. Now, for the first time, there was strong representation from among quantum theorists on this Board. E. C. Kemble served 1925-1927, J. H. VanVleck 1926-1928, G. Breit 1927-1929, R. S. Mulliken 1928-1930 and E. U. Condon 1929-1931. Thus at any one time one-quarter of the Board were theoreticians. This, combined with Tate's own appreciation of modern theoretical developments, had a marked impact on the subsequent character of Physical Review and other American research journals of physics.

During these years the total number of papers published in Physical Review rose from 189 in 1926 to 281 in 1929 with a corresponding increase in pages from about 2,150 to 2,700 per year. Approximately one-sixth of the papers published in 1929 dealt with quantum mechanics. Some of these were longer and more comprehensive then any previously published in Physical Review. For example, J. H. VanVleck's article "On σ-Type Doubling and Electron Spin in the Spectra of Diatomic Molecules," in the April issue, was forty pages long and J. C. Slater's paper "Theory of Complex Spectra" in the November 15 issue, was thirty pages in length.

This represented a conscious change of policy on the part of Tate and drew some adverse criticism from his predecessor in the post of Managing Editor. It could well have been argued, however,[9]

"that the greater length was only a reasonable manifestation of the rapidly improving quality of American physics at the time, to say nothing of the added complexity of theoretical papers occasioned by the burgeoning quantum mechanics."

Other changes were made in the Physical Review under Tate's management largely in response to the current ferment in physics and to the earlier criticisms[10] of publication delays frequently associated with that journal which might cause an author to lose priority credit. It has been said that:[11]

"It is fortunate that Tate's initial appointment as the chief editor of the Physical Review practically coincided with the 'quantum-mechanical revolution.' It would have been a calamity if the post had been filled with someone not appreciative of or sympathetic to the great developments that suddenly burgeoned in theoretical physics. Instead, he showed rare judgment and common sense in not delaying by much...refereeing noteworthy papers dealing with various applications of quantum mechanics, for America was at something of a disadvantage compared to the centers of Europe where the revolution had germinated."

In addition to minimizing refereeing delays for regularly submitted papers, Tate introduced in mid-1929 a new feature, "Letters to the Editor" for the prompt publication of brief items not at all subjected to the usual referee system.[12] At the same time, the Physical Review was made a semi-monthly publication.

Still another innovation in physics journal publication in America was accomplished in 1929 at Tate's initiative. After considerable discussion within the physics community,[13] an enabling resolution was adopted at the Annual Meeting of 1928 in favor of a quarterly publication of "supplements to the Physical Review, these to contain resumés, discussions of topics of unusual interest and, in a broad sense, material that will give valuable aid to productive work in physics and yet that cannot appropriately be published in the Physical Review."[14]

Some impetus for this undertaking came from American physicists such as E. U. Condon whose recent European experience convinced him that[15]

> "...the greatest handicap to physical research here is the lack of an adequate literature in English. The market is glutted with American textbooks for freshman and sophomore general physics but there we seem to fade out...we lack the literature which brings young men quickly into step with the research work in the various fields. We have relied too much on the German literature in spite of the fact that the education of our students in foreign languages, especially German, is very weak...
>
> "Until American physicists do recognize that their scholarship is incomplete, until they are providing themselves with their own review literature, the promoter of such a journal may have a discouraging time of it. But I hope that it can be promoted and carried through in a successful way."

Such an idea was not completely new in America. We noted in Chapter 2 that some issues of the National Research Council Bulletin were devoted to surveying broad topics. But, as Condon continuned:

> "It [a quarterly review journal of physics]" would be much more flexible... and less formidable [than N.R.C. Bulletins]."

The first issue of the new journal appeared in July, the second in October of 1929 under the title Physical Review Supplement. The contents of these original issues are of interest in revealing the identities of the authors, the topics covered and the paper lengths which give some indication of their scope of coverage. The July and October 1929 issues contained the following articles:

R. T. Birge	Probable Values of the Physical Constants	1- 73
A. H. Compton	The Corpuscular Properties of Light	74- 89
K. K. Darrow	Statistical Theories of Matter, Radiation and Electricity	90-155
E. C. Kemble	The General Principles of Quantum Mechanics I	157-215
F. L. Mohler	Recombination and Photoionization	216-227
T. H. Osgood	Spectroscopy of Soft X-Rays	228-240

These were all review type articles, meant to provide a general survey of the current state of topics with wide appeal in the physics community. In January of 1930 the new journal was renamed Reviews of Modern Physics and continues to this day to serve the valuable function of providing the reader with "perspectives and tutorial articles in rapidly developing fields of physics as well as comprehensive scholarly reviews of significant topics."[16]

Opportunities and Facilities for Education and Research

The period 1926-1929 saw increasing opportunity for the pursuit of physics by Americans in a variety of circumstances, such as research laboratories, graduate education programs and post-doctoral fellowships.

1) Laboratories

The excellence of laboratory facilities continued to be a hallmark of American science[17] whether sponsored by universities, industries or private foundations. At the University of Minnesota an entirely new physics building was opened in 1928.[18] At the Bell Telephone Laboratory C. J. Davisson and L. H. Germer were carrying on their electron scattering experiments which revealed the phenomenon of electron diffraction.[19] In 1929 the Bartol Research Foundation moved into new, well-endowed quarters on the campus of Swarthmore College.[20]

2) Doctorates in Physics Granted by American Institutions

Continuing growth in the area of graduate education during the late 1920's is evidenced by the fact that, in the three academic years 1926-1927, 1927-1928 and 1928-1929, a total of 270 doctorates in physics were awarded by American institutions.[21] Although more than 25 American universities were awarding doctorates in physics by this time, six institutions accounted for one-half of the total number. The University of Chicago led in the number of degrees awarded during this three year period with a total of 34. The California Institute of Technology was second with 26, followed by Cornell University with 24. Next were the University of Michigan, the Johns Hopkins University and the University of California conferring 18, 17 and 16 doctorates in physics respectively. The large majority of the dissertations continued to be written on experimental topics. We shall discuss in a later section those that specifically dealt with theoretical topics related to quantum mechanics.

3) Postdoctoral Fellowship Programs

With regard to coping with the new quantum mechanics the most important educational opportunities were those provided by the various post-doctoral fellowship programs. The National Research Council continued to award about two dozen grants to promising young physicists each year. The earlier restriction to American institutions was relaxed so that about fifteen of the Fellows were allowed to spend part of their fellowship tenure at foreign institutions in the years between 1926-1927 and 1929-1930. The American institutions that proved most attractive to National Research Fellows during those years were the California Institute of Technology, Princeton University and Harvard University, with 19, 15, and 12 "fellow-years" respectively.[22]

The fellowships sponsored by the International Education Board, which were described in Chapter 2, continued until the program was absorbed into the Natural Science Division of the Rockefeller Foundation that was formed in 1929. Regardless of the designation these grants benefited physicists of all nations and their specific emphasis on international study was especially welcome for the dissemination of quantum mechanics during the latter half of the 1920's.[23]

The American-Scandinavian Foundation fellowships had few physicists as recipients during this period. Very likely this was due to two facts: their stipends were relatively small, and their restriction to Scandinavian countries made them less attractive to physicists being lured to Göttingen, Leipzig, Munich and Zurich for the study of quantum mechanics.

Fortunately there was, by then, another American fellowship program which came into prominence and was especially well suited for such study. The John Simon Guggenheim Memorial Foundation established, in 1925, a broad program of fellowships "for advanced study abroad...to improve the quality of education and the practice of the arts and professions in the United States and to foster research."[24] During the years 1926-1929 there were, at any one time, as many as four American physicists studying at European centers of theoretical physics under the auspices of the Guggenheim Foundation. Guggenheim grants were usually made for six to eight month periods rather than for full academic or calendar years. Also, Guggenheim Fellowships were frequently awarded to more mature scientists rather than to recent Ph.D. recipients.[25]

The periods of unencumbered time for study and research that the various fellowship programs provided were of inestimable value not only to the individual Fellows themselves, but also to the institutions with which they were subsequently affiliated and to the future of physics in America. Especially in the cases of those who went abroad, their own horizons were significantly broadened. Moreover, their eagerness and competence did not go unnoticed within the European scientific community.

International Honors Accorded American Physicists

Not all of the American physicists who went to Europe in the late 1920's did so as students or as pilgrims to the established centers of learning. The visits of some of them were occasioned by invitation in recognition of their contributions to the world of physics.

In 1927 A. H. Compton was awarded the Nobel Prize in Physics "for his discovery of the effect named after him."[26]

By this time Compton's research interests had shifted largely to the field of cosmic rays. He had, during the previous academic year 1926-1927, gone to India in the dual capacity of lecturer at Punjab University in Lahore and of Simon Guggenheim Foundation Fellow, gathering cosmic ray data at high altitude in the Himalayas.[27]

Compton returned to America from India by way of Europe where he attended the Volta Congress in Como and the Solvay Congress in Brussels. At each of these meetings Compton delivered an address.[28]

Compton was not the only American to be invited to these two prestigious conferences held in the Fall of 1927. Irving Langmuir was also present at both of them. In addition, there were six other Americans among the 62 participants at the Volta celebration in Como[29] and two more Americans attended as institutional representatives.[30]

All of the formal presentations by the American delegates at Como featured experimental considerations and there were no American participants in the session devoted to "Le Teorie sulla Struttura della Materia e sulla Radiazioni" either as lecturers or discussants.[31] It is interesting to note, nevertheless, that theoretical work by American physicists was beginning to be cited by European speakers. In particular, mention was made of recent work by D. M. Dennison on the specific heat of hydrogen[32], by E. U. Condon on wave mechanics and the normal state of the hydrogen molecule[33], by J. R. Oppenheimer on nuclear bombardment by alpha particles[34], and by Norbert Wiener on operator theory.[34] There were, in addition, references to the Compton Effect and to Compton's suggestion of electron spin.[35] Compton's own address was delivered at the session devoted to "Esperienze sulla Struttura della Materia" and was entitled "On the Interaction between Radiation and Electrons."[36]

Although it was still primarily for their experimental achievements that American physicists gained public recognition from their European colleagues, by the late 1920's there was very strong evidence that the new generation of American physicists that emerged after World War I were not going to remain bystanders in theoretical development.

European Visitors and Immigrants to the United States

Additional international intercourse for American physicists was provided by the large number of European visitors who came to lecture in the United States between 1926 and 1929. The list which will be found in Appendix II reveals that never before had the density of foreign physicists in the United States been so high. Not surprisingly, most of these guests were associated with the new quantum mechanics, and we shall discuss their activities in greater detail in a later section.

The visits and lecture tours by established members of the European physics community continued in the tradition that dated back at least to 1872 when John Tyndall came as an apostle of science to the new world. In the intervening half century, as we have already discussed in previous chapters, dozens of European physicists had been motivated to cross the ocean by the prospect of the warm welcome and the generous stipends that awaited them and by a curiosity to see for themselves what life was like in America, as well as by the urge to share their scientific expertise with the American newcomers to the world of physics. By the late 1920's a change in attitude toward American science was occurring among European physicists which resulted in more young people coming to study for awhile with American theoretical physicists and other, more mature scientists accepting permanent positions at American universities.[37]

One young European physicist, about to leave for the United States was given the following counsel by his thesis supervisor in 1927:[38]

"When you get to America you'll say they do this wrong and their physics isn't any good...That is utterly unimportant. What is important to feel happy is not the state how it is, but the derivative. That is positive. There are only two countries at the moment where it's positive: Russia and America, where things are getting better all the time from the point of view of education in physics."

Two instances of young Europeans coming to America to study physics occurred at Princeton University where R. W. Gurney, from the Cavendish Laboratory, and E. C. G. Stueckelberg, from the University of Basel, carried on theoretical studies with E. U. Condon in the late 1920's.[39] In addition, K. F. Niessen came to the University of Wisconsin from Utrecht, as a Rockefeller Foundation Fellow to study with J. H. Van Vleck in 1928.[40]

In 1927 the University of Michigan initiated a bold program in theoretical physics which included persuading S. A. Goudsmit and G. E. Uhlenbeck to come to Ann Arbor from Leiden where they had recently completed their doctoral studies with Paul Ehrenfest.[41] As newcomers to the faculty of the physics department they joined D. M. Dennison and Otto Laporte. Dennison was returning to Michigan from a three year period of post-doctoral study in Europe. Laporte had joined the Michigan faculty one year earlier. His original training under Sommerfeld was followed by his coming to the United States in 1924 as an International Education Board Fellow to the National Bureau of Standards.[42] The new group of four young theoreticians at the instructor level was nominally headed by Professor Walter F. Colby who, together with Department Chairman Harrison Randall, were responsible for assembling the group.[43] At that time there was no other American institution which had such a sizeable group of theoretical physicsts in regular residence.

With the presence of this permanent theoretical physics group, not only were the regular offerings of the Michigan Physics Department greatly enhanced but also, perhaps more importantly for the future of theoretical physics in America, the Summer School program of the department began to blossom into the Summer Symposium in Theoretical Physics which attracted many European guest lectures.[44] Even more remarkable, however, was the fact that each year there was at least one American lecturer also. So unique and so important were these symposia that it is appropriate now to give a few details of their organization and characteristic style.[45]

In the first year, 1928, the lecturers were E. C. Kemble of Harvard University and H. A. Kramers of Rijks Universiteit, Utrecht; the following year, Leon Brillouin of the University of Paris, E. U. Condon of Princeton University, P. A. M. Dirac of Cambridge University, Karl Herzfeld, then of the Johns Hopkins University, formerly of the University of Munich, and E. A. Milne of Oxford University.

The lecturers were open not only to the students and faculty of the University of Michigan but also, without charge to holders of doctorates in physics and, at a fee of $40,[46] to students of physics from other institutions.

In describing the activities associated with the symposia, S. A. Goudsmit has written as follows:[47]

"The important thing..was not the formal lectures, but the informal atmosphere during these Summer Sessions. The lectures were held to a minimum, just one hour every morning. There were usually two distinguished lecturers present each time. The afternoon was free for discussions, excursions, picnics, games, swimming. Thus, both lecturers and visitors had a lot of leisure time which was used in a most productive way. It is no wonder, therefore, that such an arrangement attracted at one time or another all the active theoretical physicsts and many experimenters, young and old, from all over the United States. Still the number was small enough each summer so that the whole group, including our own graduate students and faculty, could easily be arranged on the steps at the rear of the Physics Building for a photograph.[48]

"Important in theoretical physics is the exchange of ideas among creative workers, mostly young men with young ideas, ideas which are only half finished, uncertain, often wrong or sterile, but occasionally leading to a deep insight into the nature of atoms and molecules. It is obvious that such raw ideas usually do not get published in the scientific journals and thus cannot be learned by reading. It is necessary to create a place and an atmosphere in which those interested and those actively engaged in research in physics can meet informally and learn from each other. The Summer Symposium in Theoretical Physics made the Michigan campus such a place."

It has also been said of these symposia:[49]

"It was possible during the ensuing year to trace in a number of articles in scientific journals ideas which had their inception in the discussions of the symposium. The influence of this meeting was national and international rather than local."

and further:[50]

"In some respects, the Michigan symposia were a new invention and as such constituted a real contribution to the technique of research in physics. Since the war, the country has seen an ever-increasing number of conferences and symposia that attempt to do that which was done here so successfully in the summers from 1928 to 1940."

There were other American physics departments which, during this same period, used guest lecturers during their summer sessions to enrich their course offerings,[51] but none of them began to approach the scope of the University of Michigan Symposia.

We have noted in earlier chapters that one of the real inadequacies of American physics during the early twentieth century was the lack of centers of theoretical activity. The University of Michigan provided the first antidote to this situation by having a permanent theoretical group within its year-round faculty and by introducing a summertime activity of a unique character. Instead of merely trying to duplicate the format found at European theoretical institutes, Michigan developed something with a characteristic American style.

Summer School Sessions had long been a part of American university life, often providing intellectual opportunity otherwise unavailable to teachers at lesser institutions, as well as to young students. The University of Michigan Summer Symposia expanded this concept by providing, in addition to the scheduled lectures, an ambience that was a unique blend of intellectual stimulation and recreation which encompassed, in a remarkably democratic fashion, some 100 members who ranged from lowly graduate students to world-famous authorities in the field of theoretical physics in a typical American state university town. Ann Arbor became "a condensation point for theoretical physicists"[52] from 1928 until 1941 when the wartime commitments of scientists of all nations made such gatherings no longer feasible.

It is interesting to note the modest budgets that were associated with these programs. Following the particularly successful symposium of 1928, the President of the University of Michigan agreed to make available sums of $5,000 per year for three years to assure the immediate continuation of the activity. This sufficed until 1931 when the Summer Symposium was made a part of the regular budget of the Physics Department and of the Graduate School. It seems that the annual cost never exceeded $7,000. Only the guest lecturers received stipends. "The rest of the visitors just came on their own to learn from the famous men and from each other."[53]

Financial Support for Physics and Physicists

Although by present day standards such financial outlays as those just described for the University of Michigan Summer Symposia seem ridiculously low, they did not seem so at the time, and they were adequate to accomplish their objective.

The same comment might be applied to the fellowship programs already discussed. In the words of one of the recipients who managed his final year abroad in 1926-1927 on an $1,100 grant from The University of Michigan, after having been supported for two years at $2,000 by the International Education Board, "You did not get very rich on that stipend and it was not what you would call plushy living but it was adequate and I was very grateful for it."[54]

In Part II of this chapter we shall discuss the job market as it applied to theoretical physics in America in the late 1920's. The large demand and small supply of physicists skilled in the new quantum mechanics made their employment choices unusually broad and sometimes quite lucrative. But even for the average physicist, including those without advanced degrees, employment prospects were favorable during the pre-Depression era of the late 1920's. According to Theodore Lyman,[55]

"The very prominent part which physics plays in the modern world must be evident to anyone who takes the trouble to think about the matter. Transportation, communication, light, and power all involve this branch of science directly and practically; while from the more theoretical side, chemistry, astronomy, geology, biology, and medicine are closely connected with physics.
The importance to society of the subject is emphasized by the demand for men well trained in this branch of science. It is impossible to meet all the requests for teachers of physics or to supply the demands of commercial laboratories, such as the General Electric Company and the American Telephone and Telegraph Company, for trained workers. It

is no exaggeration to say that we could place ten times as many graduates each year as are available at present." (My emphasis.)

Lyman's statements were made in the context of explaining the need for new research facilities to be built at Harvard University to supplement those of the Jefferson Physical Laboratory which had been built in 1884. The financing for the proposed building, known since 1947 as the Lyman Research Laboratory, provides interesting insight into the funding of physics in the 1920's.

The initial plans for the new facility called for the expenditure of more than a million dollars, about half of which was for the building itself, the rest for initial equipment and continuing endowment. The General Education Board of the Rockefeller Foundation agreed to provide $400,000 if the remaining sum could be raised by friends of Harvard University. Lyman succeeded in raising the additional funds, although at the outset, he stated, "I feel exactly as if someome has asked me to climb up the outside of Bunker Hill Monument."[56]

One detail associated with the building of the new Research Laboratory at Harvard University is of interest because of the timing involved. When Lyman had the required funds assembled in 1929 many of the donated assets were in the form of stock certificates. Lyman, who was financially astute enough to realize the precarious state of the stock market in 1929, acting entirely on his own initiative, converted the certificates into cash. When the stock market crash did come in the Fall of that year, not only was the money for the physics building safely in the bank but, with the ensuing plummeting of prices in general, the accumulated funds had a much greater purchasing power than was originally anticipated.[57] The building was completed in 1931 with enough money left over to permit the establishment of "The Physics Research Endowment Fund of 1931" with $550,000 of capital. To this day a sizeable annual income from this fund is available for the support of research in physics at Harvard University.[58]

A number of other institutions were also assisted by the General Education Board to improve their facilities and teaching programs in the natural sciences. In these instances physicsts shared the largesse with astronomers, biologists, chemists and mathematicians. The same philosophy of "making the peaks higher"[59] that had been introduced by Wycliffe Rose for the International Educations Board's selection of European recipients in 1924 was soon applied to American institutions by the General Education Board. According to Raymond B. Fosdick,[60]

"Between 1925 and 1932 the General Education Board endowed this passion for scientific investigation to the extent of 19 million dollars, strategically placed in a small number of carefully chosen institutions. In every case the criteria for choice were quality and promise."

Among the institutions receiving especially large grants were the California Institute of Technology which got 3.3 million dollars; Princeton University, 2 million; the University of Chicago and Cornell University, 1.5 million each.

General Education Board funds were pledged to institutions on a matching grant basis. Depending on the individual circumstances, the recipient institution was required to raise from other sources up to three or four times the Board's pledge to collect the full amount. Institutions which fell short of their goals could still obtain reduced support from the Board in direct proportion to the amount they did raise.[61]

At some institutions the funds raised made possible the building of new facilities, as was the case at Harvard University. At others, endowments were established for the improvement of programs already underway. At Princeton University, for example, a Scientific Research Fund of $3,000,000 was set up, the income from which paid the salaries of five research professors and furnished an endowment of $1,500,000 for research.[62]

It is interesting to note that some of the General Education Board grants were made to improve the scientific education of women in the era when it was considered appropriate to provide separately for them. In this way Byerly Hall, the science building of Radcliffe College, was constructed following an allocation of $500,000 fromt he General Education Board.[63]

The financial support for American scientific development that was provided from the Rockefeller fortune through the agency of the General Education Board during the 1920's not only was precedent setting in terms of the amount of money involved but also served as an example and stimulus to a characteristic style of American philanthropy. In particular, the concept of "conditional giving" in which the recipient institutions were required to raise even larger sums from other sources was introduced by John D. Rockefeller, Sr. and did not go uncriticized.[64] It was, nevertheless, adhered to and defended on several grounds. For example, it was pointed out that the total amounts of money accruing far exceeded anything that could be provided by the Rockefeller fortune alone and thus many more projects could be activated by "seeding." In addition, it was considered highly desirable for the recipient institutions to develop support from their local philanthropic and alumni sources. Finally, it was argued that the initial, conditional promise of Rockefeller money served as "an endorsement of the beneficiary; the value of the Board's implied approval often far outweighed its dollars."[65]

Some Professional Concerns of American Physicists

In addition to the intellectual ferment brought on by the advent of quantum mechanics, which we shall discuss in our next section, the established American physicists of the late 1920's faced other problems both internal and external to the profession. Among these were questions of how the pursuit of physics could best be carried on within the growing institutions and how to respond to voices that were raising doubts about the value of science for human destiny.

An example of a discussion related to the internal policies of the physics profession was aired in the pages of Science in 1928. This discussion reveals that American physicists continued to engage in self-criticism in the late 1920's as they had in the past.[66] In addition, we find that at least some distinguished members of the American physics profession were concerned that the pursuit of physics in this country should evolve along lines that were compatible with the broader scientific and academic environment of this country rather than in a manner which was merely imitative of existing European models.

In an article entitled "Centers of Research,"[67] S. R. Williams suggested that, as a remedy for what he saw as the paucity and mediocrity of research in some American graduate schools, each department have one leader of research who would be solely responsible for the direction and supervision of all the Ph.D. candidates in that subject. He assumed that only truly outstanding scientists would be named to such posts, in a manner similar to the policy of German universities.

Opposition to such a plan was voiced jointly by L. Loeb, K. T. Compton and R. T. Birge.[68] While admitting the continuing need to improve graduate level reserch they argued that a far preferable course was for each department to assemble "a group of able, productive men, each active in directing the research of graduate students." These authors cited the characteristics of the contemporary American scene which favored such an approach as: American institutions, in contrast to many European ones, could affort to, and indeed already did, adequately support several professorships; post-doctoral programs, such as those sponsored by the National Research Council and the Bartol Foundation, were furnishing an increasingly good supply of trained researchers; interaction among the members of the group would be more conducive to the advance of knowledge on a broad front; and, finally, the student aspirant into the field of physics would enjoy a much broader range of choice no matter which of a number of institutions he might chose for his training. They cited California, Chicago, Harvard, Johns Hopkins, Michigan and Princeton as examples of research centers of physics where such groups were already functioning productively.

Loeb, Compton and Birge then went on to say

"The chief cause of our failing in research,...we attribute primarily to the dearth in the past of able, original, well-trained men. This, in turn, is due to our rapid expansion and to disregard of the importance of research on the part of the public, resulting in the failure to support it adequately or to draw into it the best talent. It is a universal comment of distinguished foreign scientists who visit us,

'We do not understand how you are content to work with so little research assistance, so little domestic help in your homes and so little public recognition.'"

The question of public attitudes toward science came to the forefront during this period in yet another context, and elicited strong response from some physicists in publications whose readership was not restricted to the scientific community.

By way of background to this episode, we recall that in the years between the beginning of World War I and the end of the 1920's remarkable strides were made in many nations in the exploitation of basic scientific principles from physics and from other scientific fields as well. The resulting developments had considerable impact on the environment of workers, but in addition widespread changes in life style were effected through individual ownership of automobiles and radios as well as through access to moving pictures and other products of American industry. This was technology rather than science at work but then, as now, critics, especially those outside the science professions, often failed to distinguish between the two, and indeed regard attempts to do so as quibbling.

Thus it was that during the 1920's criticisms of science began to raised, not unlike those that have been voiced more stridently in recent years. "Science," as epitomized in machinery and in techniques of warfare, such as had been employed during World War I, was frequently inveighed against. Questions were raised about the wisdom with which men were using the end products of scientific knowledge and about the value for humanity of an unbridled pursuit of scientific knowledge.

A ten-year "holiday" for science was suggested in 1927 by one critic during a meeting of the British Association for the Advancement of Science.[69] The theoretical possibility of tapping nuclear energy caused one scientist to write "I trust this discovery will not be made until it is clearly understood what is involved."[70] R. B. Fosdick expressed his misgivings about the uses to which scientific knowledge had been put in a series of addresses at American colleges that were further disseminated through the publication of a book entitled _The Old Savage in the New Civilization_.[71]

For the generation of American physicists who had viewed science with a double veneration as the source of truth and beauty as well as the means to a better life for all men, such criticisms and doubts were greatly disquieting. In response, R. A. Millikan chose as his Presidential Address to the American Association for the Advancement of Science in 1929 to discuss "The Alleged Sins of Science."[72] The text of Millikan's talk was printed in _Scribner's Magazine_ the following February together with an article by Michael Pupin entitled "Romance of the Machine." Both authors reiterated their faith in human progress through science--"Crescat Scientia Vita Excolatur."[73] Millikan argued not only that science could not be held responsible for the misuses that men might make of the technology that science made possible, but also that science held the key to the solution of men's burdens. He lauded in particular the quality of life that was being extended to millions through the widespread ownership of automobiles and radios.

Pupin[74] was especially upset by European critics of American culture who equated Americanism with Machine Civilization and "sordid materialism." Pupin, who had come to America as a poor young immigrant and rejoiced in the life he found here, was doubly incensed at those critics' lack of appreciation of the idealism he had found in America and of the intrinsic beauty that he believed all machines possessed. That man should have developed such wonders based on the discovery of fundamental scientific principles of cosmic scope had, for Pupin, an almost religious significance.

It is interesting to note that criticism of science and spirited response to such criticism by physicists had already begun well in advance of the economic Depression of the 1930's which brought additional criticisms of science that we shall discuss in our next chapter.[75]

Summary

By the end of the 1920's the physicists in America enjoyed many professional advantages. The American Physical Society and publications sponsored by it were increasing in quality and quantity. Florishing intercourse with the European physics community was marked by continued mutual cordiality and increasing appreciation on their part of American scientific prospects. Good financial support was available for the training of aspirants to the profession, for the research of mature scientists and for building and equipping of research facilities. As Ehrenfest put it, the derivative was positive on almost all scores. The one cloud in the picture stemmed from the fact that there were indications it might not remain so due to the impending financial calamity and to the presence of critics who questioned the effect of scientific progress on the quality of human existence.

Part II: The American Response to Quantum Mechanics

Introductory Remarks

With the advent of quantum mechanics, first as the matrix mechanics of Heisenberg in late 1925, then in the complementary form of the wave mechanics of Schrödinger several months later, changes were ocurring in the world of physics which evoked prompt response among the members of the American physics community and profoundly influenced the subsequent activities of many individuals and institutions. American physicists did not lag behind their European colleagues in responding positively to the new approach to atomic theory despite the unfamiliar nature of the language that it required. F. C. Hoyt has recently remarked:[76]

"While the Bohr orbital theory was very slow in gaining general acceptance, particularly in this country, there never seems to have been any real doubt, here or elsewhere, that the matrix mechanics provided at least a formal way out of the difficulties and the inconsistencies in which the Bohr orbital theory had entangled itself. The precarious state of the orbital theory was particularly distressing just at the time I left Copenhagen (1925) and I recall Bohr's remark to the effect that now everything is in Heisenberg's hands."

Virtually all members of the American physics profession were eager to become at least generally informed about the new developments. A much smaller number of Americans sought to thoroughly understand them. Although the number of Americans concerned primarily with the theoretical aspects of physics remained a small percentage of the total membership of the American physics community, there was, nevertheless, during the years 1926-1929, a marked increase in the number of American physicists engaged in theoretical pursuits over that of the preceding half-decade. Whereas in the years 1920-1925 there had been only about a dozen Americans working with the old quantum theory, the number attracted to the new quantum mechanics between 1926 and 1929 was closer to sixty.[77]

In depth study of quantum mechanics drew three groups of Americans: some, such as Kemble, Van Vleck and Slater were "veterans" of three or more years of experience with the old quantum theory; others, such as Condon and Pauling, had more recently begun their study of quantum theory; while still others, such as Oppenheimer and Rabi, arrived on the scene just as the breakthrough achieved by quantum mechanics was being realized.

The pace at which quantum mechanics was developing in the late 1920's made the task of keeping abreast a formidable one. E. U. Condon recalled:[78]

"Great ideas were coming out so fast during that period (1926-1927) that one got an altogether wrong impression of the normal rate of progress in theoretical physics. One had intellectual indigestion most of the time that year, and it was most discouraging."

Similar sentiments were expressed by J. H. Van Vleck in 1927 when he wrote to Kemble:[79]

"You say you hope 'no new explosion of the first magnitude' occurs for a few months, but it seems to me that we are still feeling one wave after another of the explosion (each worse than the last) as new, progressively more involved and general formulations of the quantum mechanics keep on appearing."

Despite the difficulties involved, a goodly number of Americans sought to keep up-to-date with each development made possible by the new techniques. In order to achieve this goal a variety of means were employed. Especially desirable was the opportunity to study at one of the European centers for theoretical physics, and the availability of suitable fellowships in America at that time was a significant factor in bringing Americans into the mainstream of theoretical physics. The next most desirable course was to enjoy direct communication with the European leaders in the development of quantum mechanics who toured and lectured in the United States. Such measures, of course, suffered from restrictive boundary conditions with regard to space and time. On the other hand, while an individual could always, and in almost any location, try to keep up with the burgeoning literature in the new field on his own, that was most difficult to do. In some instances where it was geographically feasible, Americans seeking to cope with quantum mechanics banded together into small study groups for mutual support and enlightenment. In addition, the study of quantum mechanics in America was pursued by individuals holding domestic fellowships, such as those of the National Research Council, at such institutions as the California Institute of Technology and Harvard University. By means of a combination of such measures, several dozen American physicists were successful, during the period 1926-1929, in assimilating the new outlook associated with quantum mechanics into their own professional work.

Among the much larger segment of the American physics community whose research would not be directly affected by the new developments there was, nevertheless, a desire to understand in more general terms the advances that were being made possible through quantum mechanical considerations. Added to this group of physics, there were now increasing numbers of chemists and mathematicians beginning to be seriously interested in quantum mechanics and its relation to their own disciplines. The guest lecturers from abroad gave some presentations that were helpful to such audiences but, to a large extent, until printed expositions became available, there was considerable reliance on oral presentations by those few American physicists available and competent to do so. Furthermore, there was the generation of students just coming into the field of physics for whom suitable courses including the new material had to be devised. As a result, the period from 1926 to 1929 was a particularly hectic one for those who aspired to keep up with the current literature, to share their new knowledge with colleagues and students and, hopefully, to use the new methods to attack still unsolved problems.

In this section we shall consider various activities in which Americans engaged that are clearly traceable to the advent of quantum mechanics, but we delay until our next section consideration of the specific achievements attained by Americans through the use of the new language.

As we examine the dissemination of quantum mechanics among American scientists and its effect upon the profession as a whole, certain new characteristics of American physics, specific to the period of the late 1920's, will be brought to light. For example, while it was almost exclusively among the young American physicists under thirty years of age that quantum mechanics was most attractive, the older generation of American physicists, with few exceptions, came to accept the value not only of the new quantum mechanics, but also of theoretical physics, per se, to a greater extent than had been so previously. As a result, theoretical physicsts attained full professional status in America. Although still in short supply, there were by this time enough of them to bring theoretical expertise to a dozen or more university physics departments and, as individuals, they found their employment prospects definitely favorable. The emergence of a sizeable group of native American theoretical physicists also had its effect on international relations within the physics community. As we mentioned earlier, European physicists took note of the increasing quality and quantity of the theoretical physics being done by the young Americans who came to Europe to study. In addition, those Europeans who lectured in America were struck by the genuine interest and evident capability of some members of their audiences. In these ways the advent of quantum mechanics provided a significant impetus to the "coming of age" of physics in America from both the domestic and international points of view.

The Dissemination of Quantum Mechanics Among American Scientists

1) Foreign Lecturers

We open our discussion of the dissemination of quantum mechanics in America with a consideration of the role played by visitors from abroad for two reasons. The lecturers reached large and varied audiences in their travels, and it was through one of them, Max Born, that many American physicists were first alerted to "the new quantum theory."

Appendix II contains brief summaries of the itineraries of more than a dozen European physicists who visited America between 1926 and 1929. While not all of them came to lecture on quantum mechanics, most of them did. In fact, practically all the leading European developers of quantum mechanics came to America during those years. For example, Born, Dirac, Heisenberg, Hund, Schrödinger and Weyl all lectured at American universities for varying periods in the late 1920's.

In all instances the tradition of American physicsts acting as cordial hosts to the visitors from abroad continued. There was, however, a new note interjected into European-American relations in theoretical physics by the presence of the growing number of young American theorists who were ready to assimilate the most recent developments. Furthermore, in some instances, bonds of personal friendship, enhanced by common scientific interests, sprang up between guests and hosts particularly among the younger physicists. An examination of some of the details associated with the visits of three of the European guest lecturers on quantum mechanics, Born, Dirac and Heisenberg, will serve to illustrate these points.

Max Born came to lecture at the Massachusetts Institute of Technology during the latter part of the Fall Semester of 1925-1926 as part of the Physics Department's program of having at least one such guest during each academic year. It appears[80] that Born was originally expected to lecture on material related to his two recent publications, _Vorlesungen Über Atommechanik_[81] and _Atomtheorie des festen Zustandes_.[82] By a fortunate coincidence the early work on matrix mechanics had just been completed before Born left Europe. Heisenberg's first paper[83] on it had already appeared. Born's own paper with Jordan[84] on the subject was then in press. In Born's words:[85]

"...and the manuscript of a third paper by the three of us (Born, Heisenberg and Jordan) was almost completed. Though the results contained in this third paper left no doubt in my mind as to the superiority of the new methods to the old, I could not bring myself to plunge directly into the new quantum mechanics. To do this would not only be to deny to Bohr's great achievement its due need of credit, but even more to deprive the reader of the natural and marvelous development of an idea."

As a result, of the thirty lectures that Born delivered at the Massachusetts Institute of Technology between November 14, 1925 and January 22, 1926, the first nine were devoted to the old quantum theory. The tenth through the twentieth lecture dealt with the new matrix mechanics, while the final ten were on the lattice theory of rigid bodies. In his discussion of matrix mecanics Born continually introduced new material, such a Pauli's work on the hydrogen atom, as promptly as it came to his attention.

After the completion of his lectures in Cambridge, which included five delivered at Harvard University,[86] Born set out to visit other locations. His wife, who had accompanied him to the United States, had fallen ill and returned to Europe. Thus, Born recalled:[87]

"...I had to travel by myself and proclaimed the new quantum doctrine at many universities. The result was that hordes of Americans...visited Göttingen the next few years."

Among the institutions that Born visited were the California Institute of Technology, the University of California at Berkeley, the University of Wisconsin, the University of Chicago and Columbia University.[88]

During the three months that Born spent at the Massachusetts Institute of Technology he found there "an excellent collaborator"[89] in Norbert Wiener, a young mathematician.[90] Together they worked on developing an operator calculus to be applied to the continuous spectrum. Their joint effort was published in America as "A new formulation of the laws of quantization of periodic and aperiodic phenomena," and in Europe as "Eine neue Formulierung der Quantengesetze für periodische und nicht periodische Vorgänge."[91] According to Born:[92]

"In collaboration with Norbert Wiener I tried to extend the matrix theory of discontinuous energy spectra to more general systems (free particles) with continuous spectra; we developed an operator calculus which came quite close to Schrödinger's method, which was at the time not yet known to us."

Wiener gives his version of the collaboration as follows:[93]

"...Born wanted a theory which would generalize these matrices or grids of numbers into something with continuity comparable to that of the continuous part of the spectrum. The job was a highly technical one, and he counted on me for aid. There is no point in my going into the technique of a piece of work which is not only abstract but was to a certain extent a transitory stage of quantum theory. Suffice it to say that I had the generalization of matrices already at hand in the form of what is known as operators. Born had a good many qualms about the soundness of my method and kept wondering if Hilbert would approve of my mathematics. Hilbert did, in fact, approve of it, and operators have since remained an essential part of quantum theory."

Another young American who worked with Born during his stay in Cambridge was J. C. Slater, but the particular work that was accomplished suffered the fate of a "near miss." In Slater's words:[94]

"New information on matrix methods became available when Max Born arrived for his M.I.T. lectures in November, 1925, fresh from writing the first Born-Jordan paper on matrix mechanics...and at once I started working with him. He turned over to me a preprint of a later paper on the subject...and an idea almost immediately occurred to me, which I worked out during his visit of several months. In my study of classical mechanics, during the last year of my graduate work, I had become acquainted with the power of the Poisson bracket expression in classical mechanics, and I felt that this could be incorporated into matrix mechanics. I worked on this, and before Born left in January, 1926, I had produced a paper in which I showed the equivalence of the Poisson bracket and the commutator, and described how this could be used for proving various quantum-mechanical theorems. This was one of those cases where an idea came almost simultaneously to two people: before I had a chance to send in the paper for publication, the first of Dirac's papers reached Cambridge...with the same results, so I quietly allowed my effort in that direction to expire."

Born seems to have been a successful and persuasive lecturer on the new mechanics. Carl Eckart recalled that he was not attracted by the early matrix mechanics papers of Heisenberg, Born and Jordan but "in...early 1926 Born came to Pasadena and his lucid lectures aroused my interest."[95] Eckart went on to say that he was particularly stimulated by Born's presentation of the operator calculus methods recently achieved with Wiener's collaboration. Following Born's visit to Pasadena, Eckart achieved some notable results on his own which will be discussed in the section dealing with American contributions to quantum mechanics.

J. H. Van Vleck was yet another young American theoretician who established personal contact with Born during his visit to the United States. A series of letters[96] were exchanged between them on the topics of the possibility of Born including a visit to Minneapolis on his itinerary and of the accuracy of Born's citation of publications by Kemble and Van Vleck in Atommechanik. Born was not able to go to Minneapolis, but it was arranged that they should meet in Madison where Born was scheduled to lecture at the University of Wisconsin in March.

The University of Wisconsin's program of guest lecturers in physics from Europe was successful in attracting several notables associated with quantum mechanics. Some, such as Max Born and Erwin Schrödinger[97] gave a few lecturers in a short period. Others, such as Peter Debye and P. A. M. Dirac spent longer times in Madison. Debye was acting professor of mathematical physics during the spring semester of 1927.[98] Dirac came during the Spring of 1929. The succession of so many distinguished visitors made the University of Wisconsin particularly attractive to an American physicist such as J. H. Van Vleck who had been trying to cope with the rapidly developing quantum mechanics in the geographical isolation of Minneapolis.[99] Van Vleck made the transition to Madison in the Fall of 1928, and enjoyed the companionship of Dirac the following Spring. In an article included in the Dirac Festschrift, <u>Aspects of Quantum Theory</u>[100] Van Vleck has warmly described his personal and professional association with Dirac whose work was already influencing Van Vleck's own development before the semester they shared at the University of Wisconsin. Their cordial relations, enhanced by a common enjoyment of the out of doors, continued during subsequent visits that Dirac made to the United States.[101]

The Spring of 1929 also saw Werner Heisenberg coming to America to lecture on quantum mechanics, principally at the University of Chicago. He did, however, also lecture at a number of other institutions from coast to coast.[102]

Heisenberg has noted that not only did he find many Americans extremely interested in quantum mechanics but also that they were already acquainted with it.[103] In particular, he recalled long talks with Van Vleck whom he considered to be "one of those who really knew the whole thing."[104]

Heisenberg's American voyage of 1929 seems to have been a very pleasant and illuminating one for him. He has written of some of his experiences as follows:[105]

"The New World cast its spell on me right from the start. The carefree attitude of the young, their straightforward warmth and hospitality, their gay optimism--all this made me feel as if a great weight had been lifted from my shoulders. Interest in the new atomic theory was keen, and since I had been invited by a fairly large number of universities in many parts of the country, I became acquainted with many different aspects of American life. Wherever I stayed for more than a few days, I struck up new acquaintanceships that started with tennis, boating or sailing parties and quite often ended in long discussions of recent developments in atomic physics."

Following the sightseeing tradition of European visitors Heisenberg availed himself of the opportunity to visit such American attractions as Yellowstone Park where he was joined by P. A. M. Dirac. Together they continued their journey westward, returning to Europe by way of Asia.[106]

The foregoing discussion points up the fact that, by the late 1920's, the American reception accorded European visiting scientists had moved in the direction of greater cameraderie coupled with mutual respect, and less awe of "Herr Professor Doktor." In part, this may be ascribed to the youthfulness of some of the visitors but it is also indicative of the growth of Americans toward coequal status in the international community of physicists.

2) Study Abroad

In the years immediately following the advent of quantum mechanics there were more Americans studying at European centers of theoretical physics than at any previous time. Every American physicist who went to Europe between 1926 and 1929 was certain to meet several of his compatriots there not only from the physics profession but also, occasionally, from the chemistry and mathematics professions.

In the table to be found on the following three pages we have compiled a summary of information regarding the European travels of about thirty such American visitors.[107] Certain characteristics emerge from this survey, among which are the following:

a) the travellers went to Europe from a number of different institutions scattered across the United States, but the largest representations were from Harvard (6), the California Institute of Technology (5),

Americans Studying at European Centers of Quantum Physics 1926-1929

Comment: The information summarized in this table has been obtained from a variety of sources. In the interest of conserving space the following code will be used to identify the sources tapped for each particular entry: (1) Published records of funding agencies; (2) Autobiographical articles or books, listed in detail in the Bibliography; (3) Biographies or biographical articles in Festschrifts, Memorial Volumes or obituaries, listed in detail in the Bibliography; (4) American Men of Science; (5) AHQP or AIP-CHP Interviews; (6) Private communications, oral and written, between the author and scientist. In a few instances, the information presented was found in the reminiscence articles or interviews of some other scientist; such cases are designated as (2) or (5) with the name of the source appended.

The name of the American institution associated with each scientist listed is that with which he was affiliated at the time of his departure for Europe.

Name	Institution	Period	Auspices	European Locations	Sources
W. P. Allis	M.I.T.	1929-1931		Munich	(2)Slater Sommerfeld
G. Breit	Carnegie Inst.	1928		Zurich	(4)
E. U. Condon	U.of Cal.Berkeley	10/26-10/27	NRC,IEB	Göttingen,Munich	(1),(2)
D. M. Dennison	U.of Michigan	1924-1927	IEB,Mich.	Copenhagen,Zurich Munich, Leiden Cambridge	(1),(2)
Jane Dewey	M.I.T.	1925-1927	Barnard Col.	Copenhagen	(4),(6)
C. Eckart	Cal.Inst.of Tech.	1927-1928	Guggenheim	Zurich, Munich Berlin, Leipzig	(1),(4),(5)
G. E. Gibson		1927	Guggenheim	Göttingen	(1)

Name	Institution	Period	Auspices	European Locations	Sources
V. Guillemin	Harvard Univ.	1925-1927	Harvard,NRC	Munich (Ph.D.1926) Leipzig, Göttingen	(4),(6)
W. V. Houston	Cal.Inst.of Tech.	1927-1928	Guggenheim	Munich, Leipzig Copenhagen	(1),(3),(4)
F. C. Hoyt	U.of Chicago	1927-1928	Guggenheim	Berlin	(1),(4),(5)
E. Hutchisson	U.of Pittsburgh	1929-1930	Pittsburgh, private funds	Berlin	(4),(6)
E. C. Kemble	Harvard Univ.	2-7/1927	Guggenheim	Göttingen,Munich	(1),(4),(6)
E. H. Kennard	Cornell Univ.			Göttingen	(2) Hylleraas
R. J. Kennedy	Cal.Inst.of Tech.	1928	Guggenheim	Munich, Berlin	(1)
V. F. Lenzen	U.of Cal.Berkeley	1927-1928	Guggenheim	Göttingen,Zurich	(1),(4),(6) (2) Birge
F. W. Loomis	N.Y.U.	1928-1929	Guggenheim	Göttingen,Zurich	(1),(4),(6)
H. Margenau	Yale Univ.	1929-1930	Yale Univ.	Munich, Berlin	(4),(6)
R. S. Mulliken	N.Y.U.	Summer 1927		Göttingen, Zurich	(2)
H. H. Nielsen	U.of Mich.	1929-1930	Amer.Scand.	Copenhagen	(1)
J. R. Oppenheimer	Harvard Univ.	1925-1927		Cambridge,Göttingen (Ph.D. 1927)	(1),(3),(4)
		1928-1929	IEB,NRC	Leiden, Zurich	

Name	Institution	Period	Auspices	European Locations	Sources
L. Pauling	U.of Cal.Berkeley	1926-1927 (19 mos.)	Guggenheim	Munich, Zurich Copenhagen	(1),(2),(4),(5)
B. Podolsky	Cal.Inst.of Tech.	1928-1930	NRC	Leipzig	(1),(4)
I. I. Rabi	Columbia Univ.	1927-1929	Columbia Univ. IEB	Munich, Zurich Hamburg, Leipzig Copenhagen	(1),(4),(5)
H. P. Robertson	Cal.Inst.of Tech.	1925-1928	NRC	Göttingen, Munich	(1),(4)
J. C. Slater	Harvard Univ.	6/29-1/30	Guggenheim	Leipzig, Zurich	(1),(2),(4),(5)
J. A. Stratton	M.I.T.	1925-1927	M.I.T.	Zurich(Sc.D.1927)	(4),(5)Rabi,(6)
L. A. Turner	Princeton Univ.	1929-1930	Guggenheim	Göttingen	(1),(4)
M. S. Vallarta	M.I.T.	1927-1928	Guggenheim	Berlin, Leipzig	(1),(4)
J. H. Van Vleck	U.of Minn.	Summer 1926		Copenhagen,Oxford Cambridge	(2)
W. H. Williams	U.of Cal.Berkeley	1926-1927		Göttingen	(2)Birge,Condon (5)Condon
E. Witmer	Harvard Univ.	1927-1928	NRC,IEB	Göttingen	(1),(4)
C. Zener	Harvard Univ.	1929-1930	Harvard Univ.	Leipzig	(4),(6)

the University of California at Berkeley (4), and the Massachusetts Institute of Technology (4);

b) the new Guggenheim Memorial Foundation Fellowship program, inaugurated in 1925,[108] played a significant role in making these trips possible; some forty percent of the visitors were subsidized through its auspices; the International Education Board[109] and the National Research Council, both financed with Rockefeller monies, were the next strongest supporters of European study by physicists during this period;

c) the most popular European destinations were Göttingen and Munich, tied for first place, followed by Zurich, Leipzig, Berlin and Copenhagen as the next most attractive sites, with Cambridge, Leiden and Hamburg also represented.

It is appropriate to insert now some information explaining the particular nature of the attraction of the European cities just named for students of quantum physics at that time.[110]

The Institutes of Theoretical Physics at the Universities of Göttingen and Munich were presided over by Max Born and Arnold Sommerfeld, respectivly. In addition, at Göttingen there were Werner Heisenberg (1925-1926) and Pascual Jordan (1926-1927) as well as the noted experimentalist in problems of atomic physics, James Franck. Furthermore, the Mathematical Institute at Göttingen was especially strong[111] and members of it, such as David Hilbert, became interested in the ways in which quantum mechanics was requiring physicsts to use areas of mathematics hitherto unfamiliar to them.[112] All in all, the approach to quantum mechanics at Göttingen was much more mathematically sophisticated than elsewhere.[113]

In Zurich there were two institutions where fundamental work related to quantum mechanics was being carried out during this period. Erwin Schrödinger was associated with the University of Zurich from 1921 to 1927 and it was there that his wave mechanics evolved during the first half of 1926. Peter Debye was at the Eidgenössische Technische Hochschule from 1920 to 1927. Wolfgang Pauli went there in 1928. In addition, Hermann Weyl was a member of the mathematics faculty of "E.T.H." from 1913 to 1930.

Debye and Heisenberg both accepted posts at the University of Leipzig in 1927 and were joined there by Friedrich Hund in 1929.

The University of Berlin, during the years 1926-1929, had Albert Einstein and Max Planck on its faculty, but from the point of view of the new quantum mechanics, it was the arrival of Schrödinger in 1927 that was more significant.

The Institute of Theoretical Physics at Copenhagen, under the direction of Niels Bohr continued to be a gathering place not only for visitors from abroad but also for the European members of the "travelling seminar."[114]

The Cavendish Laboratory was the center of experimental activity at Cambridge, England but theoretical studies were also carried on there by R. H. Fowler and, after 1927, most notably by P. A. M. Dirac.

Paul Ehrenfest was at the University of Leiden and, not far away, at the University of Utrecht, were H. A. Kramers and L. S. Ornstein.

Until his departure for Zurich in 1928 Pauli was at the University of Hamburg where he was joined by Jordan in 1927.

It was to these European sites that dozens of American physicists were attracted soon after the advent of quantum mechanics. Within the group that we have surveyed there was a wide variety of individual experiences. Some stayed only several months, others a year and a half or more. Some were fresh post-doctoral students, a few earned their doctorates in Europe, while still others already had several years of professional experience. Some achieved significant results and international recognition for the work they accomplished during their stay overseas, while others felt outclassed in the sophisticated European scientific circles. Some were frankly homesick and returned with an eagerness to improve the state of physics in the United States.[115]

We now turn to a consideration of some details associated with the European experiences during the years 1925-1929 of one young American, J. Robert Oppenheimer, not because he was typical of the group who went abroad to study theoretical physics, but precisely because he was not. Throughout his life Oppenheimer was a unique figure among American physicists and it was during the period of our present focus that he emerged as a most promising member of the American physics profession.[116]

Oppenheimer arrived in Europe in 1925 soon after completing his undergraduate studies at Harvard University in three years. He had first come to the attention of the physics department there in the Spring of 1923, at the end of his freshman year, when he sought admission to a course in thermodynamics without having completed the normal prerequisite of the regular departmental introductory course in physics.[117] As partial evidence of his readiness for the more advanced course, Oppenheimer submitted a list of some dozen related works that he had already read. The list included treatises by Ostwald, Gibbs, Nernst, Jeans and Poincaré. The physics department granted his request[118] and two years later he was awarded his bachelor's degree in physics from Harvard summa cum laude.

Oppenheimer's first year abroad was spent at Cambridge studying with R. H. Fowler and resulted in the publication of two papers involving the use of the new quantum mechanics.[119] From there Oppenheimer moved on to Göttingen to work with Born. Oppenheimer's doctoral dissertation, written under Born's direction, was entitled "Zur Quantentheorie kontinuierlicher Spektren."[120] According to Science Abstracts:[121]

"In this mathematical paper the general theory of systems with extended spectra is first discussed, following which application is made to the hydrogen atom, the continuous Röntgen absorption coefficients, the polarization and intensity distribution of the Brems radiation, and the velocity and directional distribution of the photoelectrons. The Compton and Kramers formulae are found to be confirmed."

In addition to his thesis research, during this period Oppenheimer collaborated with Born in writing a paper dealing with the quantum theory of molecules from the point of view of wave mechanics.[122]

Born was pleased with this young American student. In February 1927 Born wrote a letter to S. W. Stratton, then President of the Massachusetts Institute of Technology, in which he noted that there were five Americans studying with him at Göttingen at that time and remarked "One man is quite excellent, Mr. Oppenheimer, who studied at Harvard and in Cambridge, England."[123] When E. C. Kemble arrived in Göttingen later that year, he was able to write back the following report on his former pupil to Theodore Lyman:[124]

"Oppenheimer is turning out to be even more brilliant than we thought when we had him at Harvard. He is turning out new work very rapidly and is able to hold his own with any of the galaxy of young mathematical physicists here."

Another physicist whom Oppenheimer had favorably impressed was J. H. Van Vleck. In recommending him for a National Research Council Fellowship, Van Vleck wrote that Oppenheimer was "a man of outstanding ability and keenness of perception in mathematical physics."[125]

Oppenheimer was awarded the Fellowship and returned to the United States in the Summer of 1927 after completing his dissertation and being awarded his doctorate by the University of Göttingen. The following academic year he divided his time among Harvard University, the University of California at Berkeley and the California Institute of Technology.[126] Oppenheimer was not yet ready, however, to settle down to an academic appointment of any permanence and decided to spend another year abroad. This time he went first to Leiden for study with Ehrenfest who then sent him to Pauli in Zurich "for discipline."[127]

When Oppenheimer returned to America in 1929 to take up a joint appointment at the University of California at Berkeley and at the California Institute of Technology, he was twenty-five years old, had already published more than a dozen significant papers on the mathematical aspects of quantum mechanics, and was a recognized member of the international community of theoretical physicists.[128]

For most American physicists who went abroad for study during this period, their European sojourns were episodes in their professional development which influenced in varying degrees their subsequent careers in the United States. An exception to this pattern is found in the case of Ralph Kronig.[129]

Following his postdoctoral trip to Europe in 1924 that we discussed in our previous chapter, Kronig returned to a lectureship at Columbia University in January 1926. A year and a half later he was awarded a fellowship by the International Education Board for study in "Denmark, Switzerland and the Netherlands."[130] In November 1927, soon after his arrival in Europe, Kronig received an invitation from Pauli to become his assistant at Zurich were he (Pauli) was about to assume the chair of Theoretical Physics at the Eidgenössische Technische Hochschule. Kronig accepted the position about which Pauli had written:[131]

"Lästige Verpflichtungen wären für Sie kaum damit verbunden; Ihre Aufgabe wäre, jedesmal, wenn ich etwas sage, mir mit ausführlichen Begründungen zu widersprechen."

Kronig has, ever since, remained in Europe. While it is highly probable that Kronig really felt more at home in Europe than in the United States since he had lived in Europe the first fifteen years of his life, he had received his undergraduate and graduate training at Columbia University and, as we shall see in our next section, was active in the dissemination of quantum mechanics among American physicists during the eighteen months that he taught at Columbia.

3) Autodidactic Groups in the United States

The frequent geographical isolation of American physicists from each other as well as from the European physics community was especially trying during the first few years after the introduction of quantum mechanics, a period that was often as perplexing as it was exciting for those who were attempting to keep abreast of the latest developments. In some instances, however, small, informal study groups were formed in areas where there existed a nucleus of interested physicists who found it helpful to share, for example, their reading of the latest issues of Zeitschrift für Physik outside the formal structure of a single academic institution. New York City and Washington, D.C. were two such sites.

Ralph Kronig was the prime mover of the group that met in New York City until his departure for Europe in 1927 but the group continued to function for several years more with a constantly changing membership.[132] The group met informally and attendance fluctuated due, in part at least, to members changing their places of employment or obtaining fellowships for study abroad. Of the original group, I. I. Rabi and Shou Chin Wang, both graduate students at Columbia University at the time, were particularly faithful members. Kronig has recently provided the author with the following summary of his recollections of the group's activity as follows:[133]

"During the period January 1926 to June 1927 I held a teaching position at Columbia University, and I gathered there the group of people to which you referred in your letter. We met in the building of the physics department, I would say about once a week, often on Saturday or Sunday afternoons. The firm core of the group consisted, besides of myself, of I. Rabi, then working on his experimental thesis about the magnetism of salts in the iron group, of S. C. Wang, still studying, and of R. A. Carrie, who had specialized in mathematics, but has later switched to history...He had then a teaching position at Barnard College. Two other persons, who attended our meetings quite often, were a band spectroscopist F. W. Loomis and an experimental physicists Cox, both from New York University...

Our club met quite informally. One of us gave an introduction on some recent paper of theoretical interest. A great deal of attention naturally was given to the publications of Schrödinger on wave mechanics, which came out in the course of 1926. Our efforts to follow the new achievements led to Rabi and myself attacking the problem of the symmetrical top. We learned at the same time a lot about hypergeometric functions. Then there were the papers by Born in which the interpretation of Schrödinger's wave function as a probability amplitude was first proposed. Until then attention had been chiefly directed to closed atomic systems, possessing definite stationary states with discrete energy values. Born's ideas now opened the possibility of handling open systems, such as occur in collision problems. Of course, the philosophic implications of the probability viewpoint gave rise to an animated exchange of opinions. In 1926 I developed a theory of dispersion of X-rays by matter, in which I proposed one of the so-called dispersion relations, allowing me to compute the index of refraction if the absorption coefficient is known as a function of frequency. This work was stimulated by the precise measurements of the index of refraction for X-rays, carried out by Professor B. Davis at Columbia University on calcite with the help of an ingenious double X-ray spectrometer. These results formed another subject for our club. The susceptibility measurements of Rabi got me interested in the theory of dielectric susceptibilities, which I considered for diatomic and symmetrical top molecules. From this arose a correspondence with J. H. Van Vleck in the Spring of 1927, he having become interested in the same subject.

The club owned jointly a coffee percolator, which, however, was employed to brew tea as stimulant for our mental exertions. When the meeting was finished, it was a regular custom to dine in a Chinese restaurant, where Wang took care that we were not given chop-suey but real Chinese dishes, and where we learnt to eat with sticks. Sometimes Rabi's wife Helen joined us."

A similar study group was formed in the Washington area from among physicists at the Carnegie Institution and the National Bureau of Standards, such as Gregory Breit and Otto Laporte. Again the membership fluctuated but, years later, Laporte recalled[134] his giving a seminar on the first three Schrödinger papers which was attended by about fifteen participants. The New York and Washington groups were aware of each other's existence and Laporte also recalled participating in one of the meetings of the New York group at the time when the pq-qp relation was the topic of discussion. Later, after Rabi returned from Europe in 1929 to a position at Columbia University he shared the responsibility for the seminar with Breit who had, by then, moved to a position at New York University.

A smaller, intramural autodidactic group functioned at the Massachusetts Institute of Technology during this period.[135] Again the members were motivated by the desire to keep up with the latest developments but found themselves unprepared for the mathematical language required. By joining forces they hoped to provide each other with mutual assistance and encouragement. In this instance three young instructors, W. P. Allis, N. H. Frank and J. A. Stratton met regularly during the academic year 1928-1929 while seeking to cope with Weyl's recently published _Gruppentheorie und Quantenmechanik_.[136]

Seminars such as those described above served an important interim function in America during the late 1920's when theoretical physics itself was in such a state of flux and the study of it was further hampered by the unfamiliar nature of the required mathematics, by the paucity of printed expositions and of formal university courses at a suitable level, and by the instability of the local membership. Welcome to participate were such diverse figures as young graduate students just being introduced to quantum theory, postdoctoral physicists just back from, or about to leave for, European study and recently arrived members of the faculty from nearby institutions. By 1930, when international travel was diminishing and more American universities could provide some expertise in quantum mechanics, the informal study group became less important.

In a few instances in the late 1920's, young men in America while not working completely alone, managed to become acquainted with the current state of quantum mechanics with such limited professional contacts that the term "autodidactic" is much more applicable to their activity than is the term "group." We shall now briefly recount the experiences of two such individuals, not only because they illustrate the phenomenon involved, but also because both young men continued their activity in theoretical physics in later years and will be mentioned again further along in our discussion.

Philip Morse had already decided that he was more interested in the theoretical than in the experimental aspects of the investigations that formed his doctoral reserch at Princeton University with K. T. Compton between 1926 and 1929 when he was awarded his Ph.D. By happy coincidence, a summer job in 1928 of a purely practical nature in electronics was found for him in Ann Arbor, Michigan just at the time when the Summer Symposia in Theoretical Physics was getting underway there. By means of attending some of the lectures by H. A. Kramers whenever time could be spared from his job, and by informal discussions in the evening with Kramers and with E. C. G. Steuckelberg, a post-doctoral student at Princeton from Switzerland who was also in Ann Arbor for the summer, Morse became sufficiently familiar with quantum mechanics to have formulated some problems using the new techniques by the end of the summer. After his return to Princeton that Fall he and Steuckelberg collaborated on solving those problems. Also remarkable is the fact that, during the academic year 1928-1929 Morse became the coauthor with E. U. Condon of the first textbook on quantum mechanics to be published by American authors without having himself ever taken a course in quantum mechanics.[137]

Also at Princeton University at this time, but as an undergraduate studying chemistry, was E. Bright Wilson. By the time he was a senior, in the Fall of 1929, he had learned of the existence of quantum mechanics, although not through any courses given within the chemistry department.[138] Rather, it is Wilson's recollection that he was introduced to quantum mechanics by George Kimball, a graduate student in the department who was also giving lectures on it to older members of the chemistry faculty. Since Condon was away from Princeton during the academic year 1929-1930 Wilson could not take a course in quantum mechanics with him. Instead he wrote a senior thesis entitled "The Elementary Principles of Quantum Mechanics," based on his own study of the new Condon amd Morse text and numerous other sources. He was aided in this endeavor by H. P. Robertson, who had recently come to Princeton from a period of study at Göttingen and Munich as a National Research Council Fellow. This study of quantum mechanics by Wilson is particularly noteworthy in view of the facts that he was an undergraduate student of chemistry at the time and it provided the initial background for his subsequent work in that field which will be discussed in our next chapter.[139]

From the above considerations it becomes clear that in America in the late 1920's, it was possible for interested individuals to become acquainted with recent developments in quantum mechanics outside of the more conventional routes of university study at home or abroad. Also, it becomes clear that it was primarily to young physicists and chemists that the new material was atractive.

4) Graduate and Postgraduate Study of Quantum Mechanics at American Universities

The preceeding discussion of the dissemination of quantum mechanics among Americans by means of autodidactic groups, study abroad and European guest lecturers should not lead the reader to believe that there were no American institutions in the late 1920's where students could take courses and do guided research in quantum mechanics. There was, in fact, a growing number of institutions where such study was possible, but the late 1920's was largely a transitional period and it was not until after 1930, as we shall see in our next chapter, that theoretical physics, including quantum mechanics, became a strong component of the curriculum at most of the major universities in the United States.

We now turn our attention to three indices of the extent to which such study was possible in the years 1926-1929, namely, courses on quantum mechanics as announced in university catalogues, Ph.D. theses involving quantum mechanical presentations, and postdoctoral research fellowships awarded for study related to quantum mechanics at American universities.

On the next page we have compiled in tabular form a list of representative institutions where formal course work in quantum mechanics was available together with the names of the faculty members responsible for those courses.[140]

Those were difficult years in which to attempt to give a course in quantum mechanics: the material was evolving rapidly; there was little help available in the form of textbook presentations;[141] physics depart-

Some Opportunities for Formal Study of Quantum Mechanics in America

1926-1929*

Institution	Faculty	Source**
Brown University	H.B. Phillips	f
University of California Berkeley	H.W. Williams	u
	H.B. Phillips, Summer 1928	f
Cal. Inst. Tech.	P. Epstein, L. Pauling	u,f
University of Chicago	C. Eckart, F. Hoyt, R. Mulliken	u,f
	W. Heisenberg, Spring 1929	
Columbia University	R. Kronig 1926-1927	u,f
	E.U. Condon, Spring 1928	f
Cornell University	E.H. Kennard 1927-1928	u
Harvard University	E.C. Kemble, J.C. Slater	u,f
	F. Hund, Spring 1929	u,f
Johns Hopkins University	G. Breit, H.C. Urey	u,f
University of Michigan	D.M. Dennison, S.A. Goudsmit	u,f
	O. Laporte, G. Uhlenbeck	
University of Minnesota	J.H. Van Vleck (until 1928)	u,f
Princeton University	E.U. Condon, H. Weyl, 1928-1929	u,f
Stanford University	J.C. Slater, Summer 1926	u,f
	J.H. Van Vleck, Summer 1927	
University of Wisconsin	P. Debye, Spring 1928	f
	J.H. Van Vleck 1928-	
	P.A.M. Dirac, Spring 1929	
Yale University	R.B. Lindsay	f

*Survey covers the academic years 1926-1927, 1927-1928 and 1928-1929.

**u indicates listing in catalogue of institution or in departmental history.

f indicates statement made by individual faculty member in American Men of Science, in a retrospective article, in AHQP or CHP Interview or in private communication with author. (Note: in the case of H.B. Phillips, information was provided by his widow.)

ment faculties were in a generally unstable state, so far as competence in quantum mechanics was concerned, due to members taking leaves of absence for study abroad[142] and/or being on the verge of succumbing to enticements from other institutions.

On the other hand, it was an exciting time with students from chemistry as well as physics eager to learn about the new developments and institutions making considerable effort to see that such instruction was available.[143] In addition, those faculty members capable of doing so welcomed the opportunity to spread the word for the new physics.

With regard to the impact of quantum mechanics on students of chemistry, in some instances they simply registered for offerings within the physics departments. It is noteworthy, however, that at the California Institute of Technology Linus Pauling introduced in 1927 a quantum mechanics course into the chemistry department. In addition, Pauling was influential in alerting chemistry students elsewhere to the importance of quantum mechanics for their discipline. In Pauling's own words:[144]

"I have no doubt that the development of quantum mechanics was of great importance to chemistry. After my return from Europe in the Fall of 1927 I had many opportunities to talk about the subject and I emphasized to students in chemistry in many universities over and over again, the importance of learning physics and quantum mechanics in order to have a good understanding of chemistry."

In 1927 Pauling had become Assistant Professor of Theoretical Chemistry at the California Institute of Technology but, beginning in 1929, he also spent two months of every year for the next five years as a Visiting Lecturer in Chemistry and Physics at the University of California at Berkeley where he gave a course in the nature of the chemical bond from a quantum mechanical perspective.[145] Chemistry graduate students at the Johns Hopkins University also were introduced to quantum mechanics through the efforts of Harold Urey, a member of the chemistry faculty there between 1924 and 1929.[146]

At some institutions the courses in quantum mechanics were brief, one-semester introductory presentations, frequently given by a visiting faculty member. The very fact that such arrangements existed, however, is an indication that the regular, established faculty recognized the necessity of promptly beginning to incorporate the new quantum theory into their curricula. We have already discussed several instances of European lecturers acting in such a capacity but it should be recognized that Americans also did so. For example, J. C. Slater and J. H. Van Vleck lectured on quantum mechanics at Stanford University during the summers of 1926 and 1927 respectively; Gregory Breit and E. U. Condon, neither of whom held regular academic appointments at the time, lectured at the Johns Hopkins and Columbia Universities respectively; and H. B. Phillips travelled once a week from Cambridge to Providence to introduce Brown University students to quantum mechanics.[147] In addition, J. H. Van Vleck, during the yars that he taught at the Universities of Minnesota and Wisconsin, journeyed to Iowa City nearly every year to give a series of lecturers on quantum mechanics at the University of Iowa at the invitiation of G. W. Stewart,[148] and J. C. Slater lectured on wave mechanics at the University of Kentucky in the Spring of 1929.[149] All of these instances make clear the fact that, through the introduction of guest lecturers, students at a large number of American universities were promply alerted to the recent breakthrough in physics that was achieved in the late 1920's.

At some institutions, such as Cornell[150] and Yale[151] Universities for example, instruction in quantum mechanics was available from regular faculty members, but only in a very limited way. On the other hand, there was an unusually favorable situation for the study of quantum mechanics at the University of Minnesota between 1926 and 1928 when John Van Vleck was professor of theoretical physics there.

I am indebted to one of those students, Robert B. Whitney for some personal recollections[152] of his experience with Van Vleck's course and for copies of the lecture notes which he took and of the assignments and examinations that were given to the students. A few attendant details of this course will be set forth, not to imply that this was typical of what American students had available to them, but rather because it is remarkable that such a course was given in Minneapolis in 1927--hundreds, if not thousands of miles away from the centers of activity in quantum mechanics.

At that time Van Vleck's course was organized into a sequence which ran through an academic year of three quarters.[153] The first section dealt with the early development of quantum theory, the latter two sections with the recent quantum mechanics as developed by Heisenberg, Schrödinger and Dirac. There were about ten students in attendance, including E. L. Hill, V. Rojansky and W. H. Brattain, all of whom were then graduate students in the physics department. Whitney recalls that they all found the course most exciting--a fact due in large measure to Van Vleck's own command of the material and his constant up-dating of it with items from recent issues of foreign and domestic journals. In this way the students were enabled to feel themselves to be at the very frontiers of contemporary physics.

While it appears likely that well over a hundred American students of physics and chemistry took introductory courses in quantum mechanics during the years 1926 to 1929, less than a dozen achieved a level of competency adequate for the writing of a doctoral dissertation which involved it. This is not at all surprising in view of the scarcity of suitable thesis supervisors (some of those listed as teaching courses in quantum mechanics were not actively engaged in research in it) and in view of how unfamiliar to the average student of physics in the 1920's were the concepts and techniques required by the new quantum mechanics.

On the next page we have tabulated the results of our investigation into the writing of such theses.[154] The first group of theses listed were entirely theoretical and dealt specifically with quantum mechanics. The second group comprised an extension of the molecular spectra studies that had been underway for some years at the University of California at Berkeley and at Harvard University. All of the theses in this group were concerned with the interpretation of observed spectra associated with specific compounds. Data were presented and considered with due recognition of recent theoretical developments. At that time, however, the quantum mechanical analysis of molecular spectra had not yet attained any degree of refinement.

While our emphasis in this section is primarily on work done at American institutions, we should not overlook the fact that at least two American students completed their doctoral studies at European

Quantum Mechanical Ph.D. Theses Written at American Universities

Prior to 1930

Name Year	Institution Supervisor	Topic Related Publication
E. L. Hill 1928	Univ. of Minnesota Van Vleck	Spin Multiplets Phys.Rev.32(1928)250-272
B. Podolsky 1928	Cal. Inst. of Tech. Epstein	Atomic Dispersion Proc.N.A.S.14(1928)253-258
C. F. Richter 1928	Cal. Inst. of Tech. Epstein	Spinning Electron of H atom Proc.N.A.S.13(1927)476-479
V. Rojansky 1928	Univ. of Minnesota Van Vleck	Stark Effect Phys.Rev.33(1929)1-15
S. C. Wang 1928	Columbia University (see Note 3.154)	Normal Hydrogen Molecule Phys.Rev.31(1928)579-586
M. Muskat 1929	Cal. Inst. of Tech. Epstein	Continuous Spectra of Hydrogen-like atoms Proc.N.A.S.15(1929)405
R. J. Seeger 1929	Yale University Lindsay and Page	Quantum Theory Critique Proc.N.A.S.17(1931)301-310 Ibid. 18(1932)303-310
F. H. Crawford 1928	Harvard University Kemble	Zeeman Effect in CO Phys.Rev.30(1927)438-457
A. Christy 1928	U.of Cal.-Berkeley Birge	TiO Band Spectra Phys.Rev.33(1929)701-729
J. L. Dunham 1929	Harvard University Kemble	Harmonic Band of HCl Phys. Rev. 34(1930)438-452
J. E. Rosenthal 1929	New York University Jenkins	BeO Band Spectra Phys.Rev.33(1929)163-168 Proc. N.A.S.15(1929)381-387

institutions during this period with theses written on quantum mechanical topics. We have already discussed[155] the fact that J. R. Oppenheimer wrote his dissertation at Göttingen under the guidance of Max Born. In addition, J. A. Stratton received his Sc.D. degree from the Eidgenössische Technische Hoshschule in Zurich in 1927 for a thesis on the wave mechanical scattering of light by hydrogen atoms, written under the guidance of Peter Debye.[156]

Quantum mechanical studies at a still more advanced level were carried on at a few American institutions by a number of postdoctoral fellows, primarily those holding National Research Council grants for domestic study.

Upon examining the tabulation on the next page of the names of the National Research Fellows[157] together with the institutions where they chose to study quantum mechanics, one is immediately struck by the preeminence of the California Institute of Technology[158] and Harvard University as focal points for those young scholars. This is not surprising, however, in view of our earlier discussion of the importance of those two institutions for the study of quantum physics in the years preceding the introduction of quantum mechanics. The inclusion of New York University and the University of Wisconsin is readily explained when we note that F. W. Loomis and R. S. Mulliken were both at New York University[159] in the years 1926-1929 and J. H. Van Vleck went to Wisconsin in 1928.

In a number of instances quite notable research was accomplished by these young Fellows, but we shall defer specific mention of them until our next section where we shall discuss in some detail the

National Research Fellowships for Study in Quantum Mechanics at American Institutions, 1926-1929

Name	Period	Institution
G. Dieke	1926-1927	Cal. Inst. of Tech.
C. H. Eckart	1925-1927	Cal. Inst. of Tech.
V. Guillemin	1926-1928	Harvard University
E. L. Hill	1928-1930	Harvard University
W. V. Houston	1925-1927	Cal. Inst. of Tech.
F. A. Jenkins	1926-1927	Harvard University
J. R. Oppenheimer	1927-1928	Harvard and C.I.T
B. Podolsky	1928-1929	University of California
O. K. Rice	1927-1929	Univ. of Cal. and C.I.T.
E. O. Salant	1927-1929	New York University
S. C. Wang	1928-1929	Univer. of Wisconsin
E. E. Witmer	1925-1927	Harvard University

contributions made by American scientists within fields opened up by quantum mechanics. In this section our focus has been on delineating the opportunities that existed in the United States between 1926 and 1929 for students to keep abreast of the latest developments in quantum mechanics with the expectation that its new outlook would become part of the perspective with which their own research would be undertaken. Thus our emphasis has been primarily on determining at what institutions and under whose guidance students could achieve this goal. Our findings in this area point, once again, to the conclusion that America had not yet achieved first rank standing in both quality and quantity. The American potential, which would be realized within the next decade, however, was clearly discernible. Many American institutions quickly sought to integrate quantum mechanics into both their physics and chemistry curricula, and at a few places the faculty was strong enough in this area to provide research guidance of very high quality.

5) Symposia, Review Articles and Books on Quantum Mechanics

In addition to the formal, detailed kind of study of quantum mechanics which we have just discussed, there were, during the late 1920's, numerous efforts made to bring some understanding of the recent developments in physics to American audiences that were scientifically literate, but which, for the most part, were not immediately affected in their own research activity by the advent of quantum mechanics.

The details associated with such wide-audience dissemination of quantum mechanics are of interest in illuminating both the kinds of audiences reached and the ways in which initiative was taken to satisfy their need.

Symposia arranged as part of the program at regularly scheduled scientific meetings offer the quickest way of reaching those in attendance, and this was done at some meetings of the American Physical Society. The list of APS Symposia, in Appendix IV, Part 5 of this thesis, reveals that during the period 1926-1929 the highest level of this type of activity in the area of quantum mechanics was achieved in 1928. In that year there were formal discussions of quantum mechanics at both the November meetings in Minneapolis and at the December meeting in New York City. In the latter case, the discussion was sponsored jointly by the American Physical and Mathematical Societies. Some idea of the interest generated by this program may be grasped by noting that a full day was devoted to it, the attendance was about 400,[160] and all of the papers presented were subsequently published in the Journal of the Franklin Institute.[161] One of the participants, John Van Vleck, has recalled[162] that the main purpose of this conference was "to brief older physicists on quantum mechanics." In this connection, it is interesting to note that two-thirds of the fifteen participants were under thirty years of age at the time.

Another example of a symposium including presentations on quantum mechanics, that took place in April 1928, is of especial interest because it was sponsored by the American Chemical Society during its regular meeting held in St. Louis. This symposium was a cooperative venture of Society's Division of Physical and Inorganic Chemistry and of the Organic Division[163] aimed at taking stock of the latest ideas on atomic structure and valence. There were about a dozen participants, including three invited physicists, K. K. Darrow, G. E. M. Jauncey and J. H. Van Vleck whose presentation was entitled "The New Quantum Mechanics." The entire program was deemed to be a great success and all of the papers were later published[164] in order that "they may be the documentary evidence of the welding in 1928 of Chemistry and Physics on a great common ground."

Publication of symposium presentations, such as those mentioned above, served the dual valuable purpose of reaching an audience beyond those in attendence and of providing some form of written, up-to-date summary which could be reread and discussed. There were, however, other review articles on quantum mechanics prepared expressly for publication in a particular journal. A few interesting examples in this category were those written by J. H. Van Vleck in the Journal of the Optical Society of America and the Review of Scientific Instruments[165] and by K. K. Darrow in the Bell System Technical Journal as part of his continuing series, "Contemporary Advances in Physics" begun in 1922.[166]

One final early example of a review article dealing with wave mechanics which appeared in an American journal is quite unusual in that it was written by none other than Erwin Schrödinger.[167] I have been unable to determine the circumstances associated with this publication. It is not identified as a translation of a European article, nor as having been based on an oral presentation in Europe or America. It summarizes, but also goes beyond, Schrödinger's articles published earlier that year in the Annalen der Physik. It was submitted from

Zurich, 3 September 1926, and appeared in the December issue of Physical Review. My conjecture is that it came about in connection with plans for the visit which Schrödinger made to America the following Spring. Aside from its content, it is significant that Schrödinger must have been convinced that there was a sufficient American audience for it to warrant his preparing a twenty-two page review article of his own work.

American publications on quantum mechanics in book form were few in number before 1930. We have already mentioned the lone entry in the textbook field, Quantum Mechanics by Condon and Morse. The only other comparable book was Max Born's Problems of Atomic Dynamics,[168] which consisted of the texts of the thirty lectures that he had delivered at the Massachusetts Institute of Technology between November 14, 1925 and January 22, 1926. Coming so soon after the introduction of matrix mechanics, it is, of course, very incomplete even by pre-1930 standards, but it is of particular interest in that it was a kind of transition volume between the old and the new quantum theories, and it was one of the first books published in either Europe or America which included an exposition of the new approach.[169]

From all of the above considerations it becomes clear that there was widespread interest among American scientists during the late 1920's in becoming informed about the new quantum mechanics. Furthermore, there were a variety of ways in which they could learn about it. The next question to which we must address ourselves, however, is whether, upon becoming informed, did they accept it or reject it?

Changes in the Outlook of American Physicists Accompanying the Advent of Quantum Mechanics

1) The Philosophical Response

Once quantum mechanics had begun to exhibit its clear superiority over the old quantum theory by virtue of the number of situations to which it could be successfully applied, there was very little overt opposition to its acceptance among American physicists as a practical tool.

This does not mean, however, that all American physicists were happy with the new directions in which their discipline was moving--directions which involved increased mathematical complexity, greater conceptual abstraction and the abandonment of classical causality.

Many of the young American quantum theorists[170] retained vivid recollections of the difficulties which some of their colleagues, especially older ones, had in adjusting to the new atmosphere. For example, E. U. Condon has related the following anecdote:[171]

"The situation reminds me of a remark made to me one spring afternoon in 1928 by Bergen Davis, a distinguished older professor of physics at Columbia, who had been struggling hard to get a real feel for quantum mechanics. Finally he said in despair: 'I don't believe that you young fellows understand it any better than I do, but you all stick together and say the same thing!'"

Condon then added "I have called this the conspiracy interpretation of quantum mechanics."

For the most part, however, American physicists seem to have accepted the innovations associated with quantum mechanics more readily than did their European counterparts. Werner Heisenberg was particularly struck by the pragmatic American approach to the subject that he encounterd during his visit of 1929. Not only did he find his American audiences keenly interested in the new theory, but also, he has recalled experiencing[172]

"...a strange feeling I had acquired during this lecture tour: while Europeans were generally averse and often overtly hostile to the abstract, nonrepresentational aspects of the new atomic theory, to the wave-corpuscle duality and the purely statistical character of natural laws, most American physicists seemed prepared to accept the novel apprach without too many reservations."

Upon discussing the matter with a young American experimentalist, Heisenberg came to the conclusion that he, with his European background, was much more firmly committed to considering various branches of physics, such as Newtonian mechanics, as "closed systems," not susceptible to amendment, than Americans seemed to be.

It should be recognized that most members of the American scientific community were not inclined to struggle with questions of the philosophical implications of quantum mechanics.[173] While there was some discussion at meetings of the American Physical Society and in the Physical Review of the significance of the Heisenberg Uncertainty Principle,[174] American scientists tended to adopt an "operational" point of view in dealing with the concepts of quantum mechanics and an open-minded, pragmatic acceptance of what quantum mechanics could accomplish.

"Operationalism" was a term that came into usage following the publication of The Logic of Modern Physics by Percy W. Bridgman[175] in which he stated that "the concept is synonymous with the corresponding set of operations."

Bridgman had begun to formulate his ideas in response to his own "disquietude" with the conceptual foundations of such areas as electrodynamics and thermodynamics,[176] but the operational perspective that he arrived at turned out to be of considerable value to his fellow physicists in coping with the mental adjustments required by both relativity and the quantum theory. In 1929 E. U. Condon and P. M. Morse wrote:[177]

"In thinking about all physics, especially in connection with the problems of quantum theory, the operational point of view stressed by Bridgman in his The Logic of Modern Physics is a most important aid."

This view has been reiterated in more recent times by a number of other scientists who have recalled Bridgman's influence on their own outlook.[178]

The requirement inherent in the Heisenberg matrix mechanics that the new theory should contain only observables was clearly appealing to Bridgman, as was the quest that convenience and simplicity be attributes of an acceptable theory. It is interesting to note, however, that Bridgman, the experimentalist, was far from satisfied with the highly mathematical aspects of quantum mechanics and he viewed them as temporary expedients which merited the warning:[179]

"There is among the younger and more enthusiastic members of the physical community a tendency to regard the present theories as final which the more sedate members must combat, even in the face of all the successes of the present theories. One of the most certain lessons of the past is that no amount of success in the youth of a theory is any guarantee of a hale and hearty old age; this is to be expected and it a consequence of the transcendence of nature."

2) Professional Recognition for Theoreticians

While the American physics profession did not veer sharply away from experimental "reality" toward overriding mathematical abstraction, it did most certainly broaden its views of what must become part of the training of future physicists. As a result, since it was no longer practical to expect all faculty members to be able to handle adequately the mathematical-theoretical aspects of their discipline, insitutions earnestly sought to add one or more theoreticians to their physics departments.

The ensuing demand for competent quantum theorists was flattering and, at the same time, somewhat disconcerting. E. U. Condon, for example, found himself to be recipient of offers from six different universities in the Fall of 1929.[180] Many of the other young American theoreticians also found themselves being wooed with what seemed at the time to be fabulous offers.[181] In addition, several American institutions turned to European physicists in seeking to fulfill their needs. In Appendix III our list of European physicists who emigrated to the United States reveals those that were successful.[182]

The change in the American attitude toward theoretical activity in physics that came about during the 1920's not only influenced the training of physics students but also made it clearly possible for a young American about to enter the field of physics around 1930 to elect to become a theoretical physicists. He could expect to receive adequate preparation at any one of a number of institutions; he would not be required to master time-consuming advanced laboratory techniques; and he could expect to find recognition and employment upon the successful completion of his studies.[183]

The question may be raised as to whether those students who chose theoretical studies in physics during those years when the ferment associated with quantum theory made it especially attractive, would have become experimental physicists if the times had been different.

For some, such as J. C. Slater and P. M. Morse, for example, it is my belief that they would have. On the other hand, there were some, such as E. U. Condon and J. H. VanVleck, who may never have felt at home in laboratory situations.[184]

It should be noted, however, that at most American institutions the new theoretical physicists of this period were not aloof from their experimental colleagues. Frequently there was fruitful collaboration among them which served to strengthen the entire American physics profession.[185]

In this way, the advent of quantum mechanics may be considered to have played a significant role in the "coming of age" of physics in America. Under its impetus highly talented young people were attracted to the profession. At the same time, the exisitng insitutions and the attitudes of the established members of the American physics profession were sufficiently flexible to provide for the influx of a sizeable number of theoreticians. Thus the whole American physics profession was strengthened and its earlier deficiency in theoretical training and activity was overcome.

<u>Summary</u>

The advent of quantum mechanics proved to be a positive factor in the process of America achieving maturity in physics. This was due largely to the readiness with which most American physicists accepted the value of the new ideas. The fact that American interest in quantum mechanics was high stimulated greater intercourse with the European physics community. As a result, European physicists gained added awareness of American potential for contributing to the development of physics and many Americans became confident that they could successfully apply the new language of theoretical physics. The advent of quantum mechanics also stimulated theoretical activity at all levels at American institutions, thereby enriching the curriculum of study and providing bona fide professional status for American theoretical physicists. In our next section, as we describe some specific instances of the successful use of quantum mechanics by Americans, added detail will emerge of the changing environment in which American physicists operated in the late 1920's.

Part III: Some Significant Achievements Related to Quantum Mechanics Made by Americans

In the years between 1926 and 1929 many papers involving the use of quantum mechanics were published by American authors.[186] In this section we shall make no attempt to catalogue all of them. Instead, a sufficient number of them will be highlighted to bring out certain characteristics that were associated with America coming rapidly to the forefront in this branch of physics. In particular, it will become clear that there were a number of young Americans caught up in the excitement of the new ideas; that they studied and worked under a variety of academic circumstances; and that the subjects they dealt with were topics of lively concern among the international physics community. This last characteristic had a two-fold result: their work was promptly appreciated but they often suffered the disappointment of a "near miss."[187] On the other hand, most importantly for the individuals themselves and for the future of physics in America, during this transitional period several young American scientists began working in particular areas that were susceptible to quantum mechanical analyses and where their contributions in the ensuing decades turned out to be outstanding.

Carl Eckart: The Equivalence of Matrix and Wave Mechanics

The fact that there were young American scientists alert and ready to take up quantum mechanics promptly has already become evident earlier in this chapter,[188] but an especially interesting example may be found in the case of Carl Eckart.[189]

Eckart went to the California Institute of Technology in 1925 as a National Research Council Fellow after taking his doctorate at Princeton University. He heard Max Born lecture on matrix mechanics in early 1926 and, soon thereafter, Schrödinger's papers on wave mechanics began to arrive in Pasadena. By the end of May 1926 Eckart reached the conclusion that the schemes--matrix and wave mechanics--were equivalent formalisms and he wrote his analysis for publication.[190] Schrödinger's treatment of the same question reached publication before Eckart's did but this does not detract from the latter's achievement.

E.U. Condon and R. W. Gurney: Nuclear Barrier Penetration

A second instance of the same quantum mechanical result being achieved almost simultaneously on both sides of the Atlantic occurred in 1928. In this instance both E. U. Condon and R. W. Gurney, a young scholar from Great Britain, who were working at Princeton University, and George Gamow, who was then studying at Göttingen, realized within days of each other that the probability interpretation of the wave function made possible the explanation of barrier penetration by nuclear particles,[191] a particularly interesting result because almost all previous application of quantum mechanics had been to atomic and molecular considerations.

Some examples of Americans whose achievements in quantum mechanics were not only unique but were very promptly appreciated internationally may be found in the cases of David Dennison, William Houston and John Slater.

D.M. Dennison: The Specific Heat of Hydrogen and Proton Spin

Dennison succeeded insolving the long standing riddle of the specific heat of hydrogen. He also showed, for the first time, that the proton has spin equal to one-half. This was accomplished just before he was to leave Europe in 1927 to return to the University of Michigan where he had received his Ph.D. three years earlier.[192] Dennison first presented his results during the final lecture in a series which R.H. Fowler had invited him to deliver at Cambridge University. At that time Fowler was just completing his book Statistical Mechanics: The Theory of the Properties of Matter in Equilibrium, in which he had discussed the unhappy state of the theory of the specific heat of hydrogen. Consequently, he was "very much excited about the new theory" proposed by Dennison since it provided a "wedding of quantum theory and of statistics of the proton." Fowler incorporated Dennison's results into his own text and arranged for them to be published soon after in The Proceedings of the Royal Society.

W. V. Houston: The Wave Mechanics of Electrical Resistivity

Houston was another young American who began his professional career just as the age of quantum mechanics was dawning. He differed from most of our other examples in this group in that his subsequent career was divided between theory and experiment.[193] Houston's entry into the domain of theoretical quantum mechanics came in 1927 when he went to Munich to study with Sommerfeld on a Guggenheim Fellowship. Sommerfeld suggested to Houston that he look into the possible correlation between the mean free path of an electron in a wave mechanical sense and the temperature dependence of metallic resistivity. Houston's success with the problem led Sommerfeld to call it "Die erste anständige Bearbeitung des Widerstandsgesetzes."

J. C. Slater: A Determinantal Method for Atomic Multiplets

One more example of an American contribution to the advancement of quantum mechanics that was promptly hailed in Europe occurred in 1929 when J. C. Slater introduced his determinantal method of treating atomic multiplets.[194]

Slater was teaching at Harvard during the late 1920's when group theory was being introduced into the extension of quantum mechanics to increasingly complicated situations, although, in reality, permutation group theory was mathematically more general than required by the quantum mechanical situations to which it was then being applied due to the restrictions imposed by the Pauli Exclusion Principle.

Slater has related that[195]

"I had what I can only describe as a feeling of outrage at the turn which the subject had taken."

He was consequently pleased to find that group theory could be avoided through the use of what came to be known as the "Slater Determinant."[196]

Upon arriving in Europe in late 1929 on a Guggenheim Fellowship, he found that his method was also welcomed especially by those European physicists who were as unfamiliar as he was with group theory. In fact, he became known as the one who "slayed" what they called the "Gruppenpest."[197]

Meanwhile, during the period 1926-1929, there were additional achievements made by Americans in extending the domain of applicability of quantum mechanics that took place at locations far removed from the European centers of such activity.

J. H. Van Vleck and S. C. Wang: Advantageous Uses of Matrix Mechanics

It is interesting to note in this connection that J. H. Van Vleck continued to use matrix mechanics in the papers that he began to bring out on the subject of electric and magnetic susceptibilities in Minneapolis and Madison for some time after most other authors had adopted a wave mechanical approach. Van Vleck found that, at least for some problems, the method of matrices was preferable to that of differential equations.

When Shou Chin Wang came to the University of Wisconsin in 1929 to work with Van Vleck as a National Research Fellow, Van Vleck persuaded him to use matrix methods in tackling the problem of the asymmetrical top, a course which Wang adopted with notable success.[198]

J. R. Oppenheimer: Autoelectric Field Currents

During the year 1927-1928 which J. R. Oppenheimer spent as a National Research Fellow at American institutions after taking his Ph.D. with Born in Göttingen, he continued to publish prolifically on quantum mechanical topics. All of the papers that he produced during this period served to advance the subject matter and to consolidate his reputation as a most promising American theoretical physicist. One of his topics, however, deserves special note and his achievement with it is well summarized as follows:[199]

"His most original contribution... was his theory of field emission, the first example of an effect due to barrier penetration (antedating the explanation of radioactive alpha decay). He developed a perturbation theory of nonorthogonal states and used it to calculate the disintegration of a hydrogen atom in an electric field. He then applied his results to the effects observed in metals. This work was done at Pasadena where Robert A. Millikan and Charles C Lauritsen were studying the phenomenon, and was the first evidence of a feature later to be so prominent in his work: his close collaboration with his experimental colleagues."

Following the initial successes achieved by quantum mechanics for the relatively simple systems such as the hydrogen and helium atoms and the hydrogen molecule, the next natural step was to investigate how the new methods could be applied to atoms with several electrons and to molecules in general. The problems associated with such analyses turned out to be quite complex, but it is noteworthy that, from the very start of this work, which would later come to be known as solid state physics and quantum chemistry, American contributions stood high. At this time it is appropriate to consider, in this connection, the work of Robert Mulliken and Linus Pauling.

R. S. Mulliken: Molecular Orbitals

Robert Mulliken's work during this period was an extension of some earlier work by Friedrich Hund and a natural outgrowth of Mulliken's own molecular spectra studies begun in the years of the old quantum theory. Within the new idiom his method of approach, which became known as that of "molecular orbitals," involved the assigning of individual sets of quantum numbers to the electrons belonging to the system under consideration, thereby bringing a kind of ordering into the electronic configuration.[200]

Linus Pauling: Shared Electron Bonds

During the same period Linus Pauling began developing his approach to the quantum mechanical nature of chemical bonds, an approach which has become known as "resonance bonding."[201] Pauling received the Nobel Prize in Chemistry in 1954 for this work. Mulliken was similarly honored in 1966.

C. J. Davisson and L. H. Germer: Electron Diffraction

Finally, we should not fail to mention another contribution to strengthening the new wave mechanics which came from the traditional American stronghold in physics--the laboratory. In 1927, Clinton Davisson and Lester Germer produced results at the Western Electric Laboratories of the Bell Telephone Company using an electron beam reflected from a crystal which clearly confirmed the wave properties of electrons. Ten years later, Davisson, the senior investigator of the team, shared the Nobel Prize in Physics for this work with G. P. Thomson of England who had achieved similar results using a different technique.[202]

We could go on and cite additional instances of respectable research results that were achieved by young American participants in the prompt development of the uses of quantum mechanics. The pages of Physical Review for 1928 and 1929, for example, contain interesting contributions from G. Breit, V. Guillemin, E. L. Hill, E. C. Kemble, R. M. Langer, F. W. Loomis, B. Podolsky, V. Rojansky and C. Zener, as well as additional valuable papers from those authors upon whom we have focused our attention. The dozen examples that we have cited, however, are sufficient to make it abundantly clear that by the end of the 1920's American scientists had entered the mainstream of contemporary physics.

Summary

As we look back now on the development of physics in American during the years immediately following the advent of matrix mechanics in late 1925 and consider which were the chief differences between this and earlier periods, we come to realize that what happened within physics in America between 1926 and 1929 was a culmination of trends that had been underway for a number of years rather than a sharp break with the past.

At home, American physicists now had a strong professional society, good and well run research publications, continued excellence in laboratory facilities and increased opportunity for theoretical study.

Internationally, there was more intercourse with European colleagues than ever before. More Europeans visited and lectured in America then in any previous four year period; a few accepted permanent positions in the United States. Many Americans were able to spend a year or more studying at European institutions and showed themselves to be thoroughly competent. In short, American and European physicists now beginning to meet on an equal level.

The advent of quantum mechanics prompted considerable interest in various sectors of the American physics community. A few dozen young members immediately sought to participate actively in its development while most of the others came to accept pragmatically not only the particular innovations associated with it but also the concept that theoretical physics was a bona fide professional classification.

The years from 1926 to 1929 were hectic and unsettled in many ways for American physicists, but during those years physics in America "came of age" in the sense that we defined that term earlier in this thesis. Full maturity still lay ahead, but we shall see in our next chapter that it was swiftly achieved within the next few years.

Chapter 3: Notes and References

3.1 All of the information about the American Physical Society presented in this section was obtained either from the Proceedings of the American Physical Society as printed in the Physical Review for the years 1926-1929 or from a recent edition of the Bulletin of the American Physical Society which annually publishes cummulative records of membership and office holders.

3.2 The data on election to Fellowship during 1928 appear somewhat confused. According to the cummulative list in Bull. Amer. Phys. Soc. 18 (1973) 942, Fellowship in A.P.S. rose from 486 in 1927 to 520 in 1928. According to H. Webb's Minutes of the 30th Annual Meeting held in December of 1928, printed in Phys. Rev. 33 (1929) 276, there was a net increase in Fellowship of 55 during 1928. The total of the lists of names of those attaining Fellowships during the various meetings held in 1928 is 63. Whatever the accurate figures are, 1928 was the year in which an unusually large number of American physicists were designated productive members of the profession. (For a graph of the growth of Membership and Fellowship in A.P.S. see Appendix IV, Part 3.)

3.3 Some others, such as J. H. Van Vleck, had already been elected to Fellowship in previous years.

3.4 The February meetings of 1928 and 1929 were held with the Optical Society of America; a joint symposium with the American Mathematical Society was held in December of 1928.

3.5 The Washington, D.C. meeting of April, 1929, as reported in Phys. Rev. 33 (1929) 1067.

3.6 Phys. Rev. 31 (1928) 301.

3.7 Phys. Rev. 33 (1929) 111; 1067.

3.8 Phys. Rev. 33 (1929) 1067; the occasion was the 1929 Spring Meeting in Washington, D. C.

3.9 Quotation taken from "John Torrence Tate" by Alfred O. C. Nier and John H. Van Vleck, typescript of an article submitted in 1974 to, but not yet published in, the National Academy of Sciences Biographical Memoirs and kindly made available to this author by Professor Van Vleck.
K. K. Darrow, in his biographical memoir "John Torrence Tate," American Philosophical Society Yearbook 1950, 325-328 paid tribute to Tate's contributions to improving the quality of American physics journals and suggested that the Physical Review should be known as "Tate's Journal" by analogy with "Poggendorff's Annalen" and other early journals whose style and content bore the marked imprint of the editor.

3.10 See Chapter 2, page 2.5.

3.11 Nier and Van Vleck, "J. T. Tate." According to Van Vleck's recollection, shared with the author in May 1975, papers from an established author, such as J. C. Slater were published without any refereeing. In other instances, Van Vleck served as an informal referee, reading papers as soon as they reached the Physical Review office at the University of Minnesota.

3.12 "Prompt publication of brief reports of important discoveries in physics may be secured by addressing them to this department [Letters to the Editor]. Closing dates for this department are, for the first issue of the month, the twenty-eighth of the preceding month; for the second issue, the thirteenth of the month. The Board of Editors does not hold itself responsible for the opinions expressed by the correspondents." Phys. Rev. 34 (1929) 161.

3.13 A file of some fifty letters from American physicists responding to Tate's request for opinions on the advisability of such an undertaking is on deposit at the Center for History of Physics, American Institute of Physics, under the designation "APS--Reviews of Modern Physics."

3.14 Phys. Rev. 33 (1929) 276.

3.15 Letter from E. U. Condon to J. T. Tate, 2 October 1928, in file described above in Note 3.13.

3.16 Statement on the inside cover of Rev. Mod. Phys. 47 (April 1975).

3.17 The statement "There is no question but what our laboratories are better now than those abroad" occurs in the Condon letter cited in Note 3.15.

3.18 The formal dedication of the new facility took place during a meeting of the American Physical Society held in the new building November 30 and December 1, 1928 with a roster of speakers that included H. Weyl, K. T. Compton, E. U. Condon, I. Langmuir, F. L. Mohler and J. R. Oppenheimer. Quantum mechanics and electron impact considerations were the principal focus of their presentations. (Phys. Rev. 33 (1929) 111).

3.19 An account of their work written by Germer appeared in an article "Low-Energy Electron Diffraction" Physics Today, July 1964, 19-23.

3.20 K. T. Bainbridge took his National Reserach Fellowship to the Bartol Foundation in 1929. According to Professor Bainbridge (Private conversation with the author, March 4, 1974) the Bartol fortune was amassed in the sugar refining industry. Howard McClenahan of Princeton University was instrumental in persuading Bartol to endow a research institute under the aegis of the Franklin Institute. The Bartol bequest, made in 1918, was subject to some years of litigation but eventually more than two million dollars came to the new enterprise. In 1924 the Bartol Research Foundation was formally dedicated during the centenary celebration of the Franklin Institute. Research sponsored by the Foundation was carried on in some buildings belonging to the Institute in Philadelphia that had been remodelled for that purpose. At the time of the move to Swarthmore the Foundation was directed by W. F. G. Swann who had assumed office in 1927.
Sources: Journ. Frank. Inst. 198 (1924) 408; 203 (1927) 866; 205 (1928) 767-829; 206 (1928) 741.

3.21 National Research Council Reprint and Circular Series, Nos. 80 (1927), 86 (1928), and 91 (1929).

3.22 Data derived from listings in National Research Fellowships 1919-1944.

3.23 The Rockefeller Foundation Directory of Fellowship Awards 1917-1950 contains lists of recipients arranged by their country of residence as of 1950. An element of confusion arises in some of the listings for American recipients, such as E. U. Condon and G. H. Dieke, who are listed as International Education Board Fellows for the same years that their names appear among the National Research Council Fellows. The fact that Rockefeller money funded both programs may have been responsible for this confusion.

3.24 Directory of Fellows of the John Simon Guggenheim Memorial Foundation 1925-1967, (New York, 1968) and John Simon Guggenheim Memorial Foundation Annual Reports 1925-1936. The quotation cited is from page 9 of the latter publication.

3.25 For example, E. C. Kemble and A. H. Compton were Guggenheim Fellows during the late 1920's.

3.26 Nobel Lectures in Physics 1922-1941, 167-193. The award to Compton was made jointly with C. T. R. Wilson for his development of the cloud chamber.

3.27 The Cosmos of Authur Holly Compton, pp. 39-43.

3.28 Electrons et Photons: Institut International de Physique Solvay 5e Conseil (Paris: Gauthier-Villars, 1928) pp. 55 - 85. Atti del Congresso Internazionale dei Fisici, 11-20 Settembre 1927 Como, Pavia, Roma: Onoranze ad Alessandro Volta nel primo centenario della morte Vol. I (Bologna: Nicola Zanichelli, 1928) pp. 161-170.

3.29 William Duane, E. H. Hall, A. E. Kennelly, R. A. Millikan, R. C. Tolman and R. W. Wood. P. W. Bridgman was also invited but could not attend. (Source: Atti del Congresso Vol. I,p. x.) A paper by Bridgman was printed in Atti del Congresso, Vol. II,pp. 239-248.

3.30 M. Pupin represented the American Physical Society, A. Trowbridge the International Education Board. (Source: Atti del Congresso, Vol. I, p. xi.)

3.31 It was during this session that Niels Bohr first publicly presented his views on the complementary nature of physical phenomena. Following Bohr's presentation was a discussion participated in by Born, Kramers, Heisenberg, Fermi and Pauli. (Atti del Congresso, Vol II, 435-598) According to Jammer (Conceptual Development of Quantum Mechanics, p. 351) Compton, Duane, Hall, Millikan and Tolman were members of the audience during Bohr's presentation.

3.32 Atti del Congresso, Vol. II,p. 334; presentation of O. W. Richardson.

3.33 Ibid.,p. 335; comment by A. Sommerfeld.

3.34 Ibid., p. 590; comment by M. Born included reference to Oppenheimer and Wiener.

3.35 Ibid., p. 588; presentation of N. Bohr. (Lorentz also took note of Compton's electron spin hypothesis. Ibid., p. 3.)

3.36 Atti del Congresso, Vol. I, pp. 161-169.

3.37 Appendix III contains a chronological list of European physicists who emmigrated to the United States together with the names of the institutions to which they came.

3.38 Paul Ehrenfest to S. A. Goudsmit, quoted by the latter in AHQP Interview, Session 2, 7 December 1963, p. 35.

3.39 E. U. Condon "Reminiscences of a Life in and out of Quantum Mechanics" Internat. Jour. Qu. Chem. Symp. No. 7 (1973) 7-22.

3.40 The Rockefeller Foundation Directory of Fellowship Awards 1917-1950.

3.41 AHQP Interviews with both Goudsmit and Uhlenbeck describe their backgrounds and the circumstances surrounding their transition into the American scientific scene. "Reminiscences of Professor Paul Ehrenfest" by G. E. Uhlenbeck, Amer. Jour. Phys. 24 (1956) 431-433, provides a detailed view of the Ehrenfest style of education. According the Uhlenbeck, Interview, p. 14, it was at Ehrenfest's insistence that two rather than one young man was lured to Ann Arbor.

3.42 AHQP Interview with Otto Laporte, Session 2, 31 January 1964, pp. 22, 23 describes his experiences at the National Bureau of Standards where he worked with the "great experimental spectroscopist" W. F. Meggers and in doing so "learned to value experiment."

3.43 Source materials for information about the blossoming of theoretical activity at the University of Michigan which provided the background for the following discussion include:
D. M. Dennison "Physics and the Department of Physics since 1900" in Research--Definitions and Reflections: Essays on the Occasion of the University of Michigan Sesquicentennial Ann Arbor, 1967 (Xerox copy obtained from the Center for History of Physics, AIP, New York City.)
S. A. Goudsmit "The Michigan Symposium in Theoretical Physics" Michigan Alumnus Quarterly Review, May 20, 1961, 178-182.
________ Letter to the author, 26 March 1970.
Center for History of Physics Interviews with David Dennison and Harrison Randall.
Meyer, Lindsay, Rich, Barker and Dennison "Physics" in The University of Michigan, available at the Niels Bohr Library.
University of Michigan Summer School Catalogues 1928-1935, also available at the Niels Bohr Library.

3.44 Michigan had begun a summer program of lectures by visiting physicists in 1923, as was mentioned in Chapter 2.
Ralph Kronig in a letter to the author, 18 June 1974, provides the following recollections from one of the pre-1928 years:
"...I spent the summer 1926 at the summer school of the University of Michigan, Ann Arbor. I went there together with the German physicist Herzfeld...At Ann Arbor H. M. S. Randall, whose interest lay in the field of spectropscopy (particularly of the infra-red) had created a very active physics department. One of the visitors that summer was the Swedish band spectroscopist Eric Hulthen...I recall that summer with much pleasure."

3.45 Goudsmit,"The Michigan Symposium", contains a list of the guest lecturers year by year in addition to much background information. The University of Michigan Catalogues not only provide information on the topics discussed by those lecturers but also reveal that, over the years, Dennison, Goudsmit and Uhlenbeck were frequently responsible for some of the lectures.

3.46 The 1929 Catalogue contains the following statement: "The privilege of attending lectures and seminars and of carrying on research in the laboratories and libraries during the Summer Session will be extended by the President of the University, on the recommendation of a chairman of a department, to doctors of philosophy or of science in this or other universities. There will be no charge except for laboratory supplies. Those desiring to avail themselves of this privilege should correspond with the President before the opening of the Summer Session."
The figure of $40 was recalled by Wendell Furry, who in a private conversation with the author in the Spring of 1970, described his experiences during the Summer of 1931 when he was still a graduate student at the University of Illinois. Furry was paid $100 by the F. W. Loomis, chairman of the Physics Department at Illinois, to bring back a set of notes on the lectures that Pauli and Sommerfeld gave at Michigan that summer. After paying the $40 fee, Furry was able to pay for modest room and board accommodations from the remainder. It is interesting to note that nine years later Furry himself was one of the invited lecturers at the Symposium.

3.47 Goudsmit "The Michigan Symposium" p. 179.

3.48 The collection of photographs deposited at the Niels Bohr Library by George Uhlenbeck contains prints of the annual group pictures as well as many informal snapshots of a number of participants. The article "Physics" by Meyer, Lindsay et al. cited in Footnote 3.44, contains a table of symposium attendance on an annual basis and separated into "Graduate Student" and "Guests". In 1929, for example, there were 59 students and 40 guests.
Note: films taken at some of these meetings are part of the documentary film "The World of Enrico Fermi."

3.49 Meyer, Lindsay, et al. "Physics"

3.50 Dennison, "Physics and the Department of Physics", p. 131.

3.51 For example, Stanford University had the following list of guest lecturers in the physics department during the summers of the late 1920's: John C. Slater, 1926; John H. Van Vleck, 1927; Harry L. Clark, 1928; Karl K. Darrow, 1929. (Information supplied by M. D. Reeves of the Stanford University Archives from the University Registers for those years.)

3.52 Phrase used to describe Ann Arbor by J. H. Van Vleck in a letter to E. C. Kemble, 9 July 1928. AHQP Microfilm 49, section 8.

3.53 Goudsmit "The Michigan Symposium," p. 181.

3.54 D. M. Dennison, "Recollections of physics and of physicists during the 1920's" Amer. Jour. Phys. 42 (1974) 1051-1056. Quoted passage occurs on page 1051.

3.55 "The Future of the Department of Physics" Harvard Alumni Bulletin, June 13, 1929, 1056-1058.

3.56 Letter from T. Lyman to E. C. Kemble, June 28, 1927; carbon copy among the Lyman Papers located in the Physics Department at Harvard University.
The General Education Board Annual Report 1929-1930 states that the final sum needed for the new physics building at Harvard University was $1,250,000 of which the Board provided $475,000.

3.57 Information on Lyman's handling of the funds was supplied by K. T. Bainbridge, in private conversation with the author on May 7, 1974; also mentioned by J. C. Slater in the manuscript of his forthcoming autobiography, kindly made available to this author by Professor Slater.

3.58 Information supplied in June 1975 by Frank L. Verdonck of the Lyman Physical Laboratory.

3.59 A favorite phrase of Wycliffe Rose repeatedly used in publications related to the General Education Board.

3.60 Adventure in Giving, p. 231. The figures following the quotation are from the same source

3.61 Ibid., pp. 140-149.

3.62 The General Education Board Annual Report 1925-1926, page 6 states: "In pursuance of the decision to aid institutions seeking to develop teaching and research in the fundamental sciences, an appropriation of $1,000,000 towards a total of $3,000,000 was made to Princeton University. The entire sum sought is to be held as endowment, the income to be devoted to advanced teaching and research in mathematics, physics, astronomy, chemistry and biology. Its facilities are, in almost all the foregoing subjects, excellent; its relative position among graduate schools has, in recent years, steadily improved; the sum will enable the University to consolidate and improve the work in a group of subjects, no one of which can be effectively cultivated in isolation from the others. The cooperative spirit of the proposed development and the high quality of the men already engaged in the enterprise promise well for its future."
Other details concerning Princeton's disposition of the funds are found in the editorial note which accompanies Shenstone, "Princeton Physics."

Note: Another $1,000,000 was given to the Princeton Chemistry Department, according to The General Education Board Annual Report 1928-1929, which accounts for the figure of $2,000,000 quoted earlier.

3.63 The General Education Board Annual Report 1929-1930. The author is particularly appreciative of this undertaking since, some years later, she benefited directly from it by being able to begin her study of physics in a beautiful building superbly equipped for undergraduate science teaching and learning.

3.64 R. B. Fosdick Adventures in Giving, pp. 133-149.

3.65 Ibid., p. 135.
Although the General Education Board continued to function until 1964, its work in medicine, natural sciences and the humanities was largely absorbed by the Rockefeller Foundation when that body was reorganized in 1928, a year which marked the end of an era. Wycliffe Rose retired; the International Education Board, having spent most of its funds, was liquidated. The "on top of this came the panic of October 1929 and the devastating years of the depression."
The General Education Board Review and Final Report 1902-1964 (New York: The General Education Board, 1964) p. 48.
Additional reference: R. B. Fosdick The Story of the Rockefeller Foundation (New York: Harper and Brothers, 1952).

3.66 See Chapter 1, pages 1.9 and 1.40 for some examples.

3.67 Science 68 (1928) 61. Williams, then a Professor of Physics at Amherst College, had studied in Berlin for two years, 1903-1905 after taking his M.A. at the University of Nebraska. He then returned to Columbia University where he received his Ph.D. in 1906. (American Men of Science, 7th edition.)

3.68 Science 68 (1928) 202-203. The fact that K. T. Compton joined Loeb and Birge in writing this reply was probably occasioned by Compton's presence in Berkeley that summer as a visiting professor in the University of California Summer Session, according to Birge History of the Physics Department, VIII (25).
I am indebted to Charles Weiner for calling to my attention the discussion published in Science which is reported on here.

3.69 The Bishop of Ripon, as cited by R. A. Millikan in "The Alleged Sins of Science" Scribner's Magazine, February 1930, 119-129.

3.70 Frederick Soddy, also quoted by Millikan but without specific citation.

3.71 R. B. Fosdick The Old Savage in the New Civilization (Garden City, New York: Doubleday, Doran and Company, Inc. 1931) It appears from the foreword that an edition of this had been published in 1928. Millikan's Address was largely a response to Fosdick's ideas.
Fosdick (1883-1969) a lawyer by profession, was active in public affairs and had served as Under Secretary of the League of Nations. He later became President of the Rockefeller Foundation, 1936-1948. (Source: Who Was Who in America 1961-1970)

3.72 Science 70 (1929) 513 identifies the Scribner's Magazine article as Millikan's Presidential Address.

3.73 The motto of the University of Chicago quoted by Millikan and translated as "Let knowledge grow, let life be enriched."

3.74 M. Pupin "Romance of the Machine" Scribner's Magazine February 1930, 130-137.

3.75 Further discussion of public criticisms of science during the late 1920's and of responses by individual physicists may be found in Kargon, The Maturing of American Science, pp. 33-36; and in C. Weiner "Physics in the Great Depression," Physics Today (October, 1970) 31-37.
Both of these authors are more concerned, however, with developments that occurred after the onset of the Depression. We shall return to this topic in the next chapter.

3.76 Letter to the author, 14 September 1972.

3.77 The numbers cited here are based on the discussion contained in the previous chapter covering the period 1920-1925 and the discussion about to be presented of Americans who published significantly on topics related to quantum mechanics, or who wrote doctoral dissertations in that field, or who held fellowships for study of quantum mechanics at home or abroad.

3.78 "The Franck-Condon Principle" Amer. Jour. Phys. 15 (1947) 365-374. The quotation used is found on page 368.

3.79 Letter, 30 March 1927, to E. C. Kemble who has kindly shared its contents with the author. Van Vleck was writing from Minneapolis in response to a letter he had recently received from Kemble, then in Munich on a Guggenheim fellowship.

The comments by Condon and Van Vleck should not be construed as reflecting peculiarly American difficulties with quantum mechanics. Victor Weisskopf has described his feelings after coming to Göttingen from Vienna in 1928 as follows:

"It was a great thing and very disagreeable too to be a graduate student at that time. It was disagreeable because too many new ideas came around. When you had barely digested the Schrödinger equation and Heisenberg's quantum mechanics, you already heard your colleagues talking about the Dirac equation and quantum electrodynamics. Learning went a little too fast; it was interesting, but discouraging."

"My Life as a Physicist," Lecture given at the Erice Summer School in High-Energy Physics, Sicily, Italy, in 1971; reprinted V. F. Weisskopf Physics in the Twentieth Century: Selected Essays (Cambridge: The MIT Press, 1972)

Egil A. Hylleraas also recalled that he found the scientific life in Göttingen "somewhat frightening" when he arrived there as an International Education Board Fellow in 1926.
Source: "Reminiscences from Early Quantum Mechanics of Two-electron Atoms" Rev. Mod. Phys. 35 (1963) 421-431.

3.80 The Archives of the Massachusetts Institute of Technology includes several letters between Born and Institute officials written both before and after Born's visit to Cambridge. In addition, Born has given some details of his experiences in the Preface to his lectures that were published as Problems of Atomic Dynamics by the Massachusetts Institute of Technology in 1926, and in his autobiography My Life and My Views. These sources, as well as others specifically cited below, provided the information on which this discussion is based.

3.81 Berlin: Verlag von Julius Springer, 1925.

3.82 Leipzig: B. Teubner, 1923.

3.83 Über quantentheoretisch Umdeutung kinematischer und mechanischer Beziehungen" Zeits. f. Phys. 33 (1925) 879-893.

3.84 "Zur Quantenmechanik" Ibid. 34 (1925) 858-888.

3.85 Problems of Atomic Dynamics (Cambridge: Mass. Inst. of Tech. 1926); Preface page ix.

3.86 Harvard University Gazette, January 4, 11, 18, 1926. Also, there are among the Lyman Papers in the Physics Department of Harvard University some half dozen letters between Lyman and Born related to the arrangement for these lectures, three of which dealt with recent developments in quantum theory, and two on the theory of solids.

3.87 The Born-Einstein Letters, p. 89.

3.88 See Sources listed in Appendix II for information on Born's visit.

3.89 Nobel Lectures in Physics 1942-1962 (Amsterdam: Elsevier Publishing Company 1964) p. 262.

3.90 Wiener had obtained a Ph.D. in mathematical logic from Harvard University in 1913 when he was not yet 19 years old. This was followed by a two-year fellowship which enabled him to go to Cambridge (England) and to Göttingen. (American Men of Science, 10th edition, 1960)

3.91 M.I.T. Jour. Math. and Phys. 5 (1925-1926) 84-98; Zeits f. Phys. 36 (1926) 174-187.

3.92 My Life and My Views, p. 36.

3.93 Norbert Wiener, I Am a Mathematician (Garden City, New York: Doubleday and Company 1956) pp. 108, 109.

3.94 "Quantum Physics In America between the Wars" Internat. Jour. Qu. Chem. Vol. I S (1967) 1-23; the passage quoted is found on page 8. The papers referred to by Slater are: the one cited in Note 3.84; M. Born, W. Heisenberg and P. Jordan "Zur Quantenmechanik" Zeits f. Phys. 35 (1926) 557-615; and P. A. M. Dirac "Fundamental Equations of Quantum Mechanics" Proc. Roy. Soc. 109 (1925) 642-653.

3.95 AHQP Interview, 31 May 1962, page 6.

3.96 AHQP Microfilm 49. In Atommechanik, on page 332, Born gave all the credit for the proposal and the working out of the crossed orbit model of the helium atom to Bohr and Kramers. In the footnote citing citing Kramer's paper in Zeits. f. Phys. 13 (1923) 312-341 Born did make reference to a Van Vleck publication in Phys. Rev. 21 (1923) 372. However, that was actually only an abstract of a paper by Van Vleck delivered at a meeting of the American Physical Society. Furthermore, it was just a criticism of unsuccessful earlier work by others attempting to account for the spectrum of neutral helium and was not specifically related to the crossed orbit model. Born had completely ignored the Phil. Mag. 44 (1922) 842-869 publication by Van Vleck which carried through the detailed calculation on the crossed orbit model as proposed by Kemble in Phil. Mag. 42 (1921) 123-133.

3.97 Science 65 (1927) 156. A symposium on quantum mechanics was held during Schrödinger's visit to Madison in which Van Vleck was invited to participate. Van Vleck has recalled a good-natured argument on that occasion between Schrödinger and himself on whether the wave function had only a probablistic significance. Schrödinger argued for a hydrodynamical interpretation.
Source: "Reminiscences of the First Decade of Quantum Mechanics" Internat. Jour. Qu. Chem. 5 (1971) 3-20.

3.98 Science 65 (1927) 58.

3.99 In "Reminiscences of the First Decade...," page 9, Van Vleck explains the reasons for his move as follows: "Wisconsin was my alma mater, Madison my native city, and additional bait was the fact that Wisconsin offered me the companionship of a distinguished European visiting professor each year."

3.100 A. Salam and E. P. Wigner, eds. (Cambridge, Eng.: Cambridge University Press, 1972) Chapter 2: J. H. Van Vleck "Travels with Dirac in the Rockies," pp. 7-16.

3.101 AHQP Microfilm 49, Section 13, tells of Van Vleck taking Dirac (who was known to be a devotee of cartoon films) to a Disney Day at the Boston Repertory Theatre in 1935. Van Vleck noted that his guest loved Mickey Mouse!

3.102 For specific examples, see Appendix II.

3.103 AHQP Interview with Heisenberg, 28 February 1963, p. 11.

3.104 AHQP Interview with Heisenberg, 5 July 1963, p. 3.

3.105 Werner Heisenberg Physics and Beyond: Encounters and Conversations (New York: Harper and Row, 1971) p. 94.

3.106 Ibid., p. 100.

3.107 It appears from references made by some of those listed in the table, such as E. U. Condon and J. C. Slater, that there were at least ten additional American scientists who went to Europe to study quantum mechanics at this time. I have, however, included the names only of persons about whom direct information is available from their own writings or from published sources such as the records of funding agencies or American Men of Science.

3.108 See previous discussion of these awards, Chapter 3, page 3.12.

3.109 See previous discussions of these awards in Chapter 2, page 2.15 and Chapter 3, page 3.12. It should also be remembered that the International Education Board subsidized the studies in quantum physics of a sizeable number of European scientists in the late 1920's such as E. Hückel, P. Jordan, W. H. Heitler, G. Gamow, S. A. Goudsmit, and F. W. London. (The Rockefeller Foundation Directory of Fellowship Awards 1917-1950.)

3.110 Biographical sketches in Kuhn, Heilbron, Forman and Allen, Sources for History of Quantum Mechanics were consulted to ascertain the institutional affiliations reported in this discussion.

3.111 We have already noted in Chapter 2, page 2.16, the decision of the International Education Board to subsidize the Mathematics Institute at the University of Göttingen, in line with the Board's policy of "making the peaks higher."

3.112 A letter from E. C. Kemble to T. Lyman, 9 June 1927, among the Lyman Papers at Harvard University, tells of a joint seminar conducted in the summer of 1927 by Max Born and members of the Mathematics Institute at Göttingen.

Dirk Struik, in private conversation with the author 29 March 1972, recalled Hilbert's keen interest in physicists' use of mathematics.

E. U. Condon, in "60 Years of Quantum Physics," Phys. Today (October 1962) 37-49, tells of Hilbert lecturing on quantum theory in the Fall of 1926 and also of Hilbert's discussions of matrix mechanics with Born and Heisenberg who had sought his help. Condon relates elsewhere, Amer. Jour. Phys. 15 (1947) 368, Hilbert's remark, made in lecture, "die Physik wird zu schwer für die Physiker."

3.113 The general intellectual atmosphere of Göttingen during the 1920's is described in Chapter 2 of Brighter Than a Thousand Suns by Robert Jungk, translated by James Cleugh (New York: Harcourt, Brace and Company, 1958) from the original Heller als tausend Sonnen (Bern: Alfred Scherz Verlag, 1956). The authenticity of Jungk's presentation is attested to by Charlotte Houtermans (Letter to the author, 20 October 1974) who was herself a graduate student in physics at Göttingen in the 1920's.

V. F. Weisskopf, in "My Life as a Physicist," pages 2 and 3 states that he believes he would have received a better training in mathematical physics with Sommerfeldin Munich than he did in Göttingen where, to some, the approach seemed too mathematically complicated and formalistic. Weisskopf recalls a visit to Göttingen by Paul Ehrenfest during which Ehrenfest advised him to distrust such an approach and to learn to distinguish between physics and formalism. In Ehrenfest's words "Physics is simple, but subtle."

3.114 A phrase used by Charles Weiner in "A New Site for the Seminar: The Refugees and American Physics in the Thirties" in The Intellectual Migration: Europe and America 1930-1960, Donald Fleming and Bernard Bailyn, eds. (Cambridge: Harvard University Press, 1969), pp. 190-228, to describe a mode of intellectual intercourse among European physicists during the 1920's and early 1930's.
A specific example of such activity occurred when E. Schrödinger went to Copenhagen to discuss his wave-mechanical approach to

quantum theory with Bohr. Their interaction is vividly described by W. Heisenberg in "Fresh Fields (1926-1927)," Chapter 6 of Physics and Beyond.

3.115 R. T. Birge quotes a letter which he received from E. U. Condon, written 22 February 1927 from Göttingen, in which Condon said "I am getting very tired of European snobbishness and will be glad to get back to America." History of the Physics Department, Volume II, p. VIII (42).

I. I. Rabi recalled some of the difficulties which he and other Americans encountered during their stay abroad with regard to being accepted into European scientific circles. Rabi also relates that he and Condon discussed, while in Europe, the need to "really do something to improve the situation ." AHQP Interview 8 December 1963 pp. 20, 28, and 29.

J. A. Stratton related, in private conversation with the author 7 Norember 1974, that there existed at the Massachusetts Institute of Technology in the late 1920's a kind of "lazzaroni"* of persons determined to change things for the better so far as modern physics at that institution was concerned. Stratton recalled his own participation in this group along with W. P. Allis, N. H. Frank, H. B. Phillips and M. S. Vallarta.
*Comment: the dictionary definition of "lazzaroni" is "a group of homeless idlers of Naples who live by chance work or begging." Its use in this context, however, stemmed from an analogous group of American scientists about 1855 who called themselves the "scientific lazzaroni" and were motivated by an "urge for a self-conscious direction of science in America." That group, comprised of Alexander Bache and a number of scientists from Cambridge, Massachusetts including Louis Agassiz, had banded together in the hope of controlling the institutional forms of science in America.
Source: Edward Lurie Louis Agassiz: A Life in Science. (Chicago: The University of Chicago Press, 1960) p. 182 ff.

3.116 Biographical material relating to Oppenheimer's early career may be found in N. P. Davis Lawrence and Oppenheimer, (New York: Simon and Schuster, 1968) and in Peter Michelmore The Swift Years (New York: Dodd, Mead and Company, 1969). Neither Davis nor Michelmore is a scientist and neither of their works contains specific information on the sources used for particular statements. Of the two, Michelmore's book contains fewer blunders but is not without error in some details. The slim volume Oppenheimer by I. I. Rabi, Robert Serber, V. F. Weisskopf, Abraham Pais and G. T. Seaborg (New York: Charles Scribners Sons, 1969) contains the addresses given at the Oppenheimer Memorial Session of the American Physical Society meeting, April 1967 and first published in Physics Today, October, 1967. The address by Robert Serber "The Early Years" deals with Oppenheimer's own writings may be found both in the Physics Today issue and in the hard-bound publication.

3.117 According to Oppenheimer's biographer, Michelmore, Oppenheimer expected to major in chemistry and consequently had taken courses in chemistry and mathematics as a freshman. His letter requesting permission to omit the introductory physics courses was addressed to E. C. Kemble, the instructor for the thermo-dyanmics course he wished to take. Kemble turned Oppenheimer's letter of 24 May 1923 over to Theodore Lyman for departmental consideration. It is presently among the Lyman Papers, 1923 Box, in the Harvard Physics Department.

3.118 Lyman's letter of Oppenheimer, 8 June 1923, relaying word of the favorable decision by the Physics Department is also among the Lyman Papers, 1923 Box. Some members of the department may have been somewhat skeptical of the basis of the action, for in the Minutes of the Physics Department meeting of 6 June 1923 (Harvard University Archives in Widener Library) note was made that permission was granted since "It appeared that Mr. Oppen-heimer, according to his own statement, had read rather widely for one of his age." (My emphasis.)

3.119 Proc. Camb. Phil. Soc. 23 (1926) 327-335 and 422-431.

3.120 Zeit. F. Phys. 41 (1927) 268-294.

3.121 Sci. Abs. 30 (1927) 463.

3.122 "Zur Quantentheorie der Molekeln" Ann. d. Phys. 84 (1927) 457-484. The analysis developed in this paper involved the "Born-Oppenheimer Approximation" to the complete solution of the wave equation for a molecule by considering the electronic and nuclear motions separately. This was directly related to problem E. U. Condon had been concerned with in the "Franck-Condon Principle" in the old quantum theory. The mathematics involved in this paper is by no means easy to follow and Condon later confessed: "I have never felt that I properly understood [it]. But Born is such a great master at seeing all physics in terms of "Entwicklungen nach einem kleinen Paramater Kappa" that I suppose it must be all right. The paper of Born and Oppen-heimer is among those difficult ones that are more often cited than read." "The Franck-Condon Principle" Amer. Jour. Phys. 15 (1947) 368.

3.123 Letter from Max Born to S. W. Stratton 13 February 1927, in the Archives of the Massachusetts Institute of Technology.

According to Oppenheimer's biographer, Michelmore, some of the European physics students at Göttingen were irritated by by Oppenheimer's personality and life style (he was clearly very wealthy), as well as by his obviously preferred position in Born's circle of aspiring physicists. See The Swift Years, Chapter 3.

3.124 Letter from E. C. Kemble to T. Lyman, 9 June 1927, Lyman Papers, 1927 Box. Kemble added "Unfortunately Born tells me that he has the same difficulty about expressing himself clearly in writing which we observed at Harvard"--a point to which we shall return shortly.

3.125 AHQP Microfilm 49. For a description of Van Vleck's personal acquaintance with Oppenheimer, see "My Swiss Visits of 1906, 1926 and 1930" Helvetica Physica Acta 41 (1968) 1234-1237. They met in Cambridge, England during the summer of 1926 when Van Vleck was enjoying a respite from his teaching duties at the University of Minnesota.

3.126 In the AHQP Interview conducted with Oppenheimer, Session 2, 20 November 1963, Oppenheimer related, page 18, that he was "invited" by Millikan to take his National Research Fellowship to the California Institute of Technology.

3.127 Ibid., p. 21. According to N. P. Davis, Lawrence and Oppenheimer, p. 25 "Young Oppenheimer's brilliance and creativity was generally recognized, but his ability to communicate his knowledge to others left much to be desired for some years to come."

3.128 Ibid., p. 24; the statement is made that in 1927 Oppenheimer had been offered positions by twelve universities, ten American and two European.

3.129 Sources of information on Kronig's activities used in this section are his article "The Turning Point," his AHQP Interview of 12 November 1962, and a personal letter to the author, 18 June 1974.

3.130 Rockefeller Foundation Directory of Fellowship Awards 1917-1950.

3.131 Quoted by Kronig in "The Turning Point" page 36.

3.132 Information related to this activity has been gathered from telephone conversations with S. C. Wang in May and June, 1974; from the AHQP Interview with I. I. Rabi, 8 December 1963, pp. 11, 25; and from the Kronig letter cited in Note 3.129.

3.133 Extensive quotation from this letter seems justified on the grounds that it provides many useful insights into the ways that this group functioned. For example, in addition to following the work being done in Europe, members of the group were stimulated to apply the new formalism to topics that were of local experimental interest.

3.134 AHQP Interview 31 January 1964, p. 24. In this interview Laporte also characterized the response of American physicists in the Washington area to Heisenberg's matrix mechanics with the phrase "all right, all right--what's a matrix?" (p. 23)

3.135 Information provided by N. H. Frank in private conversation with the author 17 December 1974.

3.136 Leipzig: S. Hirzel, 1928.

3.137 Information in this section was provided by P. M. Morse in private conversation with the author 20 October 1974. Morse explained his informal contacts with Kramers as stemming from the fact that they lived in the same rooming house that summer and the fact that he (Morse) was able to be helpful to Kramers in the matter of obtaining his accustomed glass of beer in nominally dry Ann Arbor under the Prohibition Law.
The text referred to was E. U. Condon and P. M. Morse Quantum Mechanics (New York: McGraw-Hill, 1929).
His collaborative efforts with Stueckelberg focused on the application of wave mechanics to the electronic and vibrational levels of the hydrogen molecule ion. Their results were published in Phys. Rev. 33 (1929) 932-947; 34 (1929) 57-64 and in Helv. Phys. Acta 2 (1929) 304-306.
Note: further information on Morse's career, including a list of his published articles may be found in the Festschrift In Honor of Philip M. Morse, H. Feshback and K. U. Ingard, eds. (Cambridge, Mass.: The MIT Press, 1969).

3.138 At that time the study of chemistry was pursued in quite a different manner than it is today, so it is not surprising that a topic as mathematically sophisticated as quantum mechanics was not presented to students of chemistry at Princeton. To the best of my knowledge, Linus Pauling was first to introduce a course including quantum mechanics for chemistry students. He did so at the California Institute of Technology in 1927. We shall say more about this later.

3.139 E. B. Wilson has provided the author with the information summarized in this section in a series of private conversations between January 1972 and May 1974. He also kindly loaned to the author for several days his copy of his senior thesis.

3.140 No claim is made for the completeness of this listing. It is, rather, a compilation of information from available university catalogues and, in some instances, from faculty members, as indicated in the note at the bottom of the table.

3.141 We shall have more to say in a subsequent section about the few expositions of quantum mechanics that were published in America during this period. We have already mentioned that the book by Condon and Morse, published in 1929, was the first American textbook on quantum mechanics. In the meantime, there came from Europe _The New Quantum Mechanics_ by George Birtwistle (Cambridge: University Press, 1928); _Wave Mechanics and the New Quantum Theory_ translated by L. W. Codd from A. Haas _Materiewellen und Quantenmechanik_ (London: Constable and Company, Ltd.,1928) and A. Sommerfeld _Atombau und Spektrallinien: Wellenmechanik Ergänzungsband_ (Braunschweig: Vieweg, 1929) which was translated soon after by H. L. Brose and published as _Wave Mechanics_ (New York:E. P. Dutton and Company, Inc., 1930).

3.142 Note how many of the names on this table also appeared on the list of Americans going to Europe for study during this period, on pages 3.40 - 3.42

3.143 Note on the table the frequent use in both the summer and regular sessions of visiting faculty from European and other American institutions, who were actively engaged in quantum mechanical research.

3.144 Letter from Linus Pauling to the author, 23 May 1974. In this letter Pauling also told of his introducing a course in quantum mechanics in the chemistry department of the California Institute of Technology which became the basis for the text that he co-autored some years later with E. Bright Wilson.

3.145 L. Pauling "Fifty Years of Physical Chemistry in the California Institute of Technology," _Annual Review of Physical Chemistry_ (1965) 1-14.

3.146 Prompted by the statement in the Preface to A. E. Ruark and H. C. Urey _Atoms, Molecules and Quanta_ (New York: McGraw-Hill Book Co. Inc., 1930) "Parts of the book have been used by one of us as the basis of lectures to graduate students of chemistry at the Johns Hopkins University," I wrote a letter to H. C. Urey asking whether this meant that he had taught quantum mechanics to his chemistry graduate students. Urey responded, 12 February 1975, "Yes. I did teach quantum mechanics to my graduate students at JHU and later at Columbia University." (Urey moved to Columbia University in the Fall of 1929.)

3.147 Information on the lectures delivered at Brown University was provided to the author by Mrs. H. B. Phillips, who also supplied a copy of the lecture notes that were prepared for distribution to the students. (Xerox copies of these notes have been deposited at the Center for History of Physics at the American Institute of Physics by Mrs. Phillips.) Phillips also gave this course during the Summer Session at the University of California at Berkeley according to correspondence now in Mrs. Phillips' possession. (This is confirmed in the biographical data on Phillips in American Men of Science, 7th edition, although it does not appear either in Birge History of the Physics Department nor in the Summer School Catalogue. Presumably the arrangements for Phillips' lectures were made after the Catalogue had gone to press.)

3.148 Private conversation between J. H. Van Vleck and the author, 26 March 1975.

3.149 Science 69 (1929) 451.

3.150 The Cornell University catalogues for the years 1927, 1928 and 1929 list a course entitled "Quantum Mechanics," offered by E. H. Kennard but with the designations "not given in 1926-1927" and "not given in 1928-1929." So at most, there seems to have been a course in quantum mechanics at Cornell only in 1927-1928. In the first edition of Introduction to Modern Physics, written by Kennard's colleague at Cornell, F. K. Richtmyer (New York: McGraw-Hill Book Co., 1928) there was no mention whatsoever of the recent quantum mechanics of Heisenberg and Schrödinger. In fact, Richtmyer appears to have been oblivious to the recent ferment in physics when he wrote on page 564:

"Physics of today seems quite secure, in spite of the chasm between classical and quantum theory, but perhaps some Copernicus will appear who will completely overturn our present exceedingly complex structure of physical theories and concepts and show us a beautiful simplicity in the laws of nature. If physics continues to grow at the present geometrically increasing rate, the physicist of a half-century hence will welcome such a revolution with open arms."

(I am indebted to Edward M. Purcell for calling this statement to my attention during private conversation in the Spring of 1972.) In the second edition of Introduction to Modern Physics (1934) Richtmyer introduced the concept of matter waves in his final chapter, presenting it first from an experimental point of view then following with a brief exposition of the theoretical insights of deBroglie and Schrödinger.

3.151 At Yale University there was no course entitled "Quantum Mechanics" given during this period 1926-1929. There was, however, one called "Quantum Theory and Atomic Structure" which Leigh Page had been offering for some years and which then continued to be listed but with the addition of R. B. Lindsay's name. Lindsay has since explained (Manuscript Autobiography, deposited at the Center for History of Physics of the American Institute of Physics) that he gave an introduction to quantum mechanics in the second semester of that course following Page's discussion of the older Bohr-Sommerfeld model of the atom. According to Lindsay, Page found the new quantum mechanics distasteful. This opinion seems to be borne out by Page's own statement in his book, Introduction to Theoretical Physics 1st edition (New York: D. Van Nostrand Co., Inc., 1928) when on page 581, in discussing the limitations of the Bohr theory, he wrote:

"A more promising theory has been developed in the form of a matrix dynamics by Heisenberg, Dirac, Born and Jordan, and in the form of a wave mechanics by deBroglie and Schrödinger. Unfortunately these new methods are even more of the nature of computing machines than the Bohr theory, and as yet they lack any very satisfactory physical interpretation."

(Page had already in his Preface ruled out a discussion of the recent work of Heisenberg, Dirac and Schrödinger as being beyond the scope of his book by reason of their mathematical complexity. As in the case of Richtmyer cited above, however, Page did include some discussion of wave mechanics in the second edition of his book, pages 647-653, which came out in 1935.)

Apparently in the late 1920's Page still hoped for some compromise between classical physics and quantum theory. Raymond Seeger, an advisee of Page during that period, has recently recalled that:

"...he [Page] initially set me the problem of deriving the Schrödinger equation by modifying the classical electromagnetic wave equation. Needless to say, I got nowhere [with it]." (Letter, Seeger to the author, 30 December 1974.)

Seeger also recalled that A. E. Ruark, an assistant professor at Yale 1926-1927, reported on the new quantum mechanics to the Journal Club of the physics department, a fact confirmed by Ruark in a letter to the author, 10 February 1975.

3.152 Telephone conversation between R. B. Whitney and the author, 16 November 1971. At the time when Whitney took this course he had just completed his doctoral studies in the chemistry department of the University of Minnesota where he received his Ph.D. in organic chemistry--another example of a young American chemistry student being introduced to quantum mechanics soon after its discovery. (Whitney's notes and other materials will soon be deposited at the AIP Center for History of Physics.)

3.153 Information presented in this section is based not only on Whitney's notes and recollections, but also on Erikson's History of the Minnesota Physics Department and on the Bulletin of the University: The Graduate School Announcement for the years 1927-1929, Vol. XXX, No. 72, August 30, 1927. (The latter was kindly loaned to me by J. H. Van Vleck and will soon be deposited at the AIP Center for History of Physics.)

3.154 Information presented in this table was obtained from Marckworth Dissertations in Physics and from "Doctoral Dissertations in Science" National Research Council Reprint and Circular Series, Nos. 80 (1926-1927), 86 (1927-1928) and 91 (1928-1929); plus cross referencing with Science Abstracts and, in the following instances, personal contact with the student involved: Muskat, Rosenthal, Seeger and Wang. With regard to the circumstances under which Wang wrote his thesis, he has told me (telephone conversation 20 May 1974) that he had departmental approval of and satisfaction with his work, although Columbia's professor of mathematical physics at that time, A. P. Willis, was a theoretician of the old school, according to Wang. Wang expressed his appreciation of Wills' encouragement in the Physical Review article based on his thesis, but Wang's work in quantum mechanics actually was made possible by his membership in the autodidactic group discussed on pages 3.49 and 3.50 of this thesis.

3.155 See discussion of the European activities of J. R. Oppenheimer on pages 3.45 to 3.47.

3.156 Private conversation between J. A. Stratton and the author, 7 November 1974. Stratton's thesis was published in Helvetica Physica Acta 1 (1928) 47-74 as "Streuungskoeffizient von Wasserstoff nach der Wellenmechanik."

3.157 National Research Fellowships 1919-1944.

3.158 Carl Eckart (AHQP Interview, 31 May 1962, pp. 5-9) has recalled that he went to the California Institute of Technology in 1925 hoping to work with Paul Epstein on extending deBroglie's ideas, about which Eckart had become excited while a student at Princeton. Originally Epstein did not share Eckart's enthusiam for matter waves but was later very helpful in assimilating quantum mechanics. Eckart also remarked that he found the atmosphere at Cal Tech very stimulating by virtue of the steady stream of foreign guest lecturers, citing, for example, the Spring of 1927 when Lorentz and Schrödinger lectured at successive hours with each attending the other's presentation. On the other hand, Eckart was disturbed by the long delays associated with the arrival on the West Coast of European research journals, a situation which, he said, influenced his later moving to Chicago.

We shall discuss later an almost simultaneous publication episode involving Eckart and Schrödinger, a case, interestingly, in which the delay may have been to Eckart's advantage in establishing that his paper was clearly his own, having been submitted before the arrival in Pasadena of Schrödinger's article written more than two months earlier.

3.159 It should be noted that F. A. Jenkins went to NYU after his NRC Fellowship year at Harvard and thus was available to supervise the thesis work of Jenny Rosenthal, as reported in the table on page 3.60.

3.160 Phys. Rev. 33 (1929) 275.

3.161 Jour. Frank. Inst. 207 (1929) 449-542.

3.162 AIP-CHP Interview, 28 February 1966, p. 14.

3.163 Industrial and Engineering Chemistry: News Edition, 10 April 1928, p. 6.

3.164 Chem. Rev. 5 (1928) 361-617; the quotation by G. L. Clark is taken from the introduction to this issue which was entirely devoted to these papers.

Evidence of a more personal nature of the growing rapport between chemists and physicists may be found in the 1928 correspondence between W. H. Rod e bush, a chemist, and J. H. Van Vleck (AHQP Microfilm 49, Section 12) in which Rodenbush, one of the speakers at the St. Louis meeting, comments on their recent meeting and refers to himself as "a chemist trying to cope with the new phsyics" and hence "consulting an expert."

3.165 This was a single journal at that time. Van Vleck's articles, both entitled "The New Quantum Mechanics," were published in J.O.S.A.-R.S.I. 14 (1927) 108-112 and 16 (1928) 301-306.

3.166 See, for example, Darrow's article "Introduction to Wave Mechanics" Bell Sys. Tech. Jour. 6 (1927) 653-701.

3.167 "An Undulatory Theory of the Mechanics of Atoms and Molecules" Phys. Rev. 28 (1926) 1049-1070.

3.168 Cambridge: Massachusetts Institute of Technology, 1926; a German edition Probleme der Atomdynamik followed soon thereafter (Berlin: Julius Springer, 1926).

3.169 E. C. Kemble lauded the prompt publication of the text of Born's lectures and expressed the hope that it would be "of great service in helping us American physicists to keep up with the stream of thought in a field in which we have been too prone to lag behind."
Source: Book review of Problems of Atomic Dynamics in Phys. Rev. 28 (1926) 423-424.

3.170 See, for example, D. M. Dennison, AHQP Interview, Session 3, 30 January 1964, p. 23; J. H. Van Vleck, AHQP Interview, Session 2, 4 October 1963, p. 3 and F. W. Loomis, University of Illinois Interview, 19 November 1965, p. 4.

3.171 "Reminiscences of a Life in and out of Quantum Mechanics" Internat. Jour. Qu. Chem. Symp. No. 7 (1973) 7-22; passage quoted is found on page 7.

3.172 Physics and Beyond, page 94.

3.173 Two members of the American physics community at that time, who were so inclined, have both recalled their difficulty in finding others with similar concerns.
Private conversations between the author and V. F. Lenzen, 26 October 1974 and between the author and Henry Margenau, 22 March 1975.

3.174 Among the Americans involved in those discussion were E. U. Condon, E. H. Kennard, H. P. Robertson and A. E. Ruark. A summary of their ideas may be found in Max Jammer The Conceptual Development of Quantum Mechanics," 332-335 and in Max Jammer The Philosophy of Quantum Mechanics (New York: John Wiley and Sons, 1974) pp. 169-170. See also R.B.Lindsay, Sci.Mon. 26(1928)299-305.

3.175 New York: The Macmillan Company, 1927; the quoted passage occurs on page 5.

3.176 P. W. Bridgman "P. W. Bridgman's The Logic of Modern Physics After Thirty Years," Daedalus 88 (Summer 1959) 518-526.

3.177 Condon and Morse Quantum Mechanics, p. 17; Bridgman's ideas, as expressed in The Logic of Modern Physics were also cited in 1928 by W. H. Rodebush in the course of his presentation during the American Chemical Society's St. Louis Symposium which we have discussed earlier.

3.178 See, for example, Linus Pauling AHQP Interview, session 2, 27 March 1964, pp. 9-10 and 26-27; E. U. Condon "The Development of American Physics" Amer. Jour. Phys. 17 (1949) 404-408, esp. p. 408; E. C. Kemble, F. Birch and G. Holton "Bridgman, Percy Williams" Dict. Sci. Biog. Vol. I, 457-461, esp. p. 460; and Van Vleck "Reminiscences of the First Decade of Quantum Mechanics"

3.179 "Permanent Elements in the Flux of Present-Day Physics" Science 71 (1930) 19-23; this was an address which Bridgman delivered as retiring vice-president and chairman of Section B at the meeting of the American Association for the Advancement of Science held in DesMoines, Iowa in December of 1929.
The quoted passage occurs on page 21.

3.180 Condon "Reminiscences," page 13.

3.181 At that time an academic salary of more than $3,500 would have been "fabulous." In Condon's case, he received $5,000 and the title "Professor of Theoretical Physics" for the year 1929-1930 that he spent at the University of Minnesota, according to Erikson's History of the Minnesota Physics Department. (It is noteworthy that Condon was 27 years old at the time.)

3.182 Correspondence in the Archives of the Massachusetts Institute of Technology from the year 1926, involving Max Born, Norbert Wiener and S. W. Stratton, then President of M.I.T., reveals that Born was offered a salary of $8,000 per year plus moving expenses, to take a permanent position there. It also appears that Born received attractive offers from Cornell University and from the University of Michigan, but Born decided to remain in Europe. One the other hand, Ohio State University succeeded in attracting both Alfred Landé and L. H. Thomas. Landé has credited Alpheus Smith, then chairman of the Physics Department, with realizing the need to improve the level of theoretical physics at that institution. (Letter from Landé to the author, 18 February 1972.)

3.183 In our next chapter we shall discuss the impact that the Depression had on the employment prospects of young physicists. Nevertheless, despite external circumstances, the professional status of theoretical physics was clearly established in America by 1930.

3.184 Both Slater and Morse started their work in physics in laboratory situations and seem to have switched to theory because they found it more interesting. On the other hand, Condon is quoted by Birge, History of the Physics Department, Vol. II, p. VIII (43), as having said that his early career as a graduate student in experimental physics lasted 24 hours, due to the inordinate amount of glassware that he broke. Van Vleck has stated that he was appalled by the prospect of having to pass a course in glass blowing techniques and was relieved to find that the requirement could be waived under certain circumstances.
Van Vleck "Reminiscences of the First Decade of Quantum Mechanics" Internat. Jour. Qu. Chem. 5 (1971) 3-20.

3.185 In the eyes of S. A. Goudsmit, a physicist who joined the American physics profession after receiving his training up through the doctorate level in Europe, the practical background of many young Americans gave them a decided advantage as physicists, whether experimental or theoretical, over their European counterparts. Furthermore, Goudsmit has related that, in Europe, he had been advised not to discuss his practical achievements as they were somehow unbecoming to a rising theorist. To his surprise, on reaching Ann Arbor, he found that all the physics students there were interested in, and reasonably adept at, tinkering with automobiles. It is Goudsmit's belief that the practical facility of American physicists as a whole, plus their ability to work as members of a team, was what accounted first for the success of the American cyclotron research and later for the success of the Manhattan Project.
Goudsmit "Why the Germans Did Not Get the Atomic Bomb," Morris Loeb Lecture in Physics, Harvard University, April 8, 1975.

3.186 Some idea of the numerical magnitude of American productivity in quantum mechanics for one of those years can be obtained from scanning the entries in the category "Mechanics, Quantum and Wave" that appeared for the first time in Science Abstracts 32 (1929): Thirty-three of the approximately two hundred papers listed were written by twenty-five individual American authors.

3.187 J. C. Slater, in his article "Quantum Physics in America Between the Wars," Internat.Jour. Qu. Chem. 1S (1967) 1-23, has expressed the opinion that, given the field of about 50 bright young men who entered physics during the 1920's, a decade filled with momentous discoveries, "the inevitable result of this was a great deal of duplication. Almost every idea occurred to several people simultaneously. No one had time to follow through a line of work without having someone else break in on his developments before they were finished. It is probable that any one of a dozen of the physicists whom I have mentioned, given the situation which faced physics in 1924, would have worked out the principles of quantum mechanics, if he had been free to take his own time about it. But as things were, no one had the time to do it all by himself, and wave mechanics is a composite of the work of many men. Certainly it attained a richness and variety of approach in this way which it would never have if it has been the work of one or a few isolated scientists."

3.188 See, for example, references to work done by R. Kronig, P. M. Morse, J. R. Oppenheimer, I. I. Rabi, J. C. Slater and N. Wiener on pages 3.35, 3.36, 3.46, 3.50 and 3.52.

3.189 Information summarized here is contained in Eckart's AHQP Interview 31 May 1962, pp. 6-8.

3.190 "The Solution of the Problem of the Simple Oscillator by a Combination of the Schroedinger of the Lanczos Theories" Proc. N.A.S. 12 (1926) 473-476. This paper was submitted May 31, 1926 and published July 15, 1926. A second, more detailed paper on the same topic "Operator Calculus and the Solution of the Equations of Quantum Dynamics" was sumbitted as week later and published in Phys. Rev. 28 (1926) 711-726, the October, 1926 issue. Eckart added a note in proof, dated September 2, 1926, to the effect that Schroedinger's treatment of it had by then reached Pasadena, but Eckart decided not to withdraw his paper since he had proceeded by showing that matrix mechanics was equivalent to wave mechanics, while Schroedinger had worked from the opposite point of view. At any rate there can be no doubt that Eckart's work was achieved entirely at the California Institute of Technology with the "assistance and encouragement of P. Epstein" as noted in the first paper cited above, and as explicated in the AHQP Interview, page 7.

3.191 Condon and Gurney "Quantum Mechanics and Radioactive Disintegration" Phys. Rev. 33 (1929) 127-140 was submitted November 20, 1928 and published in February, 1929, but an earlier note by these authors had been sent to Nature on July 30, 1928 and appeared on page 439 of the September 22, 1928 issue. Gamow's paper, "Zur Quantentheorie des Atomkernes," Zeit. f. Phys. 51 (1928) 204-212, was submitted August 2, 1928 and published October 13, 1928. Condon and Gurney took note of its arrival in Princeton in early November.

3.192 Dennison has provided an account of the circumstances surrounding this work in an address delivered at the APS-AAPT Meeting in Chicago, February 6, 1974 and entitled "Recollections of Physics and Physicists during the 1920's" which has since been published in Amer. Jour. Phys. 42 (1974) 1051-1056. Quotations are from page 1055.

"The Specific Heat of the Hydrogen Molecule" may be found in Proc. Roy. Soc. 115 (1927) 483-486.

3.193 Information on Houston's career has been obtained from H. E. Rorschach, Jr. "The Contributions of Felix Bloch and W. V. Houston to the Electron Theory of Metals" Amer. Jour. Phys. 38 (1970) 897-904, and from the AIP-CHP Interview conducted with Houston. The quotation from Sommerfeld occurs in the Interview and on page 898 of the Rorschach article.

Houston's results appeared in Zeit. f. Phys. 48 (1928) 449-468 as "Elektrische Leitfähigkeit auf Grund der Wellenmechanik."

3.194 "The Theory of Complex Spectra" Phys. Rev. 34 (1929) 1293-1322.

3.195 Slater Solid-State and Molecular Theory: A Scientific Biography pp. 60,61. In Chapter 8 of this work, entitled "1929, The 'Gruppenpest' and Determinantal Wave Functions," Slater gives many details of the method itself and of its reception in Europe.

3.196 It was in this paper that Slater used for the first time his celebrated, so-called, "F and G integrals" that have proved so useful in the quantum theory of atomic spectra.

3.197 This pun was incorporated into a playlet based on the Faust legend that was produced by members of Bohr's Institute in Copenhagen in 1932. An English translation of the dialogue of that skit has been published as an Appendix in George Gamow Thirty Years that Shook Physics (Garden City, New York: Doubleday and Co., 1966).

3.198 S. C. Wang "On the Asymmetrical Top in Quantum Mechanics" Phys. Rev. 34 (1929) 243-252.
Van Vleck has described some of his own work during this period and this particular episode in his article "Reminiscences of the First Decade of Quantum Mechanics."

The problem of the asymmetrical top was one that also engaged the attention of other physicists during this period, such as H. A. Kramers and G. P. Ittmann. Using a different method from Wang's, they published a three part article "Zur Quantelung des asymmetrischen Kreisels" Zeit. f. Phys. 53 (1929) 553-565; 58 (1929) 217-231; 60 (1930) 663-681.
It is interesting to note that not only was Wang's successful treatment of the topic completed before theirs, but also that Kramers and Ittmann were folloiwng Wang's work closely. They mention it in all three parts of their article, but it is especially noteworthy that in Part II, which they submitted on August 10, 1929, they referred to Wang's paper that had appeared in the July 15 issue of the Physical Review--a clear indication that in 1929 work being done in America was promptly gaining attention in Europe.

3.199 Robert Serber "The Early Years" in "A Memorial to Oppenheimer" Physics Today, October 1967, p. 35. The work refered to by Serber was published as "Quantum Theory of the Autoelectric Field Currents" Proc. N.A.S. 14 (1928) 363-365. In all, Oppenheimer had six papers published in 1928, four of them in Phys. Rev. 31 and 32, two in Proc. N.A.S. 14.

3.200 Mulliken's own account of his early work may be found in "The Path to Molecular Orbital Theory" Jour. Pure and Appl. Chem. 24 (1970) 203-215, and in "Spectroscopy, Molecular Orbitals and Chemical Bonding" Science 157 (1967) 13-24. The latter was his Nobel Prize Lecture.

Mulliken's principal papers from the 1928-1929 period were "The Assignment of Quantum Numbers for Electrons in Molecules," Part I Phys. Rev. 32 (1928) 186-222; Part II: "The Correlation of Molecular and Atomic Electron States" Ibid. 761-772; Part III: "Diatomic Hydrides" Phys. Rev. 33 (1929) 730-747.

3.201 Pauling's early papers in this field include "The Application of the Quantum Mechanics to the Structure of the Hydrogen Molecule and Hydrogen Molecule Ion and to Related Problems" Chem. Rev. 5 (1928) 173-213 and "The Shared-Electron Chemical Bond" Proc. N.A.S. 14 (1928) 359-362.

Pauling has made the interesting point in his AHQP Interview, 27 March 1964, pages 19, and 20,that American chemists were able to assimilate quantum mechanics into their thinking much more readily than their European counterparts who were less well trained in physics and mathematics and tended to specialize in a narrow field too early in their careers.

Pauling has published some remarks on his early experiences with quantum mechanics in "Fifty Years of Progress in Structural Chemistry and Molecular Biology" Daedalus, Fall 1970, 988-1014.

3.202 A very careful and interesting study of all aspects related to this work of Davisson and Germer has been made (but not yet published) by Richard Gehrenbeck as "C. J. Davisson, L. H. Germer and the Discovery of Electron Diffraction" Ph.D. Dissertation, University of Minnesota 1973. I am indebted to Dr. Gehrenbeck for his kindness in loaning to me a copy of this work.

Chapter 4

The Flourishing of Quantum Theoretical Physics in America, 1930-1935

CHAPTER 4

The Flourishing of Quantum Theoretical Physics in America 1930 - 1935

Introduction

By the early 1930's the pursuit of physics in America had clearly entered into a new phase of development. American physicists and their work were no longer outside the mainstream of the international development of the discipline. Americans relied much less heavily on European scientists than they had in the past and the European physicists who came to America, either as visitors or as immigrants, did so as professional colleagues rather than missionaries or teachers as they had in the past. By this statement we do not mean to undervalue the contributions which many of those Europeans made to the richness of the American scientific scene either in the early 1930's or in the decades ahead. Rather, we mean that the European physicists who came here could function as effectively as they did because physics in America had already attained maturity.

In any study of scientific development during the early 1930's two dominating external influences must be kept continually in mind: the severe economic depression, and the rise, in Europe, of totalitarian regimes with policies inimical to science in general and to Jewish scientists in particular. We shall have numerous opportunities in this chapter to witness the effects of these two circumstances on the growth of physics in America during this period.

While the physics profession certainly had problems to cope with during these years, it did so in a forthright manner and, in a sense, grew stronger through the effort. As individuals, American physicists continued to advance although perhaps more slowly than during earlier, more prosperous times. Experimentalists felt the constraints of restricted budgets, yet their output continued to meet or set high standards internationally. Theoreticians, who had so recently "come of age" in America, not only continued to hold their own but also became more firmly integrated into the enterprise of physics in America. In addition to satisfactory academic positions many theoretical physicists enjoyed opportunities for participation in theoretical institutes or symposia. And, most importantly, theoretical work of outstanding quality was produced in America.

As in our previous chapters we shall follow a pattern of discussion which begins with a consideration of the institutional developments within the American physics community that took place during this period. This will be followed by a description of the direction in which physics was moving in America from the internal perspective of the discipline itself. Finally, we shall refocus our attention upon those specific achievements made by American scientists that relate to applications of quantum mechanics.

Part I: Institutional Developments within the American Physics Profession

Many of the developments that took place within the American physics profession in the early 1930's can be traced either to the increasing complexity of the discipline as a whole and the consequent diversity of interests among the membership or to efforts to retain some degree of professional prosperity despite the adverse economic conditions that prevailed in society at large.

We shall see in this section that the American physics profession continued to grow in the areas of professional organizations, journal publications, educational and research facilities, even while it was experiencing some very real financial difficulties.

Professional Organizations

In the years between 1930 and 1935 membership in the American Physical Society continued to grow but more slowly than in the previous half-decade. Whereas total membership increased between 1925 and 1930 by almost 40 percent from 1760 to 2484, only 362 more members were added by 1935. Fellowship achieved a notable increase between 1930 and 1931, going from 516 to 679, but it then stayed below the 700 level until after 1935.[1]

The Society's two-year Presidential term was reduced to one year in 1932. It has been suggested[2] that the density of outstanding American physicists, worthy of this honor, had by then risen to such a degree that a one-year term was deemed appropriate. P.D. Foote, A.H. Compton and

R.W. Wood each served a single year term following the two-year tenure of W.F.G. Swann in 1931-1932.

The pattern of meetings continued to grow in number and in program. Beginning in 1930 when a three day meeting was held in Ithaca on the campus of Cornell University,[3] a summer meeting in June was added to the schedule of February, April, November and December meetings that we discussed in previous chapters.

Invited talks and symposia became regular parts of virtually every APS meeting. The list in Appendix IV, Part 5 shows the great burgeoning of this activity that took place during the early 1930's. In many instances the attendance at APS meetings was between 300 and 500[4] despite the economic "hard times."

One meeting which merits particular note was that held jointly with Section B of the American Association for the Advancement of Science in Chicago in June 1933 in connection with "A Century of Progress Exposition" - according to APS Secretary W.L. Severinghaus, "perhaps the most important scientific session in its [APS] history to date."[5] The meeting lasted six days and was attended by several visitors from Europe.[6] It included a number of symposia on diverse topics.[7] The one on the "Application of Quantum Mechanics in Chemistry" is of special interest for this thesis. J.C. Slater, a participant in that symposium, has written on his recollections of that event as follows[8]:

"The thing that impressed me - very consciously, as I look back - was not so much the excellence of the invited speakers as the fact that the younger American workers on the program gave talks of such high quality on research of such importance, that for the first time the European physicists present were here to learn as much as to instruct. One must remember that in the 1920's the first thing required of a young American graduate student of physics was to learn German so he could follow the new work in the field. Today [1968] the first thing a young physicist of continental Europe does is learn English so he can attend meetings and learn of new progress. This trend was already becoming visible in the 1930's, and I felt that the tide was turning more specifically at this 1933 Physical Society meeting than at any other one moment."

While the Physical Society remained the principal professional organization of American physicists, particularly those engaged in "pure" research during the early 1930's, there were also other American organizations for physicists primarily concerned with either the applied or the educational aspects of the discipline. Some, such as the Optical Society of America, had been in existence for a number of years.[9] Others, such as the American Association of Physics Teachers, were of more recent origin.[10]

In 1931 a new type of organization for American physicists was inaugurated with the founding of the American Institute of Physics, an event resulting from two separate sets of circumstances. While it is not possible to tell the full story of the Institute's beginnings at this time[11] it will be illuminating to examine some details associated with this action.

Karl T. Compton, first Chairman of the AIP Governing Board, stated in 1933[12]

"In one sense the American Institute of Physics is the child of the five parent national societies which have cooperated in forming it. In another sense, however, it has followed the more usual course of being born of two parents, the one financial distress and the other organizational disintegration."

The five parent national societies were the American Physical Society, the Optical Society of America, the Acoustical Society of America, the Society of Rheology and the American Association of Physics Teachers.[13]

The "financial distress" arose in connection with the cost of publication of such journals as The Physical Review and those of the other member societies. Rising costs had come about due to the increasing research productivity of members of the American physics profession. Economic hard times had aggravated the dilemma.

The "organizational disintegration" came about as the result of increasing complexity within the discipline of physics and the consequent fragmentation of the American physics community through dispersal into various societies with special interests.

American Physical Society committees were formed in response to these two predicaments, one a standing committee on the financial status of The Physical Review and the other on the organization of physics.[14] The Chemical Foundation[15] was approached by the former seeking financial help. It appears that impetus for the formation of the American Institute of Physics came from the Chemical Foundation.

Other groups of physicists and scientists closely allied to physics were also asking for help from the Foundation at that time. Francis P. Garvan, President of the Foundation, stated that[16]

"...it appeared to him that the physics groups of America were not well organized and that he would much prefer to give his support in such a way as to bring about a better organization on an eventually self-supporting basis."

Necessary steps were taken leading to the first meeting of the Governing Board of the American Institute of Physics on 3 May 1931. This is regarded as the founding date of the Institute. On this occasion it was announced that the Chemical Foundation would underwrite the cost of the Institute for at least the first year and would provide for the establishment of an office in New York City.[17]

The constitution of the AIP defines its objectives as "the advancement and diffusion of knowledge of the science and its application to human welfare" and states that the publication of scientific journals related to physics was to be an integral part of its function. We shall discuss in our next section the remarkable growth of American journals of physics that took place during the early 1930's and the role played by the American Institute of Physics in their publication.

Before leaving the topic of the launching of the American Institute of Physics we should make note of the fact that not all members of the American physics profession favored the move. Some disapproved of the public-relations aspect of the venture which seemed, to them, inappropriate for a scientific discipline. In addition, they believed that economy of publication costs could just as well have been effected by direct negotiation with an established publisher, as was done by the American mathematics community.[18]

Serial Publications

During the early 1930's the semi-monthly Physical Review continued to grow in size and prestige.[19] In 1930 and 1931 two "volumes" were issued each year, each containing about 150 research papers and requiring close to 2000 pages. By 1932 a peak was reached when four "volumes" of about 1000 pages each were issued. The size of the Physical Review leveled out somewhat in the years 1933 - 1935 at two volumes per year, each with about 1000 pages. This was not caused by a real, sharp decline in the output of the American physics profession, but rather is explainable in terms of the additional publications that were established during these years and which we shall discuss presently.

Meanwhile Reviews of Modern Physics continued as a quarterly journal of longer survey articles, a number of which gained wide-spread attention. For example, in 1932 Enrico Fermi published in Reviews of Modern Physics his memorable "Quantum Theory of Radiation," based on the lectures he had delivered at the University of Michigan Summer Symposium on Theoretical Physics in 1930,[20] and in 1935 J.H. Van Vleck and A. Sherman published "The Quantum Theory of Valence" which was later translated into Russian.[21]

Turning now to the new journals of physics that were inaugurated during the early 1930's, one is immediately struck by their number - surprisingly large in view of the economic pressures of the period and a clear indication that the discipline of physics in America was exhibiting signs of considerable vigor.

The Acoustical Society of America, which had been formed in late 1928, began publishing its own journal in October of 1929.[22] In 1930 the Optical Society of America decided that the time had come to separate the Review of Scientific Instruments from its regular Journal of the Optical Society of America. The Review of Scientific Instruments had been added to the Journal of the Optical Society of America in 1922 as a special section supported financially by The Association of Scientific Apparatus Makers of the United States of America.[23] When the separation was effected it became a monthly journal with Paul D. Foote and Floyd K. Richtmyer as Editors. It later became the organ of the American Institute of Physics, a change that will be discussed presently.

The next journal to be inaugurated was Physics,[24] "A Journal of General and Applied Physics conducted by the American Physical Society," which began monthly publication in 1931 under the editorship of John T. Tate. He gave the following reasons for launching a new journal for the publication of original research[25]:

"...the [Physical] Review has more than doubled in volume during the past six years and has become more and more the exponent of the purely introspective side of physics.... But fascinating as are the developments in atomic physics and the quantum mechanics, they do not by any means represent the whole of the science, nor are they more interesting or valuable than the original work of the greater number of professional physicists who are applying physical methods and principles to the problems of other sciences and the industries."

A feature of this journal during the early years of its publication was an Editor's Column where summaries of recent articles in the Physical Review were presented for those readers wishing to follow the developments of quantum mechanics and other theoretical topics.

In 1933 two new journals were launched. The American Physics Teacher,[26] a quarterly publication, became the organ of the recently formed American Association of Physics Teachers, and the Journal of Chemical Physics began monthly publication under the editorship of Harold Urey. Urey was assisted by a group of more than 20 associate editors drawn almost equally from the fields of physics and chemistry.[27] In an editorial published in the first issue Urey noted that, since the boundary between physics and chemistry had been completely bridged, it was appropriate to have a journal devoted to this interdisciplinary field.[28] Contributed articles were, in fact, received from authors associated with either chemistry or physics departments and many of the articles related to applications of quantum mechanics. This was a very active field during those years and the new journal, soon after its inception, followed the example of the Physical Review in providing a Letters to the Editor department for short, unrefereed reports of new work.[29]

By this time (1933) all of the journals that we have mentioned in this section were being published through the American Institute of Physics, with the standardized page size and two-column format so familiar to all readers of AIP journals today. A particularly noteworthy change occurred with regard to the Review of Scientific Instruments at that time, occasioned largely by the announcement in 1932 of the Association of Scientific Apparatus Makers that, owing to the prevailing depression conditions, it could no longer guarantee support for the Review.[30] Thereupon it was decided that the AIP should take over complete responsibility for this journal and make it the official organ of the AIP. The journal's format was altered to include all the advertising previously carried by individual physics journals. A section called "Physics News and Views" and another called "Physics Forum" for items of broad interest in the physics community were added. Initially, it was distributed free of charge to all AIP members, but this policy had to be abandoned by the end of 1934 due to high distribution costs.[31] The Review of Scientific Instruments did continue, however, on a subscription basis, as a source of news for members of the physics community until Physics Today was begun in 1948.

Solely from the point of view of journal production, it is clear that by the early 1930's the American physics profession was a well-established, productive community of physics educators, applied physicists and researchers at the frontiers of modern physics. The table below summarizes their output for a single, typical year in those journals publshed by The American Institute of Physics.

American Physics Journal Publications, 1934

Journal	No. of Articles[1]	No. of Pages[2] (approximate)
Amer. Phys. Teach.	23	190
Jour. Acous. Soc. Amer.	30	210
Jour. Chem. Phys.[3]	140	900
Jour. Opt. Soc. Amer.	66	350
Phys. Rev.	300	2000
Physics	58	410
Rev. Mod. Phys.	10	280
Rev. Sci. Inst.	108	450
Total	735	4790

[1]Brief communications, such as Letters to the Editor, have not been included.

[2]Total No. of Publication Pages less those devoted to Indexes, Book Reviews, meeting notices and other news announcements.

[3]No attempt was made to separate out those articles which were written by chemists.

Opportunities and Facilities for Education and Research

1) Research Centers

Despite the economic stringencies of the early 1930's centers of activity for research in physics continued to develop during this period in the United States. Not only did those succeed in maintaining their programs, which we discussed in earlier chapters, such as the California Institute of Technology, Harvard University and the University of Michigan but, perhaps more remarkably, a number of new facilities developed during this period. We now cite a few particular examples which are especially interesting because of the kind of physics that was carried on at those locations, the recruitment of personnel to staff the centers, and the ways in which they were funded. All of these aspects bring out the secure position that physics had achieved in America.

The Massachusetts Institute of Technology as a whole experienced a notable period of development during the early 1930's. We shall concern ourselves only with some of the details associated with its growth in the field of physics. We have already noted in our previous chapter (see page 3.95) the desire and the efforts of some members of the MIT faculty to improve that institution. One of that group, J.W. Stratton, has recalled that there was consequently great rejoicing when it was learned that Karl T. Compton had accepted the post of President of MIT in 1930.[32]

Compton believed that a knowledge of basic science was essential for modern engineering and set out to revamp the physics department in particular. A strong staff was assembled by bringing in J.C. Slater from Harvard, P.M. Morse from Princeton, and G.R. Harrison from Stanford.

They joined, among others, W.P. Allis, N.H. Frank, J.W. Stratton and M.S. Vallarta who, as we mentioned earlier, were already at MIT.

Within the next few years a number of changes related to physics were effected by this group. The George Eastman Research Laboratory was completed, undergraduate instruction in physics for all students was improved and, at the graduate level, there were initiated theoretical studies about which we shall have more to say later. In addition, close ties were established, in terms of joint meetings and exchange visits, between the faculties of physics at MIT and those at Harvard and Princeton Universities.[33]

Significant developments related to research in physics took place at the University of California in Berkeley during the early 1930's following the arrival of Ernest O. Lawrence and J. Robert Oppenheimer as new faculty members.[34] Lawrence and Oppenheimer were both under thirty years of age at the time and they brought to the Berkeley campus a fresh, young approach in their respective areas of competence. Oppenheimer's theoretical expertise soon attracted a sizeable group of doctoral and postdoctoral students about whom we shall say more in a later section. Lawrence, a man of great energy and enthusiasm, was busy building (with the aid of M.S. Livingston and others) progressively larger forms of the cyclotron which he had recently invented and for which he would receive the Nobel Prize in Physics in 1939. The building and operation of the cyclotron was an expensive proposition but the necessary funds were made available from sources outside the University which will be discussed later. The cyclotron project, later to become the Radiation Laboratory, required long hours of cooperative effort on the part of many young physicists who became part of Lawrence's "team." Lawrence's influence on American physics eventually extended far beyond the Berkeley campus as various groups at a number of institutions undertook to build their own cyclotrons.[35]

Princeton, New Jersey was the site of another set of institutional developments in physics during the early 1930's which merit our attention. In addition to changes at the University which included European physicists John von Neumann and Eugene Wigner joining the faculty and E.U. Condon returning to it, an entirely new entity emerged in 1933 with the opening of the Institute for Advanced Study.[36]

The Institute was a unique conception in the American academic framework.[37] It was not part of Princeton University, but in the early years of its existence it shared the facilities in Fine Hall with the University Mathematics Department.

The first section of the Institute to be established was the "School of Mathematics." From the start, however, it attracted permanent and visiting physicists[38] as well as mathematicians, and among the mathematicians were some, such as Hermann Weyl and John von Neumann, who had worked on the mathematical aspects of the development of quantum mechanics. In addition, when we recall the weakness in the mathematical training of physicists that had characterized the early years of the study of physics in America, the opening of the School of Mathematics of the Institute for Advanced Study must be regarded as a positive development for physics as well as mathematics in America. The Institute provided an attractive setting where physicists could become better acquainted with and more skilled in the mathematical aspects of their own discipline.

The program of the Institute had an informal structure which included lectures, seminars and study groups. Young persons who already had their Ph.D.'s (or equivalent training) were welcome to come for a year's visit, and the Institute soon became a mecca for holders of National Research and Rockefeller Foundation Fellowships.

The University of Wisconsin was another American site which attracted students of theoretical physics from Europe as well as America in the 1930's. John Van Vleck had an active group of doctoral candidates and postdoctoral fellows working there with him prior to his departure in 1934 for Harvard University. Gregory Breit, who succeeded Van Vleck at the University of Wisconsin, continued the high standard for the study of theoretical physics at that institution.

There were other American institutions, in addition to those just cited, which also improved their capabilities in physics during this period. While they may not have risen to the status of a "Center of Research," their presence clearly contributed to raising the general level of the study and practice of physics in America. One thinks, for example, of Washington University in St. Louis where the Wayman Crow Hall of Physics was opened in 1934.[39]

Thus it would seem that, although real financial difficulties did indeed exist for American physicists in the early 1930's (and we shall discuss some of them later), they did not result in physics losing ground or momentum in America.

2) Ph.D. Production

In the years from 1930 to 1935 the annual number of doctoral degrees granted by American institutions continued the upward trend that we have noted characterized the preceding decade. In 1930 there were 89 doctorates in physics awarded by 25 institutions. In 1935, 38 institutions awarded a total of 150 such degrees.[40]

Although a large fraction of the institutions involved granted only a few (three or less) doctorates in physics each year, the data clearly indicate that American students continued to enjoy ever-widening opportunities for the study of physics so far as the location of doctoral degree granting institutions was concerned. This is readily brought out by listing in tabular form those institutions which were responsible for the majority of the degrees granted in physics for the two years 1929-30 and 1934-35 (see list on next page).

Since the institutions listed in the left-hand column did not markedly alter their output of Ph.D.'s between 1930 and 1935, it is evident that the growth that did take place was largely due to the emergence of additional leaders, such as Columbia University and the Massachusetts Institute of Technology.

Leading American Institutions in the Granting of Doctoral Degrees in Physics, 1929-30 and 1934-35

1929-30		1934-35	
Institution	No. of Degrees	Institution	No. of Degrees
Cal. Inst. Tech.	11	Cal. Inst. Tech.	12
U. of Mich.	10	U. of Cal.-Berk.	11
U. of Cal.-Berk.	7	Columbia Univ.	10
U. of Chicago	7	Mass. Inst. Tech.	9
Cornell Univ.	7	U. of Mich.	9
U. of Wisc.	6	Johns Hopkins U.	7
Harvard Univ.	5	U. of Virginia	7
Johns Hopkins U.	5	Cornell Univ.	6
Total	58	U. of Wisc.	6
		U. of Chicago	5
		Harvard Univ.	5
		Yale Univ.	5
		U. of Iowa	4
		Princeton Univ.	4
		Stanford Univ.	4
		Total	104

3) Postdoctoral Fellowship Programs

In the early 1930's some changes occurred in the postdoctoral fellowship programs discussed in our previous chapters, but the concept of providing a year or more of unencumbered research time to promising young students of physics was by then firmly incorporated into the structure of American training in physics. During this period a perceptible shift occurred with respect to the Fellows' choices of institution. In general, far fewer American recipients elected to go to European locations than had done so in the late 1920's and, in fact, more European Fellows were choosing to come to America. Certain American institutions were clearly considered to provide an attractive atmosphere for the study of physics, as will become evident from a look at a few details associated with the leading postdoctoral fellowship programs of the period.

National Research Council Fellowships provided the major source of postdoctoral support for American students of physics in the early 1930's. In 1932 the stipend associated with these awards was reduced from $1800 to $1600 and remained at that level until 1940.[41] This did not, however, pose serious problems for the recipients since the cost of living by that time had dropped more than 20 percent from what it had been on the average in the 1920's when the NRC Fellowship program became established. These awards were highly desirable ones for which applications far exceeded appointments. For example, in the area of the natural sciences, which included chemistry, mathematics and physics, 399 applications were received for the year 1933-34 and fellowships were granted to 82. Only about 5 percent of those receiving doctoral degrees in science were named National Research Fellows.

In each of the years between 1930 and 1935 about two dozen young physicists were supported for study at the institution of their choice by the National Research Council. Of the approximately 90 individual recipients, 16 elected to spend at least part of their time at European institutions, but the vast majority of fellowship-years were spent at American locations. The most popular choices were the California Institute of Technology with 31 fellowship-years, Princeton University with 23, the Massachusetts Institute of Technology with 16, and the University of California at Berkeley with 14.[42]

When the Rockefeller Foundation was reorganized in 1929 it continued support of the National Research Council fellowships, but the program administered by the International Education Board came to an end. Comparable fellowships in physics were subsequently awarded by the Natural Science Division of the Foundation, created in 1929 to continue the support of natural sciences on a domestic and international basis.[43]

In the years 1930 through 1935, 59 Rockefeller Fellowships were awarded to physicists.[44] It is interesting to note that some of the European recipients of these awards chose to come to the United States during the early 1930's. Among those doing so were J.D. Cockcroft and Herbert Dingle from Great Britain, Svein Rosseland from Norway and Robert Schlapp from Scotland.[45]

Schlapp spent the year 1931-32 at the University of Wisconsin where he worked with J.H. Van Vleck on problems involving the application of quantum mechanics that we shall discuss in a later section. Another

European who made his way to Madison in the early 1930's to study with Van Vleck was William G. Penney who did so as a Fellow of the Commonwealth fund.[46] During the early 1930's Commonwealth Fund Fellows were also studying at other American institutions, such as Cornell University, Yale University, the University of California at Berkeley, the Johns Hopkins University, and the University of Michigan.[47]

Fellowships specifically awarded for study abroad continued to be available to American physicists through the John Simon Guggenheim Memorial Foundation. About a dozen such awards were made between 1930 and 1935. In the first part of this period, continental institutions continued to attract American students of physics,[48] but between 1933 and 1935 all three American physicists who received Guggenheim awards elected to go to the Cavendish Laboratory.[49]

In this section our discussion has focused primarily on the availability of postdoctoral fellowships. We defer until later consideration of the research topics chosen by the recipients.

International Contacts

Within the international scientific community of the early 1930's American physicists continued to enjoy cordial relations and increasing prestige. For example, John Van Vleck was invited to take part in the Solvay Congress on Magnetism that was held in 1930, the first American theorist to be so honored.[50] Ernest Lawrence was invited to participate in the 1933 Solvay Congress on the Nucleus.[51]

European physicists continued to visit and lecture in the United States after 1930 but there were fewer grand tours than in previous years.[52] Max von Laue and Rudolf Ladenburg visited a number of American physics laboratories in connection with the establishment, with Rockefeller funds, of an Institute of Physics at the Kaiser Wilhelm Institute in Berlin.[53] The Summer Symposium in Theoretical Physics at the University of Michigan blossomed as a major international gathering place for physicists during the 1930's. Among those who came from Europe to take part in these symposia between 1930 and 1935 were Niels Bohr, Paul Ehrenfest, Enrico Fermi, George Gamow, Werner Heisenberg, Hans Kramers, Wolfgang Pauli and Arnold Sommerfeld. These visitors shared the lecture platforms with such American scientists as Gregory Breit, Philip Morse, Robert Oppenheimer and John Van Vleck. All participants in the symposia mingled freely outside the lecture rooms, sharing sports, food and drink, and ideas about physics.[54]

As the 1930's progressed, however, European physicists turned more and more to the United States for employment in their field and as a refuge from the growing tide of European dictatorships.[55] At least two of the Michigan summer visitors later took up permanent residence in

America. George Gamow joined the faculty of the George Washington University in 1934, and Enrico Fermi came to Columbia University in 1939 after having returned several times to the Michigan Summer Symposia.[56] Other visiting physicists during the 1930's who decided to accept permanent appointments in America were James Franck, Otto Stern, and Albert Einstein.

Einstein had spent three winters at the California Institute of Technology before accepting the permanent post of professor at the newly formed Institute for Advanced Study.[57] There were some who viewed Einstein's move to the United States as having symbolic significance, as shown in the following quotation attributed to Paul Langevin[58]:

"It's as important an event as would be the transfer of the Vatican from Rome to the New World. The Pope of Physics has moved and the United States will now become the center of the natural sciences."

The migration to the United States of European physicists during the 1930's and their impact on the development of physics in America is a large topic, and one that is somewhat beyond the main focus of this thesis.[59] We do wish to stress, however, that it was only because America already had suitable established institutions and a stimulating native physics community that the newcomers could be so successfully assimilated.

Support for Physics and Physicists

We have already mentioned some of the financial sources that benefited physicists during the early 1930's - George Eastman provided the money for the new physics laboratory at the Massachusetts Institute of Technology; money from the Bamberger family made possible the Institute for Advanced Study; the Rockefeller Foundation supported its own fellowship program and that administered by the National Research Council.

In addition, the Rockefeller Foundation made other grants, such as one of $40,000 to the California Institute of Technology in the area of physics and chemistry, specifically for the study of crystal structure and the application of wave mechanics to inorganic and organic molecules.[60] Following its reorganization in 1929, the Rockefeller Foundation changed the emphasis of its support from one of capital grants to institutions toward increased funding of specific fields of interest.[61]

Ernest O. Lawrence received financial assistance from the Research Corporation, the Chemical Foundation and the Josiah Macy, Jr. Foundation toward the building of his first cyclotrons during the early 1930's.[62]

These examples of financial support for physics, however, should not be construed as indicating that the physics profession was untouched by the economic realities of the times. As a matter of fact, the Depression in the 1930's had its effects on American physicists not only in terms of the amount of money that was available for their salaries and research activities, but also in terms of the esteem in which physics was held by those outside the field. These two aspects are not, of course, independent, especially in times when money is in short supply.

In our previous chapter we noted some criticisms of science that had been raised on supposedly humanistic grounds. During the 1930's voices were raised against science and technology as causing, or at least contributing to, the worsening employment situation. According to Charles Weiner[63]

"It is difficult to determine how widespread this attitude was, because it was only occasionally articulated. It was, however, perceived as a major threat by leaders of the scientific community, because it occurred at precisely the time when scientists needed to make an effective case for increased public support."

Whether their concern was justified or not, some members of the American physics profession assumed a defensive attitude and felt that it was necessary to "promote" physics. Among those physicists sharing this point of view were members of the group responsible for the establishment of the American Institute of Physics.[64] The "Physics Forum" section of the Review of Scientific Instruments carried several editorials related to the goal of enhancing public appreciation of physicists.[65]

For his Presidential Address to the American Physical Society, delivered in December 1933, Paul D. Foote chose the topic "Industrial Physics."[66] In assessing the difficulties facing physicists, Foote remarked

"One does not require familiarity with the matrix mechanics to understand the principle of uncertainty as regards a physicist's employment during the past three years. The mathematical details of the classical dynamics may be forgotten, but D'Alembert's principle of least action nevertheless can be appreciated in the government's treatment of its scientific personnel."

With regard to the physicist's proper role in society, Foote stated

"The world owes the physicist a good living because he has become an important economic factor in a highly complex civilization. Once this fact is thoroughly impressed upon the general public, we shall have, as an incidental, all the facilities and means we require for the pursuit of our idealistic search for truth."

Later, he added

"In my opinion, this question of organized propaganda for physics and a thorough investigation of the sociological aspects of physics are the most important problems confronting our society."

Such an attitude may have been the natural outgrowth of the contemporary economic situation, but there also emerges the hint that some segments of the physics community were on their guard lest physics in America become overly theoretical and out of touch with the tangible, real world.

While attitudes may be difficult to assess in terms of depth of feeling involved and general prevalence, the effects of budgetary cuts are more readily evaluated. There is no doubt that for some years during the 1930's individual American physicists were directly affected by cutbacks in funding within universities, industrial laboratories and governmental agencies.

At many institutions reductions in salary of 10 percent or more were put into effect for the entire staff,[67] but, since the cost of living had dropped by at least that much,[68] this should not have caused undue hardship. Another means of staying within reduced budgets at some universities involved the elimination of invitations to foreign guest lecturers. For example, in 1932 the University of Wisconsin ended its commitment to such a program that had begun in the early 1920's.[69]

Reduced budgets had their impact on the research programs of experimental physicists. In 1935 F.W. Loomis wrote of the situation at the University of Illinois, where he was chairman of the Physics Department[70]

"The department, whose operating expenses have been reduced to a starvation point for over three years, suffered a financial crisis this winter and pretty nearly had to close up. It was rescued, temporarily, by the allotment of $2200 from general and engineering funds.... It is almost impossible to convey an adequate idea of the extent to which our work, both in teaching and research, has been blocked by the inability to buy necessary articles. We should have had pretty nearly to cease activity in research if it hadn't been for the equipment which was bought in our three boom years 1929-32...."

But despite all such difficulties, the pursuit of physics at American universities was by no means at a standstill, nor was the job market for physicists a total disaster. Institutions continued to seek new personnel who would bring added strength and prestige to their faculties of physics. Some physicists such as E.O. Lawrence received attractive offers from a number of institutions.[71] At the University of Illinois Loomis was able, during the "boom years" before 1933, to continue Illinois' program for up-dating the physics department which had begun in late 1929 when he was recruited to become chairman. Four promising young physicists were added to the Illinois physics department in 1930 and 1931.[72]

Nevertheless, there were real employment difficulties suffered by two segments of the American physics community--young persons not yet established in professional positions, and older scientists whose jobs were eliminated. Robert Serber has remarked that graduate students in physics during the early 1930's were in no hurry to finish their Ph.D. programs since there seemed to be no jobs to look forward to.[73] At some institutions available funds were stretched to provide at least minimal support for students who had obtained their Ph.D.'s but who had been unsuccessful in obtaining jobs.[74]

Meanwhile, the possibility of physicists finding employment at laboratories, such as those at the National Bureau of Standards, was slight. That agency suffered budgetary reductions of more than 50 percent between 1932 and 1934. As a result it was forced to lay off about one-third of its staff which had grown to more than 1000 by 1931.[75]

Although the effects of the Depression in the United States lasted until after the outbreak of World War II, gradual improvement in the employment picture began to occur by 1935. According to Charles Weiner[76]

"After 1935 the financial pinch eased and more academic jobs became available; young physicists were needed to cope with the increasing enrollments in US colleges and universities. The improvement was only gradual, however, and in the middle and late 1930's the search for employment took many physicists into work they had not previous considered (for example, into oil fields as part of industrial geophysical research teams).

Thus, in retrospect, it would seem that one effect of the Depression was to broaden the range of employment possibilities for or acceptible to physicists. At any rate, the American physics community continued to grow within the external socio-economic environment in the 1930's and, as we shall see in later sections, internal developments within the field of physics continued to make it an attractive discipline.

Before leaving the topic of employment opportunities for American physicists in the early 1930's something should be said about the question of anti-Semitism among American physicists during the early decades of the 20th century.

Comments by some authors in recent years raise the suspicion that anti-Semitic sentiments were prevalent among established members of the American physics community in the early decades of the 20th century. For example, Isador I. Rabi is quoted by Jeremy Bernstein[77] as saying

"I had no hope of getting a job [as a physicist in America in 1929]. It seemed to me that it would be very difficult to get one. Anti-Semitism was unbelievably rife in the universities and elsewhere."

Also, Nuell P. Davis' book Lawrence and Oppenheimer[78] contains the following quotation, unsupported by citation of the source:

"'New York Jews flocked out here to him [Oppenheimer] and some were not as nice as he was' said Birge. 'Lawrence and I were very concerned to have people here who were nice people as well as good students.'"

John D. Davies, writing in the Princeton Alumni Weekly, 2 October 1973, on "The Curious History of Physics at Princeton," also includes without specific reference the above quotation attributed to Birge.[79] In addition, Davies writes

"One of the exciting facets of this youthful genius [Oppenheimer] at Harvard was, according to his biographer [Davis?], that he was 'obviously Semitic.' Which brings us to another problem associated with the eventual growth of American physics. American academic institutions had been saddled with anti-Semitism....
This practice, unfair and in fact, unscholarly, had been quietly breached in some of the research fields of the physical sciences. Now the University of California decided to establish a research program in physics and against precedent, made an offer to this 'gorgeous new exotic' [Oppenheimer] for reasons which remain unclear."

Motives and attitudes are difficult to probe at any time, but especially so half a century later and about a topic on which little is apt to enter the written record. Nevertheless, although I have not made a thorough study of the issue of anti-Semitism in the American physics profession, the relevant evidence of which I am aware, leads me to the conclusion that it would be grossly unfair to say that the majority of American physicists in the early decades of this century were guilty of anti-Jewish prejudice, or to believe that a number of young Jewish physicists were unjustly denied suitable positions solely on racial grounds. Accepting such a view would require the ignoring of numerous counter-examples. There simply are too many instances of American Jews who were accepted members of the physics profession in the United States and of European Jews who were warmly welcomed by the American physics profession on either a visiting or permanent basis. Consider the following examples:

Albert Abraham Michelson was one of America's most highly respected physicists throughout his long, productive career which spanned the half-century before his death in 1931;

European Jewish physicists, Albert Einstein, Paul Ehrenfest and Max Born, received cordial receptions at the many American universities where they lectured in the years between 1921 and 1926;

Paul Epstein was recruited to the faculty of the California Institute of Technology by Robert Millikan soon after the latter became head of that institution;

Saul Dushman, born in Russia, educated in Canada, employed as a researcher by the General Electric Company from 1912, became Assistant Director of Research there in 1930;

Samuel A. Goudsmit was given an attractive position at the University of Michigan in 1927;

Ludwig Silberstein, born in Poland, trained in Berlin, came to the Eastman Kodak Company as a mathematical physicist in 1920;

Robert Oppenheimer was sought after by several American universities (see discussion on page 3.46 and in note 3.128 of Sopka thesis) following his return from his European doctoral studies with Max Born;

Isador I. Rabi was hired by Columbia University in 1929[80] and soon became one of that institution's most prestigious faculty members;

Karl Lark-Horowitz, educated in Vienna and holder of research fellowships in this country and Canada, became Professor of Physics at Purdue University in 1928 and rose rapidly to the position of Director of the Physical Laboratory and Head of the Physics Department.[81] We note, in addition, that during the 1930's, despite adverse economic conditions, the physics departments of a number of American universities, including Princeton, provided hospitable refuge for European Jewish physicists who could no longer function professionally in their homelands. One of these, Hans Bethe, recalled recently[82] some criticism of his appointment at Cornell University in 1935, not because of his racial background, but on the grounds of the number of unemployed young American phycicists at that time. Bethe found that he was very quickly made to feel welcome in Ithaca and within the American physics profession, in general, when he attended scientific gatherings or visited other institutions. Certainly the Summer Symposia on Theoretical Physics at the University of Michigan did not exhibit any trace of racial bias among its participants.

I do not mean to imply by the remarks above that there could not have been any anti-Semitic sentiments among American phycisists of the early 20th century. Undoubtedly, the physicists of this country were not all[83] noble exceptions to the general social climate of their time which did involve anti-Semitic, as well as anti-Irish, anti-Italian and other anti-minority biases. Some of the young Jews who entered the American physics profession during the late 1920's and early '30's did indeed experience real difficulty in obtaining academic employment.

They were not alone in their plight. Times were hard, positions were scarce. Since those academic institutions with openings in physics at the lower faculty levels could chose from among a relatively large pool of young scholars, they naturally were partial to those who, they believed, would function most effectively with undergraduates in the classroom. Among young Jewish physicists, as among any other sample of a population, then or now, there were some who seemed to lack such potential.[84] Such individuals may have experienced more than average difficulty in finding employment due to personal traits rather than ethnic background.

It is difficult to believe either that the growth of physics in America was slowed down by real anti-Semitic prejudice or that anti-Semitism wrecked the careers of promising young physicists in the United States. In view of the relatively large proportion of outstanding Jewish-American physicists among the post-World War II generation, it would seem that the American physics profession, as a whole, could at most be minimally charged with irrational, anti-Semitic prejudice.

Much of the evidence submitted here is admittedly anecdotal and not researched in depth (which may not be worth doing anyway); but it is quite likely that careful study would show that the American physics community was <u>far more</u> hospitable and unprejudiced than either the country as a whole or many other professional groups.

Summary

Increases in the attributes of maturity in scientific endeavor--quantity, quality and diversity of activity--characterized the progress achieved by the American physics profession during the early 1930's. In succeeding sections we shall have more to say about the specific nature of some of its scientific output. It is clear, however, that despite generally adverse economic conditions, the derivative was still positive for physics in America during the early 1930's.

In the half decade between 1930 and 1935 American physicists developed a variety of professional organizations and serial publications devoted to the pure, applied and educational aspects of their field. Opportunities for graduate and postgraduate study expanded in at least some instances. The number of institutions with strong, research-oriented faculties increased. Concommitant with these developments came a decrease in American reliance on European centers for the study of physics and an increase in the attractiveness to Europeans of America as a place to do physics.

The realities of the economic depression were felt by a number of individual physicists, but the Depression also served to hasten the emergence of a group within the American physics profession that actively sought to foster public appreciation for the value of the discipline and expanded employment opportunities for its members.

In short, by the mid 1930's physics had become an integral part of the total academic, scientific and technological life of America. In our next section we shall examine some of the internal trends that characterized the study and practice of physics in America in the early 1930's.

Part II: Some Contemporary Trends in Physics and among Physicists in America between 1930 and 1935

Introduction

The pursuit of physics in America in the early 1930's was characterized by a number of divergent trends which resulted in greater fragmentation within the physics community, but which also served to increase the community's broad strength. We touched upon some of these trends when we considered the growth of professional societies and the establishment of new research journals and when commenting upon the effects of the Depression.

In this section we turn our attention to two additional trends: the rise of interest in nuclear physics, and the consolidation of the position of theoretical study at American universities. The choice is dictated by the interactions between these two trends and the principal focus of this thesis, namely quantum physics in America. We reserve until our final section discussion of American achievements within that field.

Our omission here of specific discussion of American developments in experimental physics must not be construed as indicating either that there was any diminution of American success in that field, or that we do not place high value on it. Rather, it has been our intention to stress theoretical development over experimental achievement throughout this thesis precisely because it was in the theoretical domain that America had lagged behind, compared with both the international level of theory and the domestic level of experimentation. For America to attain a fully mature status within the world of physics it was primarily the area of theoretical activity that had to grow.

The Rise of Nuclear Physics

During the 1930's there was a sharp intensification of interest among physicists everywhere in "nuclear physics"--a term which at that time encompassed studies related to isotopes, cosmic rays, natural and artificial radioactivity and the production of high energy particles.[85]

This trend is readily explainable in the light of two separate circumstances. By 1930 it seemed to many members of the physics community that the application of quantum mechanics would routinely (or at least in principle) solve all problems associated with molecules and with the external parts of atoms. Furthermore, chemists were becoming increasingly active in working out some of the details. Hence, it was expected that the interior of the atom would provide physicists with the greatest new discoveries.[86] Secondly, 1932 turned out to be an *annus mirabilis* for such discoveries.[87] In that single year five momentous achievements occurred which related to nuclear physics. They were 1) H.C. Urey's discovery of the hydrogen isotope "deuterium"; 2) James Chadwick's demonstration of the existence of the neutron as a constituent of the nucleus; 3) Cockcroft and Walton's achievement of nuclear disintegration through proton bombardment; 4) the revelation of the existence of positrons in Carl Anderson's cosmic ray studies; and 5) Lawrence, Livingston and White's achievement of nuclear disintegrations in the target of the beam emerging from a cyclotron which was capable of generating close to 5 million electron volts.

It is of no small importance for our concern with the coming of America to the forefront in physics to note that three out of five of these achievements were made in American laboratories. And there were other Americans, in addition to those mentioned above, who were productively engaged in studies related to nuclear physics during the 1930's. To cite but a few examples: K.T. Bainbridge was working in the field of positive ray analysis; M.A. Tuve and Robert Van de Graff were engaged in high voltage generator research.

While all of the examples cited related to experimental developments, we should point out that experimental physicists engaged in nuclear study during this period were in close communication with theorists who eagerly followed the latest news from the laboratories. A particular example of such close communication occurred at the University of California in Berkeley between E.O. Lawrence and J.R. Oppenheimer. As a result, the regular meetings of "The Journal Club" at that institution became occasions of stimulating exchange for all who attended.[88]

Some quantitative assessment of the burgeoning interest in nuclear physics in America during the 1930's may be gained from consideration of the growth rates of research papers on nuclear physics appearing in the Physical Review and of doctoral dissertations written on nuclear topics. Charles Weiner has cited the following figures on these two indices[89]:

"...nuclear-physics papers, letters and abstracts increased from 8% of the publications in 1932 to 18% in 1933 and reached 32% by 1937. ...nuclear physics showed an increase in the US from 2 new Ph.D.'s in 1930 to 41 in 1939."

The rise of nuclear physics after 1930 certainly served to

demonstrate that the discovery of quantum mechanics had not "finished" physics any more than Maxwell's electromagnetic theory had done so in the late 19th century.[90] On the other hand, it has been suggested that the shift of interest to the nucleus on the part of many brilliant members of the physics profession may have been a contributing factor in the relatively slow pace of development that was experienced by applications of quantum mechanics for more than two decades after its introduction.[91]

The Strengthening of the Theoretical Component of Physics in America

The importance of theory in the study of physics was definitely accepted by the American physics community by the early 1930's. Furthermore, specific steps were taken so that Americans would have ample opportunity to be informed on theoretical development and that Americans could receive adequate training in this country to become theoreticians themselves.

In this section we shall discuss some of the measures that were taken and some of the outcomes of those measures. In keeping with the focus of this thesis, most of our examples will have direct bearing on quantum theory.

1) A New Look in Physics Faculties at American Universities

As a result of the introduction of quantum mechanics in the late 1920's, a large number of American institutions sought to insure that their physics faculties would be adequately staffed to handle the new material in particular and theoretical study in general.

We have already mentioned in an earlier section some outstanding examples of such action provided by the University of California at Berkeley, the Massachusetts Institute of Technology, Princeton University and the University of Wisconsin. The fact that other American institutions were also moving in that direction can readily be seen from the table which appears on the next page. It is interesting to note the number of institutions which employed more than one theorist, and the mixture of American and European physicists among the various institutions.

In order to bring out the kinds of courses that became available to students of physics through the presence of the faculty members listed, let us examine the situation at two universities in a little more detail.

At the Johns Hopkins University, in addition to regular courses in quantum mechanics, a Seminar in Theoretical Physics was held each year under the guidance of K.F. Herzfeld, G.H. Dieke and, later, M. Goeppert-Meyer. Between 1931 and 1935 the topics discussed in those seminars were, consecutively, Dirac's Quantum Mechanics; Group Theory and its Physical Applications; the Conduction of Electricity through Gases; and Magnetism.[92]

At Princeton University, E.U. Condon, H.P. Robertson and Eugene Wigner were in residence during the entire interval 1931-1935. In addition, P.A.M. Dirac visited there in 1931-32 and John von Neumann was a member of the department until he transferred to the Institute for Advanced Study in 1933. With such a staff a full range of theoretical courses could be offered and we find among the catalogue listings the following: Methods of Mathematical Physics; Elementary and Advanced Quantum Mechanics; Theory of Relativity; Statistical Mechanics and

Some American Physics Departments and Members of their Theoretical Staffs in the early 1930's

Brown Univ.	R.B. Lindsay
U. of Cal.-Berkeley	J.R. Oppenheimer, W.H. Williams
Cal. Inst. Tech.	P. Epstein, W.V. Houston, J.R. Oppenheimer (part time)
U. of Chicago	C. Eckart, F.C. Hoyt, R.S. Mulliken
Columbia Univ.	I.I. Rabi
Geo. Wash. Univ.	G. Gamow (1934)
Harvard Univ.	E.C. Kemble, J.H. Van Vleck (1934)
Univ. of Illinois	J.H. Bartlett and H.M. Mott-Smith
Johns Hopkins Univ.	G.H. Dieke, K.F. Herzfeld, M. Goeppert-Mayer (1932)
Mass. Inst. Tech.	W.P. Allis, N.H. Frank, J.C. Slater, J.A. Stratton, M.S. Vallarta
Univ. of Mich.	D.M. Dennison, S.A. Goudsmit, O. Laporte, G.E. Uhlenbeck
Univ. of Minn.	J. Frenkel (visiting 1930-31), E.L. Hill
Ohio State Univ.	A. Landé, L.H. Thomas
Princeton Univ.	E.U. Condon, H.P. Robertson, J. von Neumann (30-33), E. Wigner
Univ. of Wisc.	J.H. Van Vleck (until 1934), G. Breit (1935), G. Wentzel (visiting 1930)
Yale Univ.	H. Margenau

Note: We have listed only the names of faculty members capable of teaching courses related to 20th-century physics, omitting those faculty members whose theoretical expertise lay in the area of classical physics. This seems justifiable in the context of our stress on the innovations of the 1930's. With the exception of Epstein and Kemble, all of those listed had received their appointments in recent years.

Sources: Catalogues of the individual institutions and biographies of the individual faculty members.

Thermodynamics; Hydrodynamics and Elasticity; and Properties of Solid Bodies.[93]

Eugene Wigner has remarked on the atmosphere which he found at Princeton University during the early 1930's as follows[94]:

"It [the level of physics in America] was very rudimentary and very, very elementary. I felt that a great deal had to be done and often I felt that I engaged in baby talk. However, after a couple of years I realized that their interest was sincere, that they didn't want baby talk, that they wanted to learn or at least wanted me to teach the young people, and of course that made a great impression on me. I did not realize that at first. I first thought it was a sort of extravagance of the Americans that they wanted two people [himself and von Neumann] here from Berlin and perhaps it had no significance. But after a couple of years I realized that what they wanted was a transformation of the theoretical physics school into a modern, progressive, powerful school, and that of course had a great effect on me."

In retrospect, it appears extremely fortuitous that America should have achieved a readiness to take an active role in theoretical physics just at this time. Political developments in Germany had decreased the attractiveness to American students of centers of theoretical physics there which had been important in the previous decade[95] and European physicists, displaced from those centers, were able to find in the United States a congenial environment for the continuation of their own work.

2) Quantum Theory Books and Review Articles

Another indication of the increasing American concern with theoretical topics in the early 1930's may be found by examining some of the books and review articles that were written by Americans during those years and the circumstances under which they came to be written. Some books and articles were aimed primarily at the coming generation of physicists (and chemists) that was being education at that time; others were full-scale treatments of particular topics. The former group makes clear the importance that was attached to the topic for educational purposes, the latter makes clear the fact America had theoretical physicists of sufficient competence to undertake such works. Let us look at some examples of each category.

Among those of the textbook level that sought to provide students with a transition into quantum mechanics was Atoms, Molecules and Quanta by A.E. Ruark and H.C. Urey.[96] The manuscript of this book was begun in 1926 when both authors were located in the Washington, D.C. area and were members of a group of scientists in that area who met informally, drawn together by their common interest in the developments that were taking place in quantum theory. By the time the book was published in 1930 Urey had moved to Columbia University, Ruark to the Mellon Institute in Pittsburgh.

In their text the authors devoted 15 chapters to pre-1925 atomic theory, 6 chapters to wave and matrix mechanics. In the Preface we find the following statement, particularly interesting in light of American scientists' long-standing attachment to experimental results and preference for physical models:

"Orbital models are probably the nearest approach to an adequate description of atomic systems which can be secured in terms of our ordinary mechanical concepts. At the present time an earnest effort is being made to lay aside models and to focus attention on the connections between purely experimental quantities. We are in sympathy with this effort. It helps to emphasize the enduring quality of experimental facts, independent of the form in which they are expressed. On the other hand, a very large number of physicists and chemists do not wish to eliminate models from their methods of thought, nor to rely entirely on mathematical connections between their observations. In the past these models have consisted of particles moving on selected orbits, and now they consist largely of the more useful waves and nodes of the Schrödinger theory. It is only human nature to construct a new picture of a hidden mechanism as soon as an old one is discarded."

Whereas in Ruark and Urey's book quantum mechanics was introduced in a strongly physical context, J.C. Slater and N.H. Frank used a much more mathematical-theoretical approach in dealing with the topic in their Introduction to Theoretical Physics,[97] a text which evolved in conjunction with the curriculum revision that took place at the Massachusetts Institute of Technology during the early 1930's. Slater and Frank's approach was to combine material specifically needed for a discussion of applications of modern atomic theory to the structure of atoms, molecules and solids, and to chemical problems with a presentation of those mathematical topics usually associated with such classical physics topics as mechanics, electromagnetic theory, potential theory and thermodynamics. It was their conviction that the more modern topics could not be satisfactorily handled without a firm grasp of classical physics.

In 1935, Introduction to Quantum Mechanics with Applications to Chemistry, by Linus Pauling and E. Bright Wilson, was published.[98] Their association began when Wilson went to study with Pauling at the California Institute of Technology in 1931 after having received his M.A. in chemistry at Princeton University. Two years later Wilson was awarded his Ph.D. for a thesis which contained a quantum-mechanical calculation of the Lithium atom. He then remained for one more, postdoctoral year at Cal Tech before coming to Harvard University as a Junior Fellow in 1934.

Wilson had already taken at Princeton University in 1930-31 the course on quantum mechanics which was given by E.U. Condon and in which Dirac's recently published book The Principles of Quantum Mechanics[99] was used. He then took Pauling's quantum mechanics course for chemists during his first year at Cal Tech. The following year, while serving as a grader in that course, Wilson was approached by Pauling with the invitation to collaborate with him in the writing of a textbook. Wilson recalls that he wrote about one-third of the manuscript that resulted.

As indicated by the title, the Pauling and Wilson text was written primarily for chemists. But, according to the Preface, it was also aimed at experimental physicists and beginning students of theoretical physics. With a three-part audience such as this, the authors did not assume a high degree of mathematical sophistication on the part of the students and took pains to work out the mathematical details of the topics they presented. Their stated aim was "to produce a textbook of practical quantum mechanics... [and] to provide for the reader a means of equipping himself with a practical grasp of the subject, so that he can apply quantum mechanics to most of the chemical and physical problems which may confront him."

The book proved to be remarkably durable and helpful for subsequent generations of science students. Pauling has recently noted[100] "the President of McGraw-Hill told me that this book is the oldest one of their books still in print (that is, without having been revised)."

The existence of textbooks on quantum mechanics written expressly for students in the United States did not, of course, mean that Americans in the early 1930's ceased to use books on the topic written by European authors. We have already mentioned Dirac's <u>The Principles of Quantum Mechanics</u> that was widely used among advanced students.[101] Other books on quantum mechanics that were welcomed by American readers included J. Frenkel's two books on wave mechanics,[102] H. Weyl's <u>The Theory of Groups</u>

and Quantum Mechanics[103] and A. Sommerfeld's Wave Mechanics.[104] The increasing productivity of American theoretical physicists was not to separate them from their European counterparts, but rather to make them more fully integrated members of the international scientific community. This will become evident in the publications which we consider next, all of which were published abroad.

In 1932 J.H. Van Vleck's The Theory of Electric and Magnetic Susceptibilities was published as part of "The International Series of Monographs on Physics" under the editorship of R.H. Fowler and P. Kaptiza.[105] This was a field in which Van Vleck had already published a number of significant papers. In the book he provided the reader with a complete treatment of the topic from the classical, the old quantum theoretical and the quantum mechanical points of view. His own work and the contributions of others were brought together into a coherent whole; as Philip Anderson remarked later: "...even those basic ideas originated by others are illuminated and their bare bones fleshed out by Van's special point of view."[106]

"The International Series of Monographs on Physics" included a second book on theoretical physics by an American author during this period: Relativity, Thermodynamics and Cosmology by R.C. Tolman.[107] Although this topic is unrelated to the principal focus of this thesis, quantum theory, it does seem appropriate to mention the publication of Tolman's book in England as a further indication of increasing American contributions to international theoretical literature.

Tolman, a Professor of Physical Chemistry and Mathematical Physics at the California Institute of Technology, had a long history of interest in relativity and related topics, dating back to 1908 when he and G.N. Lewis presented the first American paper on Relativity Theory to a meeting of the American Physical Society.[108] Tolman had already written a book on relativity that was published in America in 1917.[109]

Another book by American authors that was published in England during the 1930's was The Theory of Atomic Spectra by E.U. Condon and G.H. Shortley.[110] This work was produced at Princeton University where, we will recall, Condon was a member of the physics department faculty. Shortley, previously a student of electrical engineering, had followed Condon to Princeton from the University of Minnesota in 1930 and, in 1933, took his Ph.D. in mathematical physics writing on "The Theory of Complex Spectra" under Condon's direction.

During the early 1930's Condon was actively engaged in the study of atomic spectra and published some related papers in the Physical Review individually and jointly with Shortley.[111] Their book was a natural and timely culmination of this activity since, as stated in the Preface,

"In this monograph we have undertaken a survey of the present status of the problem of interpreting the line spectra due to atoms. This interpretation seems to us to be in a fairly closed and highly satisfactory state. All known features of atomic spectra are now at least semi-quantitatively explained in terms of the quantum mechanical treatment of the nuclear-atom model."

A quite different group of three volumes related to quantum theory and written by an American author appeared in the German language between 1929 and 1933. These works, Elementare Einführung in die Wellenmechanik,[112] Elementare Einführung in die Physikalische Statistik,[113] and

Elementare Einführung in die Quantenmechanik,[114] were translated into German by E. Rabinowitsch from the writings of Karl K. Darrow. The first was based on an article "Introduction to Wave Mechanics," which Darrow had written for the Bell System Technical Journal[115]; the second on his "Statistical Theory of Matter, Radiation and Electricity" prepared for the Reviews of Modern Physics.[116] The third was written originally for the German translation and was subsequently published in the Reviews of Modern Physics.[117] They all bear the stamp of Darrow's facility in writing about scientific topics which we mentioned earlier[118] and which was apparently expected to appeal as much to German- as to English-speaking audiences. The first two volumes carried Vorwörter by E. Schrödinger and Max Born, respectively.

Before turning our attention to review articles on quantum mechanics it is appropriate to mention two more books that were published in the United States during the early 1930's. Both volumes presented material related to the concerns of quantum physicists, but neither was of a highly mathematical-theoretical nature. Both were part of the "International Series in Physics" inaugurated by the McGraw-Hill Book Company in 1929 with Quantum Mechanics by E.U. Condon and P.M. Morse which, we recall, was the first treatment of the topic to be published in book form by American authors.

In 1930 The Structure of Line Spectra by Linus Pauling and Samuel Goudsmit was published as, according to the Preface, "primarily a textbook for those working in the field of spectroscopy."[119] While recognizing the quantum mechanical developments of the preceding years the authors couched their presentation in terms of the older vector model representation since, they said, "It seems probably...on account of its simplicity and ease of visualization, that the vector model will continue to be used as the basis for the treatment of spectra for many years to come." Their idea was to provide "the theoretical background of the principles governing the structure of line spectra, but not an account of all observed spectra, experimental data being introduced only as illustrative examples."

Two years later Atomic Energy States as Derived from the Analyses of Optical Spectra, compiled by Robert F. Bacher and Samuel Goudsmit, became available and was hailed as a great contribution to the theoretical and experimental spectroscopist and to the physical and mathematical chemist[120] as well as to investigators in all areas of atomic physics, such as the theories of magnetism and electron impact. This was the first such compilation to appear since 1922. Many advances had, of course, been made in the meantime. "Consequently, Bacher and Goudsmit's tables are the first to be adequately correlated with theory." Furthermore, "the appearance of their volume [was] particularly timely because the last two or three years may be said to mark the closing of the frontier in the understanding of extranuclear atomic spectra...the basic principles and modus operandi are now quite completely known."[121]

The Reviews of Modern Physics, begun in mid-1929, continued in the early 1930's as a disseminator of survey-type articles related to quantum mechanics and its applications. Between 1930 and 1935 about 18 such articles were published, written principally by American authors already established in the field, such as G. Breit, E.U. Condon, D.M. Dennison, C. Eckart, E.C. Kemble, P.M. Morse, R.S. Mulliken, J.C. Slater and J.H. Van Vleck. There were a few, however, by European authors,[122] the most notable of which was E. Fermi's "Quantum Theory of Radiation," based on the lectures he had delivered at the University of Michigan Summer Symposium in Theoretical Physics in 1930.[123]

At a less advanced level than most of those appearing in the _Reviews of Modern Physics_ were the review articles which Karl Darrow continued to write for the "Contemporary Advances in Physics" series in the _Bell System Technical Journal_. Some of these dealt with quantum mechanical topics, but a large fraction of them now focussed on nuclear topics, which is not at all surprising in view of the rise of nuclear physics in the early 1930's.

3) Symposia in Theoretical Physics

Symposia, especially in a field such as theoretical physics, can provide the participants with valuable opportunities for exchanging ideas informally and for learning about the latest, not-yet-published advances in the field. When they bring together participants from varied geographical locations they help to overcome some of the feeling of isolation such as American theoretical physicists continued to experience in the 1930's as they worked in the many institutions as indicated on page 4.38. That was a time when the world population of theoretical physicists had not yet reached an unmanageable number from the point of view of most of them actually coming together to talk and get to know each other on a personal basis.

The Summer Symposia in Theoretical Physics that were held each year at the University of Michigan were by far the most successful such undertakings at the time anywhere in the world. We have already discussed[124] the beginnings of this activity and some of the aspects of its flourishing during the 1930's. These symposia differed from others that we shall discuss in this section in the following ways: their duration, which was eight weeks; their mixture of participants, which included invited lecturers, established practising theoreticians from Europe as well as America, and young graduate students and recent Ph.D.'s; and the variety of discussion topics which, during the early 1930's, included quantum mechanics and nuclear physics.

S.A. Goudsmit has portrayed some of the international atmosphere of the Michigan summer gatherings in the following words[125]:

"One of the summers that Fermi was with us we later called the Italian Summer, because several of his colleagues from Rome came through Ann Arbor and added to the general liveliness. We also had a Dutch Summer, when Paul Ehrenfest, from Leiden, came with two of his assistants and attracted several other physicists and astronomers of Dutch origin from various universities over here. A number of visiting physicists met their future wives at these summer meetings and have pleasant memories of our symposia for this reason as well. George Gamow, who had received permission to leave Russia to attend the 1934 Symposium, stayed on in this country and took a position at George Washington University."

The annual pictures taken during each of these gatherings show a congenial looking group which continued to number about sixty even during the worst Depression years, thanks to the modest costs involved.[126]

There can be little doubt of the important role played by these symposia in furthering theoretical activity among American physicists and in making Americans part of the international community of theoretical physicists--two essential aspects of the "coming of age" of physics in

America.

When compared with the University of Michigan Symposia in Theoretical Physics, those symposia which were part of American Physical Society meetings during the years 1930-1935 would seem to be of only minor significance.[127] On the other hand, we must not ignore them since they do provide some measure of the position held by theoretical physics within the entire American physics community. Not surprisingly, a number of those held during that period, such as the ones at Pasadena in June 1931, at Boston in December 1933, at Ann Arbor in June 1934 and at Minneapolis in June 1935, were on nuclear physics. The principal ones related to quantum mechanics were those held at Chicago in June 1933 which we have already discussed[128] and at Pittsburgh in December 1934 when the topic was "Group Theory and Quantum Mechanics" and the participants were G. Breit, J.H. Van Vleck and E. Wigner.

Symposia sponsored by the American Physical Society during the 1930's frequently included visiting European scientists as participants. They also frequently included both experimentalists and theoreticians--a further indication of the ready communication that existed between these two groups of American physicists. It is noteworthy in this context that even at the Michigan Symposia experimentalists were sometimes invited to lecture, as in 1934 when A.H. Compton lectured on cosmic rays and E.O. Lawrence lectured on the artificial disintegration of atomic nuclei.[129]

Our final example of American symposia in theoretical physics represents yet another, totally different kind of undertaking. Begun late in the period under present consideration, these meetings were to gain increasing importance in the late 1930's.

In April 1935, as a joint venture of the Department of Terrestrial Magnetism of the Carnegie Institution and the George Washington University, a series of Theoretical Physics Conferences were inaugurated, funded by grants of $1000 each from the two cooperating institutions.[130]

In the early 1930's President Cloyd Marvin of the George Washington University embarked on a program to improve the intellectual level of that institution's offerings, especially in the sciences. With the advice of M.A. Tuve of the Department of Terrestrial Magnetism, Marvin arranged attractive appointments for George Gamow and Edward Teller in 1934 and 1935 respectively.

The Theoretical Physics Conferences, as conceived by Tuve, were to be patterned along the lines of the Solvay Congresses. Attendance was by invitation only. A small number of "experts" were asked to carry on informal discussion of a particular topic for three days.

The following is taken from a report on the first of these conferences[131]:

"A Conference on Theoretical Physics to coordinate current theoretical researches was held in Washington April 19 to 21, 1935....Invitations were extended to a small representative group of active theoretical physicists to discuss informally those current problems which they judged most important. This initial meeting was devoted to certain theoretical problems in the rapidly expanding field of nuclear physics. Devoted solely to the clarification of the current status of the subject and to discovering the profitable directions for immediate attack, these sessions were subsequently evaluated by those present as uniquely effective in advancing the progress of their own researches....
At this initial Conference about 35 representatives of 20 universities and research-organizations were present."

It is of interest, in the light of the focus of this thesis, to note that the second Washington Conference on Theoretical Physics devoted its attention to the topic "Molecular Physics." It attracted a sizeable number of those American physicists who had been working in quantum mechanics and a number of recently arrived colleagues from Europe.[132] Most of the subsequent conferences in this series, which continued until 1946, dealt with topics related to nuclear physics. Particularly noteworthy was the one held in January 1939 during which "fission became a public matter."[133]

4) Theoretical Ph.D.'s and Postdoctoral Fellowships; the Rise of American Centers for Theoretical Study in Physics

As a natural result of the widespread distribution of theoreticians among the physics faculties of American institutions in the early 1930's, which we have discussed above (see page 4.36), a number of young persons were able to write doctoral dissertations on theoretical topics and, in many instances, to continue their studies at the postdoctoral level.

In the years from 1930 to 1935 approximately 60 quantum theoretical Ph.D. theses were accepted by American universities.[134] During that same period more than 20 National Research Fellowships were awarded for theoretical postdoctoral study related to quantum mechanics.[135] While some of these National Research Fellows went to Europe for all or part of their study,[136] the majority remained in the United States. Thus we can use these two indices, the production of theoretical Ph.D.'s and the attraction for National Research Fellows, to help us recognize which American locations were the rising centers for theoretical study in physics in the early 1930's.

In the discussion which follows we shall describe a number of settings in which young persons studied theoretical physics at an advanced level in America during the early 1930's. We shall not yet concern ourselves with specific results of their studies, reserving such considerations until the final section of this chapter. We begin by describing in alphabetical order those locations where the largest numbers of established and budding theoreticians were concentrated in the United States during this period. Following these we shall mention some other sites which merit attention not by numbers but by virtue of their future individual potential and as indicators of the new pervasiveness of American theoretical activity.

California: Berkeley and Pasadena

During the early 1930's, principally as the result of the influence of J. Robert Oppenheimer, a number of young persons went to California to study theoretical physics. One of them, Robert Serber, has recalled Oppenheimer's appeal as follows[137]:

"Oppenheimer as a teacher: Oppenheimer's fascinating personality played a major part in his unique powers as a teacher. I can cite my own experience, the impact of my first meeting with him. In 1934, I received a PhD from John H. Van Vleck at the University of Wisconsin and a National Research Fellowship that I intended to spend at an eastern university. On the way east from Wisconsin I stopped in at Ann Arbor for a month at the summer session. Oppenheimer was there and after I heard him lecture and spent some time with him I reversed my direction and went to Berkeley. Upon arriving I discovered that most of the National Research Fellows in theoretical physics were already there.

By this time Oppenheimer's course in quantum mechanics was well established. Oppie (as he was known to his Berkeley students) was quick, impatient and had a sharp tongue, and in the earliest days of his teaching he was reputed to have terrorized the students. But after five years of experience he had mellowed (if his earlier students were to be believed). His course was an inspirational as well as educational achievement. He transmitted to his students a feeling of the beauty of the logical structure of physics and an excitement about the development of physics. Almost everyone listened to the course more than once; Oppie occasionally had difficulty in dissuading students from coming for a third or fourth time. The basic logic of Oppenheimer's course in quantum mechanics derived from Pauli's article in the Handbuch der Physik. Its graduates...carried it, each in his own version, to many campuses.

Oppie's way of working with his research students was also original. His group consisted of eight or ten graduate students and about half a dozen postdoctoral fellows. He met the group once a day in his office. A little before the appointed time the members straggled in and disposed themselves on the tables and about the walls. Oppie came in and discussed with one after another the status of the student's research problem while the others listened and offered comments. All were exposed to a broad range of topics. Oppenheimer was interested in everything; one subject after another was introduced and coexisted with all the others. In an afternoon they might discuss electrodynamics, cosmic rays, astrophysics and nuclear physics."

Oppenheimer's influence on his students extended beyond physics and the normal academic setting. Serber continues with these comments:

"Dinner in Frisco: Oppie's relations with his students were not confined to office and classroom. He was a bachelor then and a part of his social life was intertwined with ours. Often we worked late, continued the discussion through dinner and then later at his apartment on Shasta Road. When we tired of our problems or cleared up the point at issue, the talk would turn to wider realms of the intellect, of art, music, literature and politics. If the problems were going badly we might give up and go to a movie. Sometimes we took a night off and had a Mexican dinner in Oakland or went to a good restaurant in San Francisco. In the early days this meant taking the Berkeley ferry and the ride across the bay. The ferries back to Berkeley did not run often late at night, and this required passing the time waiting for them at the bars and night clubs near the ferry dock. Frequently we missed several ferries."

Oppenheimer held appointments at both the University of California at Berkeley and the California Institute of Technology in Pasadena. Since the two institutions had different academic calendars it was possible for Oppenheimer to go to Pasadena for a period of several months after the close of the Spring Semester at Berkeley.[138] It was customary for many of his Berkeley students to make the annual trek with him. In addition, in Pasadena there were other students of theoretical physics such as Milton Plesset,[139] who worked with Oppenheimer.

Students who wrote their doctoral dissertations officially under Oppenheimer's guidance in the early 1930's at the University of California were Melba Phillips (1933), Glen Camp (1935) and Arnold Nordsieck (1935). In addition, there were three other graduate students, Harvey Hall (1931), J.F. Carlson (1932) and Leo Nedelsky (1932) who worked with Oppenheimer during this period while "nominally" under the guidance of William H. Williams.[140]

The National Research Fellowships awarded to Hugh Wolfe (1929-31), Wendell Furry (1932-34), Frederick Brown (1934-35), Edward Uehling (1934-36) and Robert Serber (1934-36) all carried the designation "University of California and California Institute of Technology"--a definite clue to the Fellow's membership in the Oppenheimer circle.[141]

Some of the young persons mentioned above, such as Leo Nedelsky and Melba Phillips, extended their stays at the University of California through appointments as instructor or research associate in the Physics Department.[142]

A number of papers dealing principally with quantum electrodynamics were published during the early 1930's under the joint authorship of Oppenheimer and individual members of his theoretical group.[143] But they were not the only ones in California working in quantum theoretical physics during those years. At the California Institute of Technology, for example, William Houston was engaged in theoretical and experimental studies related to crystals and he supervised the doctoral theses of two students, Clyde Crawley and Lorenz Huff.[144]

Cambridge, Massachusetts: Harvard University and the Massachusetts Institute of Technology

Cambridge, Massachusetts was the site of an active and expanding community of theoretical physicists during the early 1930's. We have already mentioned the changes that occurred in the Physics Department at the Massachusetts Institute of Technology following the arrival of John C. Slater from Harvard in 1930. (See pages 4.12, 4.13.) Edwin C. Kemble continued as the senior theoretician in the Harvard University Physics Department where he was joined in 1934 by John Van Vleck. (We recall that Van Vleck had been Kemble's first graduate student and in 1922 wrote the first theoretical thesis using quantum theory that was accepted by an American university.) Wendell Furry joined the Physics Department at Harvard in 1934 as an Instructor upon the completion of his National Research Fellowship for study with J.R. Oppenheimer.

A number of doctorates in theoretical physics were awarded by each of these institutions and several postdoctoral fellows elected to take their awards to either or, in some cases, both of these institutions.

At Harvard University the following graduate students completed their theses under Kemble's direction in the early 1930's: James H. Bartlett (1930), Clarence Zener (1930), Montgomery H. Johnson Jr. (1932), Eugene Feenberg (1933), Hubert James (1934) and Julian Knipp (1935).[145] In 1935 Richard Present, who had been working with Kemble, was the first Harvard graduate student to complete his dissertation with Van Vleck. Another student Malcolm Hebb, who had begun his studies with Van Vleck at the University of Wisconsin and had moved to Cambridge in 1934, completed his thesis in 1936.[146]

Meanwhile, at the Massachusetts Institute of Technology, Nathan Rosen (1932), J.P. Vinti (1932), M.F. Manning (1934), H.M. Krutter (1935) and R.D. Richtmyer (1935) completed doctoral theses in theoretical physics.[147]

Young physicists who came to study theory in Cambridge upon the completion of their doctorates elsewhere included Robert Bacher from the University of Michigan and William Hansen from Stanford University who chose to take their National Research Fellowships to the Massachusetts Institute of Technology. In addition, George Kimball, a chemist, came from Princeton University to MIT as a National Research Fellow to study the application of quantum mechanics to chemistry in the years 1933-35. Physicists George Shortley from Princeton University, Theodore Sterne from the University of Cambridge, England, and Lloyd Young from the University of Michigan elected to divide their National Research Fellowship years between Harvard and MIT.[148]

The University of Michigan

In view of our previous discussions (see pages 3.17-3.19, 4.48) of the summer symposia in theoretical physics that were held yearly at the University of Michigan and the relatively large staff of theoreticians with regular appointments on the physics faculty there, it is not surprising to find that this institution also drew young persons wishing to study theoretical physics as regularly enrolled graduate students.

We recall that during the early 1930's there were four theoretical physicists among the faculty of the University of Michigan. David M. Dennison, Samuel A. Goudsmit, Otto Laporte and George Uhlenbeck. Under their tutelege the following student wrote doctoral dissertations related to quantum theory: Arthur Adel (1933), Robert Bacher (1930), Edward Baker (1930), Russell Fisher (1931), Sherman Gerhard (1933), David Inglis (1931), Harold Koenig (1933), Morris Rose (1935), Edwin Uehling (1932), Ta-You Wu (1933), and Lloyd Young (1930).[149] Nathan Rosen went to the University of Michigan as a National Research Fellow for a year following the receipt of his doctorate from the Massachusetts Institute of Technology in 1932. In addition, spectroscopic investigations, a large part of the experimental program carried on at the University of Michigan, were frequently carried out in close collaboration with the theoreticians on the faculty.[150]

Princeton, New Jersey

We have already mentioned (see pages 4.14 and 4.15) the opening of the Institute for Advanced Study in Princeton in 1933 and noted the fact that its School of Mathematics attracted, as postdoctoral fellows, students of theoretical physics such as Carl Eckart, Boris Podolsky, Nathan Rosen and Clarence Zener. In addition, National Research Fellows Benedict Cassen (1930-32), Leon Linford (1930-32) and Vladimir Rojansky (1931-32) chose to take their fellowships for study related to quantum mechanics at Princeton University.[151]

During the early 1930's the Physics Department at Princeton University included theoreticians E.U. Condon, H.P. Robertson, John von Neumann (until 1933 when he transferred to the Institute) and Eugene Wigner. During this period theoretical doctoral dissertations were accepted by the Princeton Physics Department from George Shortley and Frederick Seitz.[152]

Seitz' thesis was written under Condon's guidance but Seitz also worked with Wigner during this period most notably on a quantum mechanical calculation of the cohesive energy of metallic sodium.[153]

Another promising young American physicist who studied mathematical physics with Wigner during the 1930's was John Bardeen. His thesis, for which his doctoral degree was awarded in 1936, dealt with the theory of the work function of metals. (In 1935 Bardeen became a Junior Fellow in the Society of Fellows at Harvard University where he continued his theoretical work with John Van Vleck. Van Vleck had introduced Bardeen to quantum mechanics in the late 1920's at the University of Wisconsin where Bardeen was associated with the electrical engineering department.[154])

Further opportunity for contact with other theoretical physicists were enjoyed by Princetonians annually when week-long visits were exchanged between themselves and their counterparts from Harvard University and the Massachusetts Institute of Technology.[155]

<u>The University of Wisconsin</u>

During the years prior to his departure for Harvard University in 1934, John Van Vleck acted as the mentor for a number of doctoral and postdoctoral students of theoretical physics at the University of Wisconsin. As noted earlier in this chapter (see pages 4.19, 4.20) Commonwealth Fellow William Penney and Rockefeller Foundation Fellow Robert Schlapp came from England and Scotland respectively to study theoretical physics with Van Vleck. In our next section we shall have more to say about their collaboration. Another of Van Vleck's postdoctoral fellows during this period was Albert Sherman, a chemist who received a National Research Fellowship to study the application of quantum mechanics to chemistry following the completion of his doctoral studies at Princeton in 1933.[156]

A remarkable number of graduate students worked with Van Vleck during his years at the University of Wisconsin in the early 1930's. Theoretical doctoral dissertations were completed by six students between 1930 and 1934: John Atanasoff (1930), Amelia Frank (1934), Alfred Goble (1933), Olaf Jordahl (1933), Neill Whitelaw (1933) and Robert Serber (1934). In addition, Robert Merrill, Morris Ostrosky and John Stehn later received doctorates at the University of Wisconsin for theoretical work that had been begun under Van Vleck's direction. Van Vleck's influence at the University of Wisconsin also extended to students of experimental physics, such as Glenn Havens, Robert Janes and Ragnar Rollefson.[157]

Van Vleck's departure from the University of Wisconsin did not signal the end of theoretical activity in the Physics Department since he was replaced by Gregory Breit who continued the strong program in theoretical studies that he had been engaged in at New York University since 1929.

Additional Opportunities for Study of Theoretical Physics at Other American Sites: New York City, the Johns Hopkins University, the Universities of Chicago, Illinois and Minnesota, Brown University

The institutions discussed thus far in this section accounted for more than three-fourths of the doctorates in theoretical physics that were conferred in America during the early 1930's and attracted more than four-fifths of the National Research Fellows wishing to study theoretical physics during that same period. Among the other American institutions where theoretical physics was studied at an advanced level in the years between 1930 and 1935 a few more merit particular attention at this time.

We have already mentioned in connection with Gregory Breit's move to the University of Wisconsin in 1934 the fact that he had been at New York University during the immediately preceding years. Breit had joined the faculty at New York University in 1929 following a five-year period at the Carnegie Institution in Washington. Under Breit's guidance five students at NYU wrote theses related to quantum theoretical physics. They were Brooks Brice (1930), Louis Granath (1931), Newton Gray (1933), Irving Lowen (1934) and Lawrence Willis (1933).[158] In addition, Montgomery Johnson went to New York University as a National Research Fellow for two years immediately after receiving his Ph.D. at Harvard University in 1932 and John Wheeler, another National Research Fellow, chose to go to NYU for a year after completing his doctorate at the Johns Hopkins University in 1933.[159] Jennie Rosenthal, one of the few American women studying theoretical physics at that time, was a research associate at NYU between 1931 and 1933.[160]

The informal seminar in theoretical physics that was begun in the mid-1920's by physicists in the New York area (see page 3.49) flourished in the 1930's under the joint guidance of Breit and I.I. Rabi who had returned to Columbia University as a member of the faculty in 1929. According to Rabi, those seminars became "the thing to come to" to such an extent that it occasionally was necessary to cut the group down to manageable size by making the discussion so advanced that the "pikers" would be discouraged from attending.[161]

At the Johns Hopkins University the presence of Karl Herzfeld, Gerhard Dieke and Maria Goeppert-Mayer, all of whom had come to this country from European institutions,[162] made possible study in modern theoretical physics for some young persons during the early 1930's. Jennie Rosenthal was a National Research Fellow there in 1930-31[163] and graduate students Arthur Lewis (1930), John Mauchly (1932), Richard Lee (1933) and John Wheeler (1933) wrote doctoral dissertations on topics related to quantum theory.[164]

Doctoral theses accepted by the University of Chicago during the years 1930-1935 continued to be based almost exclusively on experimental work despite the fact that Carl Eckart, Frank Hoyt and Robert Mulliken were members of the physics faculty there and gave a number of courses related to the new developments in quantum theory. Forty-two Ph.D. degrees in physics were awarded by the University of Chicago during those years. The only theoretical dissertations among that group that I have been able to identify were those written by Niel Beardsley (1932) and Beryl Dickinson (1935), both of whom worked with Eckart.[165] Although Mulliken himself was busy with theoretical considerations during those years, his graduate students were all engaged in experimental investigations.

Mulliken has commented recently[166] that after his arrival at the University of Chicago in the fall of 1928 he sought development of "the best possible experimental facilities for obtaining...molecular spectra" in the new physics building that was under construction. Plans were made but implementation was slow. Mulliken remarks "...there was a long delay. The absence of a high-resolution grating dampened my enthusiasm for taking part personally in the detailed analysis of molecular spectra, and influenced me toward more theoretical developments. The experimental work continued, however, in the hands of an active group of graduate students and postdoctoral fellows."

Work on the spectra of specific molecules was carried on at Chicago by S. Bloomenthal, W.O. Crane, P.S. Deslaup, J.S. Millis and P.G. Saper, all of whom thereby earned their Ph.D. degrees, and by Fellows H.T. Byck and J.L. Dunham who had come to Chicago from Johns Hopkins University and Harvard University, respectively. We mention these students, although they were not quantum theorists, for two reasons, namely, 1) their work was prompted by interest in the interpretation of their results in the light of recent developments in quantum mechanics, and 2) their work emphasizes once more the close collaboration between theoretical and experimental investigators that characterized so much of the American endeavor in physics during that period.

At the University of Illinois also, interest was keen on modern physics from both an experimental and theoretical point of view. F.W. Loomis, Chairman of the Physics Department, whose work relating to quantum theory had begun back in the "old" days, had students, including Polykarp Kusch,[167] working with him on the study of complex band spectra. In addition, under the guidance of James H. Bartlett, who joined the University of Illinois faculty in 1930, five theoretical theses were written by physics students in the years between 1932 and 1935. Bartlett's graduate students during that period were Wendell Furry (1932), Clarence Ireland (1932), Frederick Brown (1933), John Gibbons (1933) and Richard Cook (1935).[168]

The physics department at the University of Illinois also accepted a theoretical doctoral dissertation from Harold Mott-Smith who was already a member of the faculty and had been engaged in physical research since 1922.[169] His thesis, completed in 1933, was supervised by R.D. Carmichael of the Illinois mathematics department.[170]

Finally, we mention a few more theoretical doctoral theses in physics which were written in the early 1930's and which are of particular interest in emphasizing the growth and development of the American physics community's capacity for providing advanced graduate-level training in theoretical physics.

At the University of Minnesota in 1933 Wilfred Wetzel wrote a quantum theoretical dissertation under the guidance of Edward Hill.[171] Wetzel thus became the first of the third-generation graduate students descended from Edwin Kemble, as shown on the chart of dissertation genealogy on page 4.68.

R. Bruce Lindsay, another American physicist who had begun his research within the framework of the old quantum theory, became associate professor of theoretical physics at Brown University in 1930. By this time his own research interests tended to focus upon the philosophical implications of the new quantum mechanics and upon the theoretical aspects of acoustics.

Under his guidance, however, the following graduate students wrote dissertations related to the quantum theory of the early 1930's: Frederick White (1934) and Hugh Donley (1935).[172] Donley's thesis and those of three other Brown graduate students in later years (A.O. Williams, 1937; R.L. Mooney, 1938 and W.J. Yost, 1940) were concerned with applying to specific atoms the current Hartree method of self-consistent wave functions, a topic in which Lindsay was particularly interested; he had been the first to attempt to treat the old Bohr-Sommerfeld atom in a self-consistent manner. (See discussion on page 2.56.)

The above discussion of more than 60 quantum theoretical dissertations[173] written at 13 different American institutions between 1930 and 1935 gives clear evidence of 1) the growing capability of the American physics profession to provide its young aspirants with theoretical training; 2) the increased recognition in America of the validity of the professional designation "theoretical physicist"; and 3) the participation by American physicists in the application of the new techniques inherent in quantum mechanics.

I have attempted to include in this presentation all the theoretical theses written in the years 1930 to 1935 at the institutions discussed. I am aware of two or three more isolated theoretical theses which have not been listed. They were written at other institutions and neither the institutions involved nor the students themselves seemed representative of the emerging strength of theoretical physics in America.

Before leaving the topic of the production in the United States of theoretical Ph.D. theses in physics during the early 1930's it will perhaps be informative to introduce some additional perspectives on the data already gathered.

For example, an annual comparison of the number of theoretical theses with the total number of American theses in physics, as shown on the graph below, makes it clear that the vast majority of students wrote dissertations that were based on experimental work.

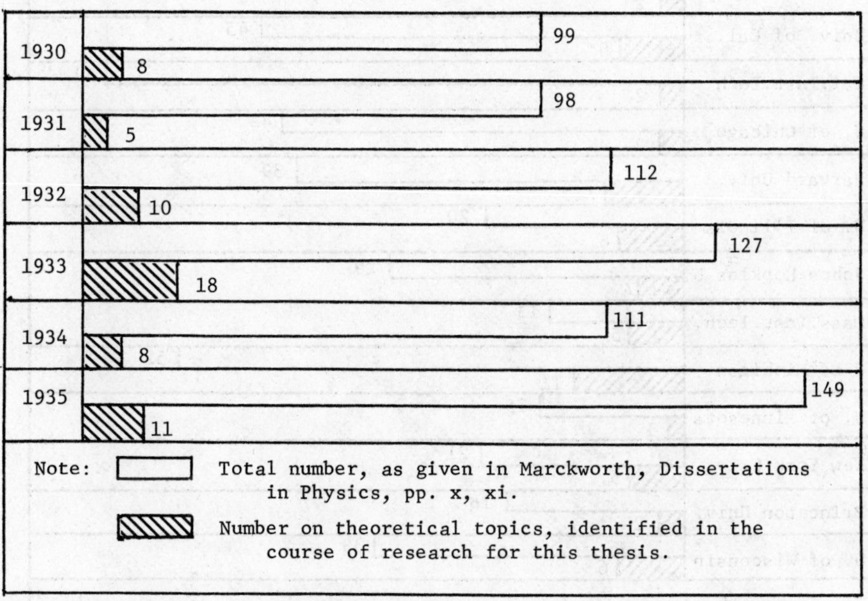

In only one year, 1933, did the number of theoretical theses approach 15 percent of the total. Taken over the entire six-year period, the proportion of theoretical theses is slightly less than 9 percent. Obviously, the new, post-quantum-mechanical generation of American students of physics did not desert the laboratories in great numbers. But this is what would be expected. There were still many more experimentalists than theoreticians supercising garduate students in physics at American universities in the 1930's.

The next graph shows, however, that among the more than a dozen American institutions then accepting theoretical doctoral dissertations in physics, the proportion of such theses, in some instances, was above 15 percent.

Institution	Total	Theoretical
Brown Univ.	7	2
Univ. of Cal.	45	6
Cal.Inst.Tech.	59	2
U. of Chicago	42	2
Harvard Univ.	39	7
U. of Illinois	20	6
Johns Hopkins U.	29	4
Mass.Inst.Tech.	13	5
U of Michigan	52	11
U. of Minnesota	15	1
New York Univ.	21	5
Princeton Univ.	18	2
U. of Wisconsin	33	6

Note: [open bar] Total number of physics doctorates granted by individual institutions between 1930 and 1935, as given in Marckworth Dissertations in Physics, pp. x, xi

[hatched bar] Number on theoretical topics, identified in the course of research for this thesis.

On the other hand, it is evident that some institutions in these years were under-represented in the proportions of theoretical dissertations. The California Institute of Technology and the University of Chicago, for example, seem to have continued to place prime emphasis on experimental physics where they had a well-established reputation of excellence. Nevertheless it is clear from the above figures that by 1935 theoretical physics was definitely established as an integral part of education in physics in the United States.

Qualitative analyses of the data related to American theoretical dissertations in physics, such as those attempted above, must be recognized as being of limited significance--the time span covered is short, the actual numbers of such theses per institution are small. More crucial, however, is the fact that such analyses reveal nothing about the individuals involved nor about the distribution of research effort and resources in special cases. Hence we next present a chart of "dissertation genealogy" for a sizeable subset of the American writers of theses related to quantum theory.

The chart on the next page shows vividly the influence of Edwin C. Kemble on the production of American theoretical doctoral dissertations. During the years from 1922 to 1935, from a total of about 85 such theses accepted by American institutions, 26 were written by students who either worked with Kemble or with one of Kemble's former graduate students. In reality, Kemble's influence on the emerging American community of theoretical physicists during this period was even greater than indicated by this chart which does not show that there were other students

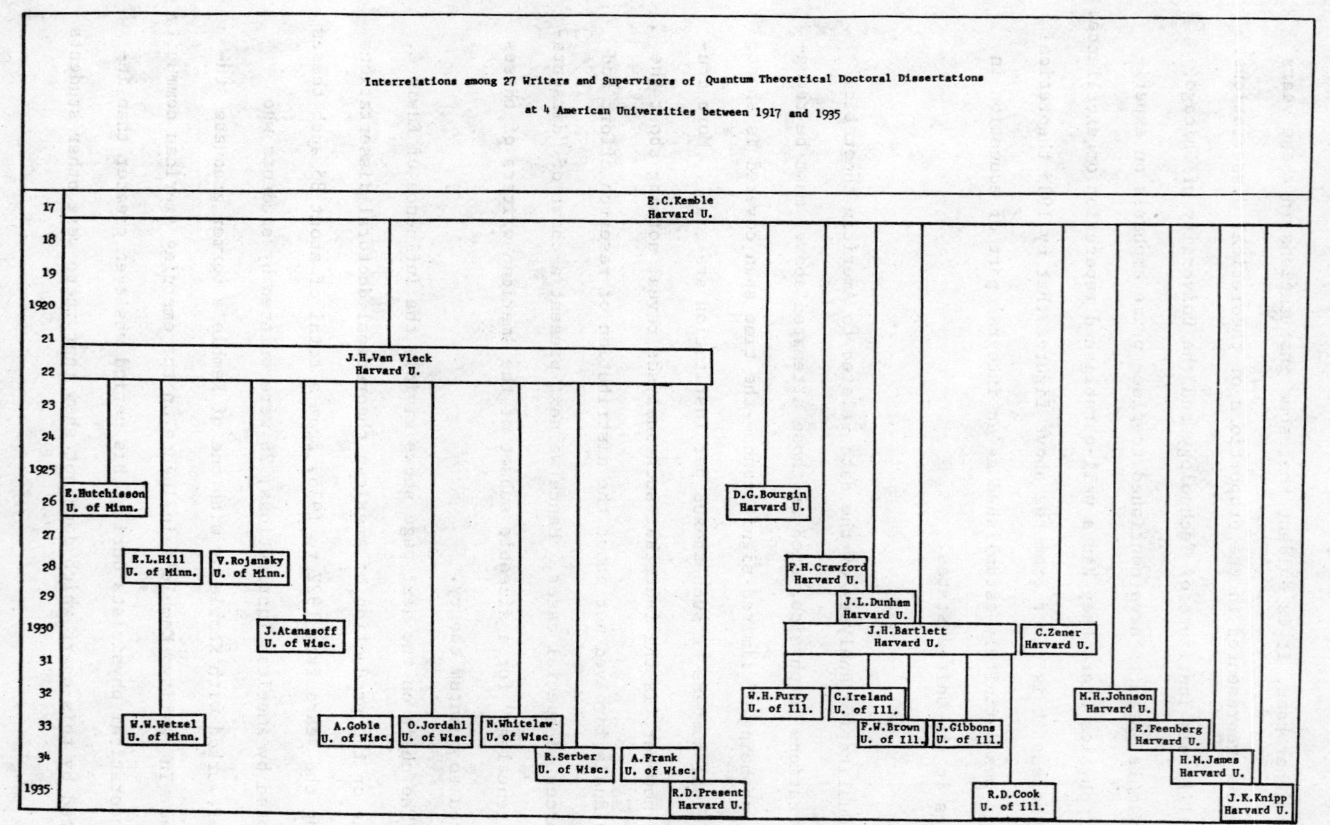

Interrelations among 27 Writers and Supervisors of Quantum Theoretical Doctoral Dissertations at 4 American Universities between 1917 and 1935

and post-doctoral fellows who studied quantum theory at Harvard with Kemble. Robert Mulliken and John Slater have both paid tribute to Kemble for providing their introduction to the formal aspects of quantum theory.[174]

The foregoing discussion of quantum theoretical doctoral dissertations in physics written by American students during the early 1930's should not be construed as indicating that all of the writers of those theses went on to become outstanding theoretical physicists. Some did and are still in the forefront of their chosen areas. Some developed other interests, perhaps due to the pressure of finding suitable employment. Others became part of what might be called "the second tier" of physicists on whom much of the solid strength of the American scientific enterprise in physics relies.[175]

The production of Ph.D. level research is inextricably tied in with the institutional environment and the availability of a specific thesis supervisor. One reason for providing such a detailed discussion as given above was specifically to highlight the fact that, by this time--the early 1930's--America had many institutions where students could study theoretical physics at an advanced level, and a considerable number of native American physicists who were capable of sponsoring theoretical doctoral research.

The charts on the following pages summarize certain aspects of the professional careers and the academic affiliations for a group of particularly outstanding American theoretical physicists who rose in the field

The Institutional Affiliations of Some Leading American Quantum Theorists between 1920 and 1935

1920 21 22 23 24 1925 26 27 28 29 1930 31 32 33 34 1935

G.Breit — J.H.U. * Harv. Univ. U. of Minn. Carnegie Institution Carnegie Inst. New York University U.of Wisc.

E.U.Condon — (Univ. of California) U.of California * Col. U. Princeton Univ. Univ. of Minn. Princeton University

D.M.Dennison — (Swarthmore College) Univ. of Michigan * University of Michigan

C. Eckart — (Washington U.-St.Louis) B.S. 1922, M.S. 1923 Princeton Univ. * Cal.Inst.of Tech. University of Chicago

E.C.Kemble — Harvard University Harvard University

P.M.Morse — (Case Inst. of Technology) Princeton University * Mass. Inst. of Technology

R.S.Mulliken — Univ.of Chicago * Harvard University Harv. U. N.Y.U. N.Y.U. Univ.of Chicago U. of Chicago U. of Chicago

J.R.Oppenheimer — (Harvard College) * Harv. U. C.I.T. University of California California Institute of Technology

L.Pauling — (Oregon St. College) California Institute of * Tech. California Institute of Technology

J.C.Slater — Harvard University * Harvard University Mass. Inst. of Technology

J.H.Van Vleck — Harvard University * University of Minnesota U. of Wisc. University of Wisconsin Harv. U.

1920 21 22 23 24 1925 26 27 28 29 1930 31 32 33 34 1935

Legend: ( ) Undergraduate Study
* Ph. D. Awarded
////// European Study and Travel

Sources: American Men of Science
Autobiographical Materials: Articles, Interviews, etc.

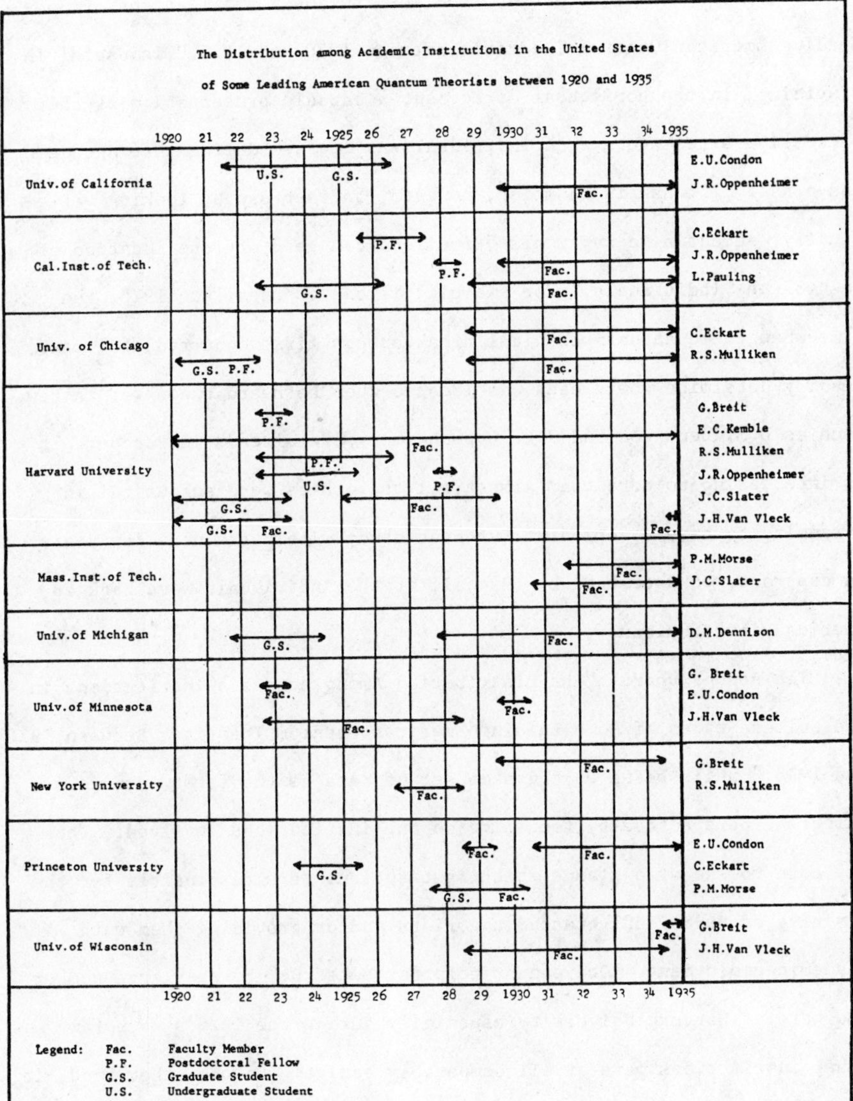
The Distribution among Academic Institutions in the United States
of Some Leading American Quantum Theorists between 1920 and 1935
1920 21 22 23 24 1925 26 27 28 29 1930 31 32 33 34 1935
Univ. of California
U.S.
G.S.
Fac.
E.U. Condon
J.R. Oppenheimer
Cal. Inst. of Tech.
P.F.
P.F.
Fac.
Fac.
G.S.
C. Eckart
J.R. Oppenheimer
L. Pauling
Univ. of Chicago
Fac.
Fac.
G.S. P.F.
C. Eckart
R.S. Mulliken
Harvard University
P.F.
Fac.
P.F.
U.S.
P.F.
Fac.
G.S.
G.S.
Fac.
Fac.
G. Breit
E.C. Kemble
R.S. Mulliken
J.R. Oppenheimer
J.C. Slater
J.H. Van Vleck
Mass. Inst. of Tech.
Fac.
Fac.
P.M. Morse
J.C. Slater
Univ. of Michigan
G.S.
Fac.
D.M. Dennison
Univ. of Minnesota
Fac.
Fac.
Fac.
G. Breit
E.U. Condon
J.H. Van Vleck
New York University
Fac.
Fac.
G. Breit
R.S. Mulliken
Princeton University
G.S.
Fac.
Fac.
G.S.
Fac.
E.U. Condon
C. Eckart
P.M. Morse
Univ. of Wisconsin
Fac.
Fac.
G. Breit
J.H. Van Vleck
1920 21 22 23 24 1925 26 27 28 29 1930 31 32 33 34 1935
Legend: Fac. Faculty Member
P.F. Postdoctoral Fellow
G.S. Graduate Student
U.S. Undergraduate Student

during the 1920's and early 1930's.

The first of these charts, "The Institutional Affiliations of Some Leading American Quantum Theorists between 1920 and 1935," is useful in providing, in the horizontal direction, a capsule presentation of the successive sites where each individual was located during that decade and a half. It has in addition, however, another use by looking vertically; selecting a particular time, one can read off the location of each of the individuals. For example, if one chooses late 1925, the time when the quantum mechanical breakthrough first occurred, one can readily determine where each was located when he heard the news. A few, such as D.M. Dennison and J.R. Oppenheimer were already in Europe. It is interesting to note that almost all the others were successful in arranging to have a period of European study within the next few years. In contrast, we note that by 1934 all of the individuals were back at American institutions.

The second chart, "The Distribution Among Academic Institutions in the United States of Some Leading American Quantum Theorists Between 1920 and 1935," while based on the same set of data as the first, provides a different perspective by focussing on the institutions involved. Thus we are able to see at a glance which institutions were responsible for the training of individual quantum theorists and/or providing them with academic employment. One cannot help but be struck by the large number who were at Harvard University especially during the 1920's when few other institutions were at all comparably active. On the other hand, it is equally noteworthy that by the early 1930's the supply of American quantum theorists was well distributed among a sizeable number of widely dispersed academic institutions.

Summary

Through consideration of two major trends in physics and among physicists in America in the early 1930's we come to better appreciate the intellectual atmosphere in which physicists studied and carried on research in the United States during that period. We cannot help but notice the contrast between that period and the previous decades with regard to America's position within the world of physics. The rise of nuclear physics brought new excitement to the discipline for physicists everywhere, and Americans participated at the forefront in developing the growing body of knowledge in this field.

In this thesis we have naturally stressed the relation of work in quantum mechanics to the growth of theoretical physics in America. We would be remiss, however, if we neglected to point out that there were additional areas of physics which contributed to increased theoretical activity by American physicists during the 1920's and 1930's. For example, technological advances within the field of communication, made possible by electronic developments, brought renewed interest in problems related to sound and electromagnetic phenomena. More versatile equipment gave impetus to a more sophisticated mathematical-theoretical approach to understanding the operation of the multitude of new devices that were rapidly coming into use.[176]

It was, however, true that work on quantum theory by Americans played a particularly important role in bringing Americans into the main-stream of physics at a crucial time for the discipline itself and for the growth of the American physics profession. The strengthening of the theoretical component of physics in America was part of the overall growth of physics in the United States which made possible a greater role for American physicists within the international community at large. This growth further resulted in America becoming a more attractive place of work for physicists from other countries to come to as visitors or as immigrants. To be sure, the foreign physicists who came to America contributed greatly to the subsequent development of physics in America. It is, however, incorrect to maintain that it took foreigners to make physics "come of age" in America. Rather, it took the already mature state of physics in the United States to provide a creative environment and new home for colleagues from abroad.

Part III: Areas in which American Quantum Theorists Made Noteworthy Contributions, 1930-1935

Introduction

By the early 1930's quantum mechanics had become recognized as the tool which would provide solutions to a wealth of problems associated with physics and chemistry that had proved to be stumbling blocks during the preceding decades. This did not mean, however, that a golden age had dawned wherein everything suddenly fell into place in those sciences. It turned out that the calculations associated with many of the possible applications of quantum mechanics involved formidable complexity. In the days before the development of high speed computers many long hours of tedious work might be required to carry through a particular calculation.

Quantum mechanics did not always provide quick and easy solutions to the old dilemmas, but it did bring a new perspective to a number of diverse topics. As a result, work was underway in the years between 1930 and 1935, reformulating the theory associated with such topics as atomic and molecular spectra, magnetism, solid state and chemical bonding. Quantum-mechanical considerations were also being applied in the rapidly emerging field of nuclear physics. Theoretical development in these areas was coordinated with the working out of specific applications which could be cross-checked with experimental results. These activities occupied many members of the world-wide community of physicists and the representation of Americans among them was high.

In this section we shall present examples of particular areas where

Americans achieved notable results which contributed to the advance of the discipline and firmly established the precedent of American participation in the front ranks of theoretical physics. We make no attempt to provide a complete catalogue of all the research done in the United States during the early 1930's that related to quantum mechanics; we do present examples selected to illustrate the number and diversity of individuals and topics involved. We do not attempt to discuss these examples in great detail since most of them involve considerable complexity of language as well as concept. It is hoped that the references provided with each will be useful to the reader who desires more information.

Before turning to these specific applications of quantum mechanics perhaps it would be well to remark upon the attitude which most American quantum physicists of the 1930's seemed to bring to their work. With very few exceptions they continued to exhibit the pragmatic acceptance of the new modes of thought that we commented upon in the previous chapter (see pages 3.67-3.69). Little attention was paid to any philosophical dilemmas that might have surfaced with the advent of quantum mechanics. Instead, most emphasis was placed on seeing what could be accomplished by means of the new techniques. Victor Lenzen, Robert Lindsay and Henry Margenau were among the few American physicists at that time who sought to probe the more philosophical aspects of quantum mechanics itself and the impact of quantum mechanics on the way in which physicists needed to view their own activity.

The final four chapters of Lenzen's book <u>The Nature of Physical Theory: A Study in Theory of Knowledge</u>,[177] completed in 1931, were devoted to an examination of quantum mechanics and its ramifications. In the Preface Lenzen stated that the stimulus for his book came primarily from philosophers. He added, however, that it was also written for those physicists who are interested in "methodology and systematic construction." Lenzen, a member of the physics department at the University of California at Berkeley, has remarked that there were few physicists nearby with whom he could discuss his philosophical concerns.[178]

Some three thousand miles away from Lenzen's home territory, R.Bruce Lindsay at Brown University in Providence, Rhode Island and Henry Margenau at Yale University in New Haven, Connecticut were at work on their book Foundations of Physics,[179] in which fundamental problems of physics were discussed from a philosophical point of view. Margenau has recently made the following comments about this publication[180]:

"Our book, written in 1933-35, was regarded by the publisher and his referees as a curiosity, indeed an anomaly because of its philosophical flavor. It was accepted as a 'high flyer', considered of doubtful commercial promise. In fact, the Wiley people discarded the plates after the first printing, expecting the sales to be limited. When later printings were called for the photo-offset process, undesirable and rare in those days, had to be used, until it was reproduced by Dover."

More than one quarter of this book dealt specifically with quantum mechanics, and the philosophical implications attendant upon the acceptance of both quantum mechanics and relativity permeated the entire discussion.

Aside from their collaborative efforts in preparing this book[181] both Lindsay and Margenau were publishing papers during the early 1930's about the philosophical aspects of physics in such journals as The Scientific Monthly, The Monist and Philosophy of Science.[182]

The paper critical of quantum mechanics written in America during the first half of the 1930's that received the most attention was "Can

Quantum-mechanical Descriptions of Physical Reality Be Considered Complete?" by A. Einstein, B. Podolsky and N. Rosen.[183] As might be expected from the name of the first author, if not from the phrasing of the title, their conclusion was "No." This paper was another item in the Bohr-Einstein debate over the validity of quantum mechanics. We shall not attempt to summarize its contents nor recount its background, tasks which have been done in great detail by Max Jammer and others.[184] It is interesting to note, however, that the two junior authors were both young men who had done their graduate work in physics at American institutions[185] and were fellows at the Institute for Advanced Study at Princeton during 1934-35. Podolsky had already coauthored another paper with Einstein and Tolman,[186] pointing out quantum-mechanical difficulties with describing past as well as future events, and Rosen had coauthored a paper with R.M. Langer on the mathematical requirements that must be placed on the Shcrödinger Ψ-function.[187]

Many responses to the Einstein-Podolsky-Rosen paper were written; for the purposes of this thesis it seems most appropriate to mention those of Edwin C. Kemble and Wendell H. Furry, both members of the Harvard University Physics Department at the time. Kemble argued that quantum mechanics must be applied only to a system which is regarded as a statistical ensemble, not to one composed of individually identifiable particles, as Einstein and his young colleagues had done.[188] Furry pinpointed the essence of the Einstein-Podolsky-Rosen dilemma as being due to a contradiction between quantum mechanics and the assumption by those authors that a system free of mechanical interference necessarily has independent real properties.[189] This was by no means the end of the discussion of the

Einstein-Podolsky-Rosen paper, a discussion which continued well beyond the time span with which this thesis is concerned.

While I have not studied the question extensively, it seems that most of the American authors involved with quantum mechanics in the early 1930's were pleased to accept it because of its ability to give experimentally verifiable results. One such person, Hubert James, whose work during this period we shall presently discuss, has described his feelings as follows[190]:

"I remember as among the happiest and most exciting days of my life the period in which we saw the numerical results come out better and better as we added more and more terms to our calculation."

Some of the quantum-mechanical topics to which American scientists turned their research efforts during the early 1930's have already been mentioned in our earlier sections dealing with books and doctoral dissertations related to quantum mechanics, written during that period. We now present a more detailed examination of several areas where valuable results were accomplished by Americans. It must be recognized, however, that few of them could be more than good beginnings which would lay the groundwork for subsequent decades of development. In this connection we recall the following assessment of the situation regarding quantum mechanics that was made by P.A.M. Dirac in 1929[191]:

"The underlying physical laws necessary for the mathematical theory of a large part of physics and the whole of chemistry are thus completely known, and the difficulty is only that the exact application of these laws lead to equations much too complicated to be soluble."

Atomic and Molecular Spectroscopy

It is appropriate to begin our discussion of quantum-mechanical applications with a look at those areas where, it might be said, "it all began," since it was spectral studies that provided most of the impetus for the growth of quantum theory from its beginning at the turn of the century.

Atomic spectroscopy was the first topic to be handled with essentially complete satisfaction by quantum mechanics, a task in which there was considerable American participation. We mentioned in our previous chapter (see page 3.75) John C. Slater's paper "The Theory of Complex Spectra," published in 1929 and in which he introduced the powerful tool that became known as "the Slater Determimant." Edward U. Condon has credited this work as the source of his interest in atomic spectra,[192] a topic which he pursued during the early 1930's at Princeton University. (It is appropriate to note that during the period when this theoretical work was being done on atomic spectra Princeton University continued its well-established program in its experimental aspects in the hands of Allen G. Shenstone and Henry N. Russell.)

Condon's theoretical study was aided by his graduate student, George Shortley. Together they wrote the book The Theory of Atomic Spectra, published in 1935, which became the classic reference in this field (see earlier discussion of this work on page 4.44), and which provides a ready source of information on the many scientists who contributed to its development. It is not feasible in this thesis to go into detail about the work of the individual Americans involved. It is, however, noteworthy that the following received multiple (i.e., three or more) citations for their theoretical contributions in the Index of Names contained in this book: R.F. Bacher, J.H. Bartlett, G. Breit, E.U. Condon, C. Eckart, D. R. Inglis, G.H. Shortley, J.C. Slater and J.H. Van Vleck, all names with which we are already familiar in this presentation.

Molecular spectroscopy, more complicated than its atomic counterpart due to the number of particles involved and the presence of vibrational and rotational modes of motion which give rise to additional lines, nevertheless proved amenable to quantum-mechanical techniques in the years between 1930 and 1935. Robert S. Mulliken was by far the most prolific of American authors who dealt with this topic, publishing some 40 papers on molecular structure derived from spectral study during that half decade. Initially Mulliken's papers in this series appeared in <u>The Physical Review</u>, but the later ones will be found in <u>The Journal of Chemical Physics</u>.[193]

The principal feature of Mulliken's analyses was the use of the concept of molecular orbitals.[194] This method, which Mulliken developed beyond the initial suggestion by Friedrich Hund, made possible the assigning of specific quantum numbers, consistent with the Pauli Exclusion Principle, to each of the electrons associated with the molecule under consideration, a system that was picturesquely refered to by J.H. Van Vleck as a "housing arrangement for electrons"[195] and that became highly valuable not only to spectroscopists but also to chemists concerned with the structure of molecules and with problems of valence. We shall have more to say about this in our next section.

Quantum-mechanical analyses of specific molecules based on their spectra were carried on at a number of institutions during the early

1930's. One that merits particular notice was the University of Michigan with its long-established laboratory for the study of infra-red spectroscopy, where theoretical members of the physics department worked in close collaboration with the experimentalists. A particularly interesting molecule studied at the University of Michigan during this period was that of ammonia, which has a pyramidal structure consisting of a triangular base formed by the three hydrogen atoms and an apex as the location of the nitrogen atom. Wave mechanics predicts that such a structure would show unusual spectral characteristics due to a kind of oscillation between the two sides of the hydrogen-atom plane by the nitrogen atom. In 1932 D.M. Dennison and J.D. Hardy were able to report the existence of just such a phenomenon.[196] Dennison and his theoretical colleague, George Uhlenbeck, published a discussion of the shape and dimensions of the ammonia molecule later that same year.[197] In addition, Philip Morse and Nathan Rosen, at the Massachusetts Institute of Technology, determined, by a different method, the average distance involved in the nitrogen atom's oscillation.[198] Their results agreed with those of Dennison and Uhlenbeck.

A valuable text which gives a picture of the state of molecular spectroscopy in the mid-1920's is <u>The Optical Basis of Chemical Valence</u> by Ralph Kronig,[199] who, we recall, had done his graduate work at Columbia University in the early 1930's and later became a professor of physics at the University of Groningen. In this book, as in the previously noted <u>Theory of Atomic Spectra</u> by Condon and Shortley, American authors are very well represented among the research citations.

Theories of Valence

Two[200] distinct quantum-mechanical approaches to the topic of chemical valence were developed during the early 1930's, each helpful in its own way, yet neither so satisfactory as to displace the other from usage. As mentioned earlier, exact solutions to chemical problems by means of quantum mechanics, although possible in principle, were impossible in practice due to mathematical complexity. Hence the progress that was made rested on the use of particular simplifying approximations. Without attempting to go into the details of the two most widely used methods here, we can present some of their distinguishing features and recount some of the circumstances of their development.

One method resulted from the application by Linus Pauling and John Slater of ideas originally proposed by W. Heitler and F. London for a quantum-mechanical treatment of the hydrogen molecule[201] and hence became known as the Heitler-London-Pauling-Slater (or H-L-P-S) method. Pauling and Slater were not collaborators in this enterprise. Slater, we recall, had become chairman of the Physics Department at the Massachusetts Institute of Technology in 1930; Pauling was professor of theoretical chemistry at the California Institute of Technology. Each worked independently of the other and their initial publications on this topic appeared almost simultaneously.

Slater's principal papers on valence theory, "Directed Valence in Polyatomic Molecules"[202] and "Molecular Energy Levels and Valence Bonds,"[203] were published in The Physical Review in the issues of 1 March and 15 September, 1931. Pauling's "The Nature of the Chemical Bond. Application of Results Obtained from the Quantum Mechanics and from a Theory of Paramagnetic Susceptibility to the Structure of Molecules" appeared in the Journal of the American Chemical Society of 6 April 1931.[204] This was the first in a series of seven papers written by Pauling in the next two years under the general title "The Nature of the Chemical Bond." The early ones were published in the Journal of the American Chemical Society,[205] the others in the Journal of Chemical Physics.[206] Pauling's Nobel Prize in Chemistry, awarded in 1954, was given for his development of these concepts.

The second quantum-mechanical method of describing the nature of chemical valence developed from work by F. Hund, J.E. Lennard-Jones, R. S. Mulliken and others. It is sometimes refered to as the H-LJ-M method. This involved the concept of molecular orbitals which we have already discussed in the previous section dealing with Mulliken's work related to molecular spectroscopy. It is interesting to note that the Nobel Prize which Mulliken received for his contributions in this field was awarded for chemistry rather than for physics.

The H-L-P-S model for chemical valence viewed the individual electrons in a molecule as localized to particular atoms in atomic orbitals which had definite shapes and formed "pair bonds." Thus the concept of "directed valence" was introduced. In order to account for the shapes of some molecules certain of the atomic orbitals must be regarded as "hybridized" between the s and p states, for example. Associated "resonance energies" can be calculated for particular molecules and, in many cases, could be shown to agree with experimentally determined values.

The H-LJ-M model for chemical valence viewed the molecule as containing "uncorrelated, itinerant electrons whose paths are called molecular orbitals."[207] Some of these orbitals were more localized than others. Mulliken calls these the "chemical orbitals" since they are the ones fundamental to the formation of chemical bonds.[208]

Both the H-L-P-S and the H-LJ-M models involved the concept of "paired electrons," in the sense of their spins being opposite or anti-parallel, in the formation of stable bonds. This notion is not too far removed from ideas inherent in the old, pre-electron spin, valence theory of Lewis and Langmuir. (See discussion page 1.51.) Incidentally, G.N. Lewis wrote about his post-quantum-mechanical thinking on the matter in an article entitled "The Chemical Bond" published in 1933.[209]

While these two models, the valence bond and the molecular orbital, each developed independently of the other and became regarded, to some extent, as rivals, it would be a mistake to search for a solution which would label one "right" and the other "wrong." In 1935 J.H. Van Vleck wrote[210]

"The two procedures represent different approximations to the solution of a complicated secular equation. The method of molecular orbitals permits factorization into one-electron problems, but at the expense of adequate cognizance of the terms due to electron repulsion, which are too fully recognized in the H-L-P-S procedure."

More recently Mulliken stated[211]

"The answer [to the question of how it is possible for both methods to be useful and helpful] lies in the fact that both methods need a considerable amount of correction for electron correlation before their descriptions become accurate. The fact that they differ so strongly from each other is explained by noting that they lie as it were on opposite sides of an accurate description, which then lies between them."

Van Vleck followed closely the development of the two different methods in the early 1930's. He wrote a series of articles "On the Theory of the Structure of CH_4 and Related Molecules"[212] in which he utilized both approaches, and another article "The Group Relation Between the Mulliken and Slater-Pauling Theories of Valence"[213] in which he showed that they were, in fact, intimately related and hence could be expected to give the same formal valence rules. In addition, he and Albert Sherman, a post-doctoral student at the University of Wisconsin at the time, prepared a lengthy survey and comparison of the two, entitled "The Quantum Theory of Valence" which was published in Reviews of Modern Physics in 1935.[214]

Although it is beyond the scope of this thesis to consider in detail the impact of quantum mechanics on the discipline of chemistry, it is appropriate to mention that the American physicists R.S. Mulliken and J.C. Slater were recognized from the start as being principal contributors to the newly emerging field of quantum chemistry, along with the American chemists Linus Pauling, and Henry Eyring of the Chemistry Department of Princeton University.[215]

The James and Coolidge Treatment of the Hydrogen Molecule

Since we have stressed so often the near impossibility of the complete solution of most problems by quantum-mechanical methods, particularly in the early days of its use, it is of special interest to look at the remarkably successful way in which the hydrogen molecule was handled by Hubert M. James and Albert S. Coolidge. James at that time was completing his doctoral studies in physics at Harvard University under the guidance of E.C. Kemble, and Coolidge was a member of the faculty of the Chemistry Department at that institution. Coolidge had completed his own doctoral studies ten years earlier and had recently been attending Kemble's course on quantum mechanics. The story of their collaboration can best be told in James' own words which permit us to gain valuable insight into the background of their joint effort and into the circumstances attending its completion--elements from the "private" aspect of science not usually available.[216]

"I remember that Dr. Coolidge had audited Professor Kemble's course on wave mechanics the year before I took it. Soon after that Professor Kemble agreed to direct my Ph.D. thesis work, and suggested that I apply the Heitler-London method to calculation of the binding energy of the molecule Cl_2. At the same time Coolidge was carrying on calculations on H_2O. Both of these calculations were complicated and laborious, and at Professor Kemble's suggestion we checked each others work. Coolidge published his work as "A Quantum Mechanics Treatment of the Water Molecule," Phys. Rev. 42, 189 (1932), and acknowledged my checking, but I was not able to return the favor, since the Heitler-London method (with the customary approximations) turned out to be utterly inadequate for the treatment of Cl_2. (I never published a word about the results of that hard years labor, and am surprised to see, on looking back, that I didn't even mention it in my thesis.) After I had convinced Kemble and myself that I had not simply made an error in my calculations, I proposed that I treat a much simpler problem, that of the molecule Li_2, avoiding the usual approximations in the application of the Heitler-London method. My results showed that all calculations made up to that time on molecules with inner shells were unreliable, some apparently good results being due to cancellation of several serious approximations. This work formed the first part of my thesis, and was published as the first part of my paper on Li_2, J. Chem. Phys. 2, 704 (1934).

I completed this work in the summer of 1932, while I was on vacation at my home in West Virginia. The result brought me to look for a better method than that of Heitler and London for the treatment of molecules, and I naturally did this in the context of the simplest typical molecule, H_2. The idea of applying to H_2 a treatment analogous to that of the helium atom by Hylleraas came to me in the bathtub - in keeping with the tradition of Archimedes, but with less evident relevance. I made some general notes on the project while I was at home, but I did not have facilities there to begin the calculation. I was familiar enough with molecular calculations to realize how much labor would be involved, and my earlier cooperation with Coolidge had made evident the great advantages of collaboration in such complex numerical calculations. When I returned to Cambridge in the fall of 1932 I discussed with Professor Kemble the possibility of undertaking this project in collaboration with Coolidge. He agreed to this, and himself made the arrangement, specifying (though I had not brought up the point) that my name should appear first on a joint paper that might come out of the work. From that time on Coolidge and I shared alike in all aspects of our joint research, alternating the name that was to appear first on successive papers without any attempt to evaluate our relative contributions to the work.

Inclusion of collaborative work as part of a Ph.D. thesis is not common, and called for comment in my thesis. I include for your convenience a copy of page 100 of my thesis, which deals with this point.

Coolidge and I always tried, so far as possible, to check each others calculations by other methods than simple repetition. Some thirty years later our tables and intermediate results were used by Kolos and Wolniewicz in debugging the computer programs they used in extending our work, and I, for one, was rather proud of the fact that they found no errors in them. Our collaboration was always pleasant, and it continued until we felt that we had employed the variational method in all cases that would repay the effort. For example, we assessed the work involved in an accurate treatment of the ground state of the beryllium atom, and concluded that the interest of the result would not justify the labor. (We started in with hand-powered desk calculators, and the motor-driven ones we later used were primitive by todays standards.) With the development of high-speed digital computers the field was opened up again, but I had other interests and did not return to it.

It did not seem at all remarkable to me at the time that I was collaborating with a chemist. As an undergraduate I had been a chemistry major, with a minor in mathematics. I went to Harvard expecting to work toward a Ph.D. in chemistry, but I switched to physics on the first day there, when it became evident that my interest in modern physics and mathematics made this departmental affiliation more appropriate. I don't remember the details, but I have the impression that my post-doctoral stay at Harvard was supported (at least in part) through the Chemistry Department. I have always worked near the borderline between chemistry and physics.

I might add one other comment. The labor involved in our initial calculation on H_2 was much exaggerated in rumors that had wide circulation at the time; I heard the report that we had spent three years on it. Actually the calculations can not have started before late September, 1932, and we sent off a preliminary report of results to the Physical Review in March, 1933. After my struggles with the Cl_2 molecule this seemed rather simple. I remember as among the happiest and most exciting days of my life the period in which we saw the numerical results come out better and better as we added more and more terms to our calculation."

The degree of success achieved by James and Coolidge may be grasped by a comparison of their calculated value of 4.697 ev with the experimentally determined one of $4.73 \pm .04$ ev for the binding energy of H_2. (Previous attempts by other authors at a quantitative treatment of the ground state of the hydrogen molecule by quantum-mechanical methods had yielded values for the binding energy which lay between 3.13 and 4.10 ev.[217])

Magnetism and Crystal Field Theory

Soon after the advent of quantum mechanics John H. Van Vleck began applying it to the study of electric and magnetic susceptibilities. His work in this field continued into the 1930's. We have already discussed his book on that subject, published in 1932, by which time he was recognized as an authority in the field.[218] (See page 4.73.)

The years immediately following that event were particularly fruitful for Van Vleck and his young doctoral and postdoctoral students at the University of Wisconsin. For example, William Penney and Robert Schlapp made, at Van Vleck's suggestion, studies of the effect of a crystalline field on the paramagnetic susceptibilities of rare earth compounds and of salts containing certain ions of the iron group. They published their results jointly in The Physical Review.[219] At about the same time Van Vleck succeeded in solving the puzzle associated with the large discrepancy in anisotropy between certain nickel and cobalt salts.[220] Van Vleck has called this his "favorite paper" because "it involved only rather a simple calculation, and yet it gave consistency and rationality to the apparently irregular variations in magnetic behavior from ion to ion."[221]

In the course of their research Penney and Schlapp introduced a crystalline potential model. Van Vleck later developed their idea by extending it to include incipient covalency.[222] This work by Van Vleck was the basis for what became known as "ligand field theory," a concept that has proved to be of great importance for the chemists' understanding of covalency and bonding in salts involving ions of the transition group of elements.

Solid State

The application of quantum mechanics to materials in the solid state was one of the most challenging and difficult ones faced by physicists in the 1930's. We have just noted some contributions to crystal theory made by physicists at the University of Wisconsin whose principal focus was magnetic behavior. Other physicists at other sites were developing new approaches from other perspectives. Among the important results achieved at American institutions we will cite but a few examples.

At Princeton University Eugene Wigner and Frederick Seitz developed a "cellular" method for studying the cohesive energy of metallic sodium.[223] At the Massachusetts Institute of Technology John Slater extended the Wigner-Seitz method to the computation of the excited bands of electrons in a metal.[224] Slater dealt specifically with the element sodium. Several of his graduate and postdoctoral students applied his methods to other substances.[225]

Quantum Field Theory, Electrodynamics and Nuclear Physics

The topics discussed thus far in this section have all represented the more or less successful application and extension of the new quantum-mechanical techniques to physical and chemical problems of long standing. The new school of theoretical physics that grew up around J. Robert

Oppenheimer at the University of California in Berkeley devoted most of its research endeavor to the challenges inherent in the emerging areas of quantum electrodynamics, quantum field theory and nuclear physics. In the early 1930's all of these topics were in their infancy and progress was slow for the most part. Nevertheless, some important research results were achieved by members of the Oppenheimer school.

We have already mentioned (see Note 4.139) the collaboration of Oppenheimer and Plesset on the subject of pair production by gamma rays. Among other important research results achieved by Oppenheimer and members of his theoretical group we cite but a few examples.

In 1930 Oppenheimer showed[226] that the positive particles in Dirac's "hole" theory could not be protons, but must be positive particles with the same mass as that of the electron. In the same paper Oppenheimer also calculated the time involved in the mutual annihilation of such particles and antiparticles. We recall (see page 4.34) that the existence of just such positive particles was established two years later by Carl D. Anderson of the California Institute of Technology.

In 1934 Oppenheimer and one of his postdoctoral students, Wendell Furry, published a paper and a subsequent Letter to the Editor with (what sounds to us now as) the curious title "On the Theory of the Electron and Positive."[227] Robert Serber has provided the following summary of this collaboration[228]:

"[They] formulated the Dirac theory as a field theory, essentially in its modern form. The charge renormalization and the vacuum polarization effects were pointed out although the problem of gauge invariance remained as a difficulty. The vacuum polarization effects were declared to be observable, with a warning that other radiative corrections existed for electrons."

Serber also notes "Similar considerations were being made by Dirac at about the same time."

In all Oppenheimer published, singly or jointly with one of his students, more than twenty papers or notes on topics were at the frontier of physics in the years between 1930 and 1935.

Note

The Summary of this chapter has been combined with the Conclusion of the thesis which is placed after the Notes and References on pages 4.93 to 4.127.

Chapter 4: Notes and References

4.1 See Appendix IV, Part 3 for a graph showing the growth of APS Membership and Fellowship. I have discovered no ready explanation for the marked rise in Fellowship that took place in 1931. On scanning the names of those so designated (Phys. Rev. 37 (1931) 1011, 1676; 38 (1931) 1916; 39 (1932) 173) one is surprised to find the names of many physicists whose work we discussed earlier in this thesis, and who, one might expect, would have already attained APS Fellowship. For example, among those listed are S. Goudsmit, G. Uhlenbeck, L. Pauling, D. Dennison, F.R. Bichowsky, I.I. Rabi, and S.C. Wang. Perhaps 1931 was the year in which the Society made up for some omissions during the unsettled years of the late 1920's.

4.2 Comment made by J.H. Van Vleck during a private conversation with the author. I have been unable to find any discussion in the minutes of the APS meetings of the change in the Presidential term of office. Appendix IV, Part 4 contains further information on APS presidents.

4.3 Phys. Rev. 36 (1930) 372.

4.4 Attendance figures are included in the Minutes of most of the meetings listed in Appendix IV, Part 5.

4.5 Minutes of the meeting, Phys. Rev. 44 (1933) 313.

4.6 See Appendix II for the names of several visitors from Europe.

4.7 See Appendix IV, Part 5 for symposium details.

4.8 Slater, "Quantum Physics in America Between the Wars," Physics Today (January 1968) 43-51; quoted passage on p. 43.

4.9 The Optical Society of America held its first annual meeting in December 1916, according to Jour. Opt. Soc. Amer. 1 (1917) 45.

4.10 The American Association of Physics Teachers held its second annual meeting in December 1932, according to the Am. Phys. Teach. 1 (1933) 21.

4.11 Among the possible sources of information concerning the founding and early years of the American Institute of Physics I have investigated and drawn upon the following:

1) AIP annual files deposited at the Institute's Center for History of Physics. Unfortunately, however, these files were incomplete as of February 1, 1976. At that time folders for the years 1931, 1932 and 1933 were missing. A notation was found indicating that they were in the possession of Henry A. Barton, retired Director of AIP. Dr. Spencer Weart, Director of the Center, when apprised of this fact, stated that efforts would be made to secure their return in the near future.

2) Karl T. Compton, "The American Institute of Physics," Rev. Sci. Inst. 4 (1933) 57-58.

3) H.A. Barton, "The Story of the American Institute of Physics," Physics Today 9 (January 1956) 56-66.

4) AIP-CHP Interview with H.A. Barton conducted by Charles Weiner, 3 and 16 March 1970.

5) "AIP 40 Years," Physics Today 24 (June 1971) 29-44, a pictorial essay on highlights of the growth of AIP arranged by decade with materials assembled by the staff of the Niels Bohr Library.

4.12 Reference 2 of the previous Note.

4.13 The AAPT was in the process of being formed when plans for the AIP were completed, hence it was arranged to have the AAPT included as a "founding member society." Details concerning the formation of AAPT and its early relations with both APS and AIP may be found in the AIP-CHP Interview with Homer Dodge, first President of APPT, conducted by Donald Shaughnessy, 23 January 1963.

4.14 These are the names of the Committees as given by Compton, reference 2, Note 4.11. Barton, reference 3, Note 4.11, speaks only of a "Committee on Applied Physics" as being instrumental in bringing about the founding of AIP. I have not yet sought clarification of this point from records of the American Physical Society. It appears, however, from the various sources listed in Notes 4.11 and 4.13, that the following members of the physics profession were particularly active in promoting the formation of AIP: K.T. Compton, P.D. Foote, George Pegram and F.K. Richtmyer.

4.15 "The Chemical Foundation is a corporation organized at the suggestion of the Alien Property Custodian, by members of the American

Dyes Institute, the American Manufacturing Chemists Association, and other gentlemen engaged in various branches of the chemical industries to buy from the Alien Property Custodian and hold for the chemical industries and for the country at large, the German-owned United States chemical and allied patents taken over by the Alien Property Custodian under the amendment of November 4, 1918 of the 'Trading with the Enemy Act'....

"Its charter provides that after the redemption of the preferred stock, the free net earnings of the Corporation shall be 'used and devoted to the development and advancement of chemistry and allied sciences in the useful arts and manufacturers in the United States.'"

From the Prospectus of The Chemical Foundation, Inc. printed in A. Mitchell Palmer and Francis P. Garvan, _Aims and Purposes of The Chemical Foundation, Inc. and the Reasons for its Organization_ (New York, 1919).

Garvan, the Alien Property Custodian, became President of the Foundation at the time of its founding in 1919.

4.16 As quoted by Compton in reference 2 of Note 4.11.

4.17 Information in this and the next paragraph comes from Barton's account in reference 3 of Note 4.11. I have not yet read the AIP Constitution. A copy presumably would be found in the missing 1931 AIP file mentioned above.

4.18 In private conversation with the author, 28 August 1975. J.H. Van Vleck shared his recollections of such criticisms being voiced by C.E. Mendenhall, who was at that time chairman of the physics department at the University of Wisconsin.

4.19 See, for example, the comment by W. Heisenberg (AHQP Interview Session 3, 11 February 1963, page 5), "I certainly know that I should say after 1930, I have read the _Physical Review_ quite regularly. Then it started to become the leading journal, but not much before that time."

4.20 _Rev. Mod. Phys._ _4_ (1932) 87-132.

4.21 _Ibid._ 7 (1935) 168-228. Reference to the Russian translation is found in Van Vleck "Reminiscences of the First Decade of Quantum Mechanics," _Internat. Jour. Quan. Chem._ _5_ (1971) 17. Unfortunately the specific literature citation is not known at this time.

4.22 Jour. Acous. Soc. Amer. 1 (1929-30) 5-8 gives the history of the Acoustical Society to date. Its organizational meeting was held at the Bell Telephone Laboratories, December 27, 1928. Its membership included representatives from university and industrial laboratories and soon numbered about 500. Harvey Fletcher of the Bell Laboratories was elected first president of the new organization whose purpose was "to increase and diffuse the knowledge of acoustics and promote its practical applications."

4.23 J.O.S.A.-R.S.I. 6 (1922) 1.

4.24 This journal was renamed the Journal of Applied Physics in 1937 and has continued as such ever since.

4.25 Physics 1 (1931) 69.

4.26 The present name of this (now monthly) journal, the American Journal of Physics, was adopted in 1940.

4.27 Among those initially serving in this capacity were: E.U. Condon, D.M. Dennison, G.B. Kistiakowsky, L. Pauling, C.P. Smyth and H.S. Taylor. Another member, R.H. Fowler, deserves special note since he was a British physicist. This journal was clearly open to European contributors. In the very first issue there was an article by Peter Debye, submitted from Leipzig. Some months later there was one coauthored by Fowler and J.D. Bernal from the University of Cambridge.

4.28 Jour. Chem. Phys. 1 (1933) 1. In a later section we shall discuss the area of quantum chemistry where the two disciplines of physics and chemistry were particularly closely united.

4.29 Ibid., 288.

4.30 Barton, "The Story of the AIP," pp. 57, 58. Rev. Sci. Inst. 4 (1933) 1 contains an editorial detailing the plans and advantages of "The New Review."

4.31 Rev. Sci. Inst. 6 (1935) 209-211 spelled out the current financial plight of the publication and announced a regular subscription rate of $3.00 per year and a reduced rate of $1.50 per year for members of the Founder Societies of the Institute.

4.32 Private conversation with the author, 7 November 1974.

4.33 This discussion of developments at the Massachusetts Institute of Technology is based on the following sources:

Reports of the Physics Department in annual MIT Presidential Reports;

Slater, _Solid-State and Molecular Theory_, Chapter 21;

G.R. Harrison, "Karl Compton and American Physics," _Physics Today_ November 1957, pp. 19-22;

Private conversations with MIT faculty members N.H. Frank, G.R. Harrison, J.A. Stratton, Dirk Struik and with Mrs. H.B. Phillips.

4.34 This discussion of the influence of Lawrence and Oppenheimer on the study of physics at the University of California is based on the following sources:

The Oppenheimer biographical references cited in Note 3.116;

Herbert Childs, _An American Genius: The Life of Ernest Orlando Lawrence_ (New York: E.P. Dutton and Co. Inc., 1968);

Birge, "History of the Physics Department," Vols. III and IV;

Private conversation with V.F. Lenzen, 26 October 1974.

4.35 In private conversation with the author, 4 March 1974, Kenneth T. Bainbridge recalled Lawrence as a very impressive physicist who was remarkably generous in helping others to build and operate their own cyclotrons.

4.36 Details of the founding and early years of operation of the Institute for Advanced Study are found in the following _Bulletins of the Institute for Advanced Study_: No. 1, December 1930; No. 2, February 1933 and No. 3, February 1934.

4.37 The Institute for Advanced Study was made possible financially by a $5,000,000 gift in 1930 from Louis Bamberger and his sister Mrs. Felix Fuld. (The Bamberger name is well known in New Jersey as associated with the leading Newark department store.)

Abraham Flexner was the first Director of the Institue and no doubt was influential in developing its philosophy of operation. Flexner had a long association with educational enterprises and was a champion of high standards of excellence. In his book, Universities: American, English, German (New York: Oxford University Press, 1930) which was based on three lectures given at Oxford University in 1928 at the invitation of the Rhodes Trust, he was scathingly critical of the existing educational institutions in America. He particularly deplored what he saw as a naive faith in education as a servant of society and the lack of adequate academic standards associated with such entities as schools of education and correspondence courses.

4.38 Albert Einstein became a Professor at the Institute in 1933. Paul Dirac was Visiting Professor in 1934-35. Among the visiting fellows that year were Carl Eckart, Boris Podolsky, Nathan Rosen and Clarence Zener.

4.39 The occasion was celebrated by a meeting of the American Physical Society held in the new facility, according to Phys. Rev. 47 (1935) 196.

4.40 Data for these years are available in the following publications:

1929-30 N.R.C. Reprint and Circular Series No. 95;
1930-31 Ibid., No. 101;
1931-32 Ibid., No. 104;
1932-33 Ibid., No. 105;
1933-34 Doctoral Dissertation Accepted by American Universities, No. 1, Donald Gilchrist, ed. (New York: H.W. Wilson Co., 1934); identified as "compiled for the National Research Council and the American Council of Learned Societies";
1934-35 Ibid., No. 2.

Note: Marckworth Dissertations in Physics, pp. x and xi, has a chart summarizing the annual degree production, 1930-59, arranged by institution, plus the total for each year and each institution. That set of data does not entirely agree, year by year, with that obtained from the sources listed above. The total number, however, for the six year period 1930-1935 inclusive, is 696 according to Marckworth, 690 according to the others, a discrepancy of less than one percent.

4.41 The figures quoted in this paragraph are taken from Rand, "The National Research Fellowships," Sci. Mon. 73 (August 1951) 71-80.

4.42 Data derived from information in National Research Fellowships, 1941-1944, pp. 24-39.

4.43 Rockefeller Foundation Annual Report, 1929.

4.44 Based on information compiled from the Rockefeller Foundation Directory of Fellowship Awards 1917-1950 by the American Institute of Physics-Niels Bohr Library, and made available to me by Gerald Holton.

4.45 Rockefeller Foundation Directory of Fellowship Awards 1917-1950.

4.46 Penney wrote to Van Vleck 24 July 1974 the following information about his obtaining that Commonwealth Fund Fellowship:

"Kronig advised me to try for a scholarship to take me to the United States and said that my first choice should be to work with you. The Secretary of the Commonwealth Fund told me much later that nobody had ever applied to go to Wisconsin to study physics before I applied, and they thought it was worth a try!"

4.47 Commonwealth Fund Fellows 1925-1937 (New York: The Commonwealth Fund, 1938).

4.48 J.H. Van Vleck, for example, spent seven months in Europe in 1930 on a Guggenheim Fellowship. According to the plan of study that he submitted with his Fellowship application (AHQP Microfilm No. 49) Van Vleck hoped to broaden his outlook on theoretical physics through visits with Bohr in Copenhagen. Born in Göttingen, Debye, Heisenberg and Wentzel in Leipzig, Kramers in Utrecht, Sommerfeld in Munich and Pauli in Zurich.

Van Vleck has included some recollections of this European sojourn in his papers "Reminiscences of the First Decade of Quantum Mechanics" and "My Swiss Visits of 1906, 1926 and 1930."

4.49 K.T. Bainbridge, R.B. Brode, and S.K. Allison each went to the Cavendish Laboratory as Guggenheim Fellows, according to Annual Reports of the John Simon Guggenheim Memorial Foundation for the years 1933, 1934 and 1935.

4.50 Le Magnetism: Rapports et Discussions du Sixieme Conseil de Physique Tenu a Bruxelles du 20 au 25 Octobre 1930 sous les auspices de l'Institut International de Physique Solvay. (Paris: Gauthier-Villars et Cie, 1932).

Van Vleck's address, entitled "Sur les susceptibilities des ions samarium et europium," appears on pages 55-64.

4.51 Structure et Proprietes des Noyaux Atomiques: Rapports et Discussions du Septieme Conseil de Physique tenu a Bruxelles du 22 au 29 Octobre 1933 sous les auspices de l'Institut International de Physique Solvay (Paris: Gauthier-Villars, 1934).

4.52 See Appendix II for numerous examples.

4.53 Birge, History of the Physics Department, Vol. III, pp. X(7) and (8).

4.54 Goudsmit, "The Michigan Summer Symposia in Theoretical Physics," the film "The World of Enrico Fermi."

4.55 See list of some emigrés in Appendix III.

4.56 Laura Fermi described their initial American experience at the University of Michigan in 1930 and her husband's growing attraction to America in Atoms in the Family (Chicago: The University of Chicago Press, 1954) Chapter 8, "A Summer in Ann Arbor," pp. 76-82.

4.57 Frank, Einstein, pp. 265-270.

4.58 Jungk, Brighter Than a Thousand Suns, p. 46. I have been as yet unable to establish the authenticity of this quotation.

4.59 Charles Weiner's article "A New Site for the Seminar: The Refugees and American Physics in the Thirties" contains a careful discussion of this topic. A broader view of the subject, including all the affected professions, is given by Laura Fermi in Illustrious Immigrants, (Chicago: University of Chicago Press, 1968)

4.60 Rockefeller Foundation Annual Report of 1932.

4.61 Rockefeller Foundation Annual Reports of 1932, and 1933. Warren Weaver, who became director of the Natural Sciences Division of the Rockefeller Foundation in 1932, has written of his early experiences in that capacity and the philosophy with which he administered his position in Scene of Change, Chapter 5 "The Rockefeller Foundation," pp. 58-75.

4.62 R.T. Birge, History of the Physics Department, Vol. IV, p. XI (21). Childs, An American Genius, pp. 167, 176 and 226.

4.63 "Physics in the Great Depression," p. 35.

4.64 In private conversation with the author, 15 January 1975, George R. Harrison explained the impetus for his undertaking to write Atoms in Action: The World of Creative Physics (New York: Wm. Morrow and Co., 1937), a project begun with the encouragement of H.A. Barton and K.T. Compton and with a $1000 advance from the American Institute of Physics.

It was hoped that Harrison could do for physics what Edwin E. Slosson had done for chemistry in his book Creative Chemistry: Descriptive of Recent Achievements in the Chemical Industries (Garden City, N.Y.: Garden City Publishing Co. Inc., 1919).

An earlier edition of Slosson's book seems to have been issued in 1917. A posthumous edition, revised by Harrison E. Howe, was issued by The Century Co., New York in 1930 as part of a series "The Century Books of Useful Science."

Slosson's book was considered to have been very effective in building public awareness of and appreciation for the role of chemistry in daily life. Julius Stieglitz, Professor of Chemistry at the University of Chicago and former President of the American Chemical Society, with whom Slosson had done his graduate work, wrote in his Introduction to the Revised Edition

"In the thirteen years that have passed since Slosson's 'Creative Chemistry' was first published in 1917, a profound change has taken place in the public recognition of the role of chemistry in the life of our nation....Our political leaders, our public-spirited citizens, all centers of education are alive now to the power of chemistry - in sharp contrast to the days of 1917! The man to whom more than to any other single indvidual, this country owes its awakening to the almost infinite beneficent possibilities of the science of chemistry is the late Edwin Emory Slosson. His 'Creative Chemistry' has educated our whole people, from our senators down to the thousands of eager pupils in the high schools of town and country, in the meaning, the power and the opportunities of chemistry in problems of health, of agriculture, and industrial wealth."

American Physical Society President, Paul D. Foote in "Industrial Physics," Rev. Sci. Inst. 5 (1934) 64, stated "We need more books on physics like Slosson's Creative Chemistry."

H.A. Barton, retired Director of the American Institute of Physics, also remarked in his AIP - CHP Interview, 16 March 1970, p. 39, on the effectiveness of Slosson's book in the area of public relations for the chemistry profession.

Biographical note: In his lifetime, Edwin Emory Slosson (1865-1929) was known as a chemist, educator and popularizer of science. He was Professor of Chemistry at the University of Wyoming from 1892 until 1921 when he became Director of Science Service in Washington, D.C., a post which he held until his death. (Source: Dict. Amer. Biog.)

4.65 See, for example, K.T. Compton "Physics in National Planning," Rev. Sci. Inst. 5 (1934) 235-236;
H.A. Barton, "Support for Science," Ibid., 339-340 and
_________ "Science in the Public Press," Ibid., 386.

4.66 Ibid., 57-66. The passages quoted are found on pp. 57, 60 and 63. Foote was at that time Director of Research at the Gulf Oil Co. where he had gone in 1929 from his previous position at the National Bureau of Standards.

4.67 Among those institutions where salary reductions were introduced were the California Institute of Technology (Source: Millikan Autobiography, p. 246), the University of California at Berkeley (Source: Birge History of the Physics Department, Vol. IV, p. XI (12); the University of Minnesota (Source: Erikson "History of the Minnesota Physics Department," reports for the years 1932-33 and 1933-34); the University of Illinois (Source: Interview with F.W. Loomis 19 November 1965, p. 6).

4.68 Figures for the cost of living indicator, compiled by the Bureau of Labor Statistics and included in Rand, "The National Research Fellowships," give the 1930 value as 119.4. By 1933 it had dropped to 92.4.

4.69 Van Vleck, "Reminiscences of the First Decade of Quantum Mechanics," p. 9.

4.70 Annual Report of the Physics Department, 1934-35, as quoted by G.M. Almy in "Life with Wheeler [Loomis] in the Physics Department, 1929-1940," an address delivered on the occasion of Loomis' retirement in 1957. A xerox copy of this address was kindly made available to the author by Maynard Brichford, Archivist at the University of Illinois.

This passage is also quoted by Weiner in "Physics in the Great Depression" and he notes that Almy's manuscript is in the Niels Bohr Library at the American Institute of Physics.

4.71 Northwestern University is an example of one institution which sought to lure Lawrence away from the University of California at Berkeley with an offer of the Chairmanship of the Physics Department and a salary of $6000 a year--an offer made in 1930 when Lawrence was 29 years old. The prospect of Lawrence's possible departure from the University of California prompted that institution to immediately promote him to a full professorship with a salary of $4500. Some years later (during the winter of 1935-36) Harvard University tried to persuade Lawrence to accept a new position there as dean of a new graduate division of science and engineering.

Source: Birge, History of the Physics Department, Vol. III, pp. IX (13)-(16); Childs, An American Genius, pp. 154-156, 231, 241.

4.72 G.M. Almy and J.H. Bartlett in 1930, P.G. Kruger and H.M. Mott-Smith in 1931, according to Interview with Loomis, p. 6.

4.73 CHP-AIP Interview with Robert Serber, 19 February 1967.

4.74 Letter from F.W. Loomis to E.H. Williams, June 1936, quoted by Almy in "Life with Wheeler," p. 4, tells of trying to divide $1800 among five or six worthy young unemployed physicists at the University of Illinois.

4.75 Cochrane, Measures for Progress, pp. 310-311, 320-322. 1931 had been a "banner year" at the National Bureau with a budget of $4,000,000 and several active research programs.

4.76 Weiner, "Physics in the Great Depression," p. 36.

G.R. Harrison, in private conversation with the author, 9 September 1975, provided an interesting sidelight on a relationship between work in physics at an American institution and the Depression. Harrison was able to secure the assistance of 143 WPA (Works Progress Administration) workers, starting in about 1935, in analyzing data which were later published as The Massachusetts Institute of Technology Wavelength Tables (Cambridge: MIT Press, 1939). It is Harrison's recollection that only about a dozen of these workers were spectroscopists. The others' previous employment had ranged from trombone player to drugstore clerk.

4.77 The New Yorker, 13 October 1975, p. 86.

In private conversation with the author on 5 April 1976 Professor Rabi recalled that, while there were many instances of good, or even excellent, work being done by individuals, the general intellectual environment at American universities in the early decades of the 20th century was not at a very high level, at least when compared with the leading European universities. Faculty members, for the most part, had similar social backgrounds and departments tended to function as "little families." The resulting "clubby" atmosphere was one in which first or second generation Jews would be out of place. Few young Jews of that era elected to follow scientific careers. (The medical and legal professions were far more attractive to them.) Gradually change was effected as the university intellectual level improved and the faculty members' interests broadened until ability became more important than personality in the consideration of candidates for academic positions. In Rabi's estimation, his experience in the physics department of Columbia University, in regard to promotions and general acceptance, was superior to that of some of his Jewish colleagues in other departments.

4.78 New York: Simon and Schuster, 1968, p. 83.

This book was very unfavorably reviewed by Frank Oppenheimer, Robert's brother, in Physics Today, February 1969, 77-80. The reviewer took particular issue with the quoted remarks attributed to Birge, saying

"Raymond Birge, former head of the physics department at the University of California at Berkeley, appears as an anti-Semite (presumably because he did not take to the 1930 version of hippies) despite all evidence that I have to the contrary."

4.79 In a letter of J.H. Van Vleck, kindly shared with me, H.D. Smyth on 17 April 1974, commented on several inaccuracies in Davies' ariticle. With regard to anti-Semitism at Princeton, Smyth stated

"Also I have found no hint of anti-Semitism in the [physics] department or elsewhere in this university though obviously I cannot guarantee its absence."

4.80 The position of lecturer at $3000/year at Columbia was offered to Rabi by radiogram while he was still in Europe and, he believes, upon the recommendation of Werner Heisenberg who was lecturing in the United States at that time, according to Bernstein in the account cited in Note 4.77.

4.81 Biographical data cited in this discussion have been checked with appropriate editions of American Men of Science.

4.82 Private conversation with the author, 12 November 1975. Bethe remarked that his two Jewish grandparents on his mother's side were more than enough to account for his dismissal from German universities under Nazi control.

4.83 At the Symposium honoring the 50th Anniversary of the Discovery of Electron Spin during the American Physical Society meeting in New York, held on 2 February 1976, S.A. Goudsmit related that when he recommended Robert Bacher for a position at Cornell University upon the completion of his doctoral work at the University of Michigan in the early 1930's, R.C. Gibbs asked whether Bacher was Jewish or not, apologizing at the same time for raising the issue but saying that [F.K.] Richtmyer insisted upon knowing.

In a letter to the author, 27 February 1976, Professor Goudsmit stated

"...it is a fact that I never met an anti-Semitic physicist in America when Jewish physicists had trouble getting jobs at certain universities and industrial laboratories. It was the fault of the higher administration, not of the colleagues. Thus the subject you address yourself to, anti-Semitism among physicists, was either absent or of minimal effect. But Jewish physicists did suffer from administrative prejudices."

Bacher is not Jewish. He did get a position as instructor at Cornell University in 1935, a time when Richtmyer was Dean of the Graduate School there. (Note also discussion Phys.Today,Dec.76,p.42)

4.84 J.R. Oppenheimer, for example (who had no difficulty getting a job), was said to have been "quick, impatient and had a sharp tongue, and in the early days of his teaching he was reputed to have terrorized the students." R. Serber (an admirer of Oppenheimer) Physics Today, October 1967, p. 38.

A reliable source has recently recalled that a few of the young Jewish physicists of that era among his acquaintance were unusually soft-spoken and inarticulate when compared with many of their fellow students. In addition, their own brilliant ability in grasping difficult material quickly seemed to make them insensitive to the difficulties encountered by less gifted students.

4.85 I am indebted to Wendell Furry for emphasizing to me the scope of "nuclear physics" in the 1930's during private conversation 7 March 1974.

4.86 Among physicists who were active at that time and who have expressed opinions along these lines are:

O.M. Corbino in "New Goals for Experimental Physics," Minerva 9 (1971) 530-538, an English translation by F. Segré of an address delivered in September, 1929;

L. Nordheim, AHQP Interview, 30 July 1962, p. 22;

M.A. Tuve, private conversation with the author, 24 January 1975;

E.U. Condon, "Reminiscences of a Life in and out of Quantum Mechanics," p. 18.

4.87 For a discussion of the impact of these events see Charles Weiner "1932 - Moving into the New Physics," Physics Today (May 1972) 40-49; for a discussion of the broader aspects of the rise of nuclear physics see Exploring the History of Nuclear Physics. C. Weiner, ed. AIP Conference Proceedings No. 7 (New York: AIP 1972), a report of conferences held in 1967 and 1969 at the American Academy of Arts and Sciences.

4.88 Childs, An American Genius, p. 145 and recollections of Victor Lenzen, shared with the author in private conversation 26 October 1974.

4.89 "1932 - Moving into the New Physics," p. 47. Weiner states that these data were assembled by Henry Small at the AIP Center for History and Philosophy of Physics.

4.90 K.T. Bainbridge, in private conversation with the author 4 March 1974, recalled hearing the statement made at an APS meeting in the late 1920's that physics was now "finished."

At the APS Symposium mentioned in Note 4.83 I.I. Rabi spoke of the feeling that he found in Europe after the advent of quantum mechanics that "physics will be over in six months--at least physics as we have known it" among such physicists as Max Born and Werner Heisenberg.

4.91 J.C. Slater, Solid State and Molecular Theory, pp. 3, 4.

4.92 Johns Hopkins University Circular, School of Higher Studies of the Faculty of Philosophy - Announcement of Courses 1931-32, 32-33, 33-34 and 34-35.

4.93 Princeton University Official Register - The Catalogue Issue, 1931-32, 1932-33, 1933-34 and 1934-35.

4.94 AHQP Interview, Session 3, 4 December 1963, p. 19.

4.95 Two American scientists who have commented upon the deteriorating situation which they encountered in Germany in the 1930's are R. S. Mulliken, "Molecular Scientists and Molecular Science: Some Reminiscences," and J.C. Slater, Solid-State and Molecular Theory, pp. 158, 159.

4.96 New York: McGraw-Hill Book Co. Inc., 1930.

Information relating to the circumstances under which this book was written was obtained from the Preface to the book itself, from the AHQP Interview with H.C. Urey, Session 1, 24 March 1964, pp. 25, 26 and from autobiographical material provided to me directly by A.E. Ruark, and originally prepared for the AIP Center for History of Physics.

The Ruark and Urey text was the second volume of a new series of textbooks published by McGraw-Hill, the "International Series in Physics," begun in 1929 with F.K. Richtmeyer as Consulting Editor. Condon and Morse's Quantum Mechanics, discussed in our previous chapter, was the initial volume; Slater and Frank's Introduction to Theoretical Physics, about to be discussed, was one of the subsequent volumes of a long series of books that were to become mainstays in the libraries of students majoring in physics in America. The familiar, matching pea-green bindings that became a kind of trademark of the series, are still frequently seen. It is interesting to note this particular development of physics textbooks for upper-level American students in light of the criticism previously leveled on the lack of suitable American physics textbooks beyond the elementary level. (See letter from E.U. Condon to J.T. Tate, quoted on page 3.9 of this thesis.)

4.97 New York: McGraw-Hill Book Co. Inc., 1933.

Information relating to the circumstances under which this book was written was obtained from the Preface to the book itself, from Slater, Solid State and Molecular Theory, p. 192, and from private conversation with N.H. Frank, 17 December 1974.

4.98 New York: McGraw-Hill Book Co. Inc., 1935.

Information relating to the circumstances under which this book was written was obtained from E.B. Wilson in private conversation, 13 January 1972.

4.99 Oxford: The Clarendon Press, 1930.

4.100 Letter to the author, 23 May 1974.

4.101 In private conversation with the author, 7 March 1974. Wendell Fury recalled using (and finding very difficult) Dirac's "masterpiece" while a graduate student at the University of Illinois.

4.102 Wave Mechanics: Elementary Theory (Oxford: The Clarendon Press, 1932). Wave Mechanics: Advanced General Theory (Oxford: The Clarendon Press, 1934).

Frenkel was known personally to members of the American physics community since he had come from the Soviet Union to be Guest Professor of Theoretical Physics at the University of Minnesota in 1930-31, according to Erikson's "History of the Minnesota Physics Department." The first of the above books was written during this period when, according to the Preface, Frenkel had "the opportunity of presenting the subject to a highly qualified American audience."

During his stay in America Frenkel participated in APS meetings and published papers in the Physical Review, a particularly noteworthy one being "On the Transformation of Light into Heat in Solids," Phys. Rev. 37 (1931) 17-44; 1276-1294.

Slater has commented "...Frenkel was a great favorite with me as well as with Americans in general. He was a physicist of the type I preferred, writing clearly and comprehensibly: His books on wave mechanics, written in English, were among the most useable of all those that were published in the early days" (from Solid State and Molecular Theory, p. 158).

4.103 London: Methuen and Co. Ltd., 1931; reprinted by Dover Publications Inc., 1951. H.P. Robertson prepared this translation from the second German edition of Gruppentheorie und Quantenmekanik (1930).

4.104 New York: E.P. Dutton and Co. Inc., 1930; translated by H.L. Brose from Atombau und Spektrallinien: Wellenmechanischer Ergänzungsband (Braunschweig: F. Vieweg und Sohn, 1929).

4.105 Oxford: The Clarendon Press, 1932.

4.106 Philip W. Anderson, "Van Vleck and Magnetism," Physics Today (October 1968) 23-26; quoted passage from page 24.

4.107 Oxford: The Clarendon Press, 1934.

4.108 Goldberg, "The Early Response to Einstein's Special Theory of Relativity, 1905-1911."

4.109 The Theory of the Relativity of Motion (Berkeley: University of California Press, 1917).

4.110 Cambridge: The University Press, 1935.

4.111 Phys. Rev. 35 (1930) 1342-1346; 36 (1930) 1121-1133; 37 (1931) 1025-1043.

Sources of information on Shortley include Marckworth's Dissertations in Physics; and comments contained on Condon, "Reminiscences of a Life in and out of Quantum Mechanics," pp. 17, 18.

4.112 Leipzig: Verlag von S. Hirzel, 1929; 2nd edition, 1932.

4.113 Leipzig: Verlag von S. Hirzel, 1931.

4.114 Leipzig: Verlag von S. Hirzel, 1933.

4.115 Bell Sys. Tech. Jour. 6 (1927) 653-701.

4.116 Rev. Mod. Phys. 1 (1929) 90-155 (called Supplements to Physical Review at that time, but presently bound and shelved as Reviews of Modern Physics).

4.117 Ibid. 6 (1934) 90-155.

4.118 See Chapter 2, p. 2.36.

4.119 According to the AHQP Interview with Pauling, Session 2, 27 March 1964, pp. 11, 12, 14, 15, 35, the collaboration between Pauling and Goudsmit which resulted in the writing of this text began during the late 1920's when Pauling and Goudsmit were studying in Copenhagen. Both became interested in the topic of hyperfine structure.

The early chapters of The Structure of Line Spectra were based on Goudsmit's doctoral dissertation, as translated by Pauling.

4.120 Book review by A.G. Shenstone, Rev. Sci. Inst. 4 (1933) 178.

4.121 Book review by J.H. Van Vleck, Ibid. 103.

4.122 N.H. Frank, in private conversation on 17 December 1974, has provided me with the following details of the background to the paper which he coauthored with Arnold Sommerfeld, "The Statistical Theory of Thermoelectric, Galvano- and Thermo-magnetic Phenomena in Metals," Rev. Mod. Phys. 3 (1931) 1-42;

Frank was studying with Sommerfeld in Munich at the time (early 1930) and was pleasantly surprised when Sommerfeld suggested that Frank take over the burden of preparing a paper on the topic which Sommerfeld had promised John T. Tate he would write for the Reviews of Modern Physics.

Sommerfeld inserted the following note when the article was published:

"I wish to emphasize that the parts of this report as far as they carry further than the results of my paper, Zeits f. Phys. 47, 1, 1928, are due mainly to Dr. Frank, on whom the credit and the responsibility therefore fall."

(It is not clear whether Sommerfeld was more concerned with acknowledging his young colleague's work or with dissociating himself from any errors or deficiencies it might contain.)

4.123 Rev. Mod. Phys. 4 (1932) 87-132.

4.124 See pages 3.17-19 and 4.18.

4.125 "The Michigan Symposium in Theoretical Physics," p. 181.

4.126 According to the official publication of the University of Michigan on the Summer Symposium for the year 1933, participants could be housed in a fraternity house by paying $20 for the entire session. It was also stated that "Good board can be obtained at a very moderate cost."

4.127 Appendix IV, Part 5 contains a list of APS Symposia Topics and Panelists up through 1935.

4.128 See page 4.4.

4.129 From materials related to the history of the University of Michigan at the Niels Bohr Library of the American Institute of Physics.

4.130 My principal source of information on the background of these conferences is Merle A. Tuve, who kindly shared with me, in private conversation 24 January 1975, his recollections concerning them.

There is also some discussion of this topic in Laura Fermi, Illustrious Immigrants: The Intellectual Migration from Europe, 1930-1941, pp. 181-2.

4.131 The Carnegie Institution of Washington Yearbook, 1935, p. 246. I am indebted to Dr. Louis Brown of the Department of Terrestrial Magnetism for providing me with xerox copies of the pages from each of the Yearbooks which dealt with the Theoretical Physics Conferences, and for arranging my visit to "DTM" during which I met with M.A. Tuve.

4.132 Annual group photographs were taken of the participants at these Conferences, a full series of which hang on the wall at the Department of Terrestrial Magnetism but without complete sets of identifications.

E.C. Kemble graciously shared with me his copy of the 1936 photograph. Full identification has been established for that year and the information has been passed on to the Niels Bohr Library along with the negative and a print made in recent years from Kemble's 1936 print. Kemble has also given to NBL two original copies of informal "action" pictures, taken during the 1936 meeting.

4.133 Letter to the author from Louis Brown, 20 November 1974; also referred to in Fermi, Illustrious Immigrants, p. 182.

4.134 N.R.C. Reprint and Circular Series Nos. 95 (1929-30), 101 (1930-31), 104 (1931-32), 105 (1932-33);

Doctoral Dissertations Accepted by American Universities, Nos. 1 (1933-34), 2 (1934-35), Donald Gilchrist, ed. (New York: H.W. Wilson Co., 1934, 1935), "compiled for the National Research Council and the American Council of Learned Societies."

4.135 National Research Fellowships, 1919-1944.

Most of the Fellowships for study related to quantum mechanics naturally were awarded in the category of "physics." It is interesting, however, to note that three were given as chemistry awards, two as mathematics.

The chemistry awards went to George Kimball who had received his Ph.D. in chemistry at Princeton University in 1932, to John Kirkwood who had received his Ph.D. in chemistry at the Massachusetts Institute of Technology in 1929, and to Albert Sherman who had received his Ph.D. in chemistry at Princeton University in 1933.

The mathematics awards went to Benedict Cassen who had received his Ph.D. in physics at the California Institute of Technology in 1939, and to Neal McCoy who had received his Ph.D. in mathematics at Iowa State University in 1929.

Note: the listings in National Research Fellowships include such biographical information as where the recipient obtained his doctoral degree and his professional occupation subsequent to his fellowship tenure up to 1944, the year the pamphlet was published.

Quantum mechanics became increasingly attractive to chemists during the 1930's and we shall have more to say about their involvement with it in our next section. On the other hand, while quantum mechanics required more sophisticated mathematics than most American physicists of the time felt at home with, the American mathematics community, for the most part, took little interest during the 1930's in quantum mechanics or any other area of applied mathematics. D.G. Birkhoff was one notable exception to this attitude for which he chided his fellow mathematicians in his address "Fifty Years of American Mathematics" delivered in 1934. Birkhoff himself published some papers that were concerned with the relationship between quantum mechanics and asymptotic series. (Those papers and the address referred to may be found in George David Birkhoff: Collected Mathematical Papers.)

4.136 Philip Morse took his NRC Fellowship to the Universities of Munich and Cambridge in 1930-31;

Milton Plesset spent the second year, 1933-34, of his Fellowship at the Institute for Theoretical Physics in Copenhagen;

Albert Sherman went to Cambridge for the second year of his Fellowship, 1934-35;

John A. Wheeler also chose Copenhagen for his second year of study, 1934-35;

Charles T. Zahn spent the year 1932-33 at the Physical Institute in Leipzig as an NRC Fellow;

Clarence Zener also went to Cambridge for part of the second year of his Fellowship.

In addition, Ernst Stueckelberg, a Swiss physicist who had been studying at Princeton (see Chapter 3, page 3.52) was awarded a National Research Fellowship for a year of study, 1930-31, at the University of Munich. (He then went back to Princeton as assistant professor of physics for two years before returning permanently to Switzerland in 1933.)

4.137 "The Early Years," Physics Today (October 1967), 38.

4.138 Birge, History of the Physics Department, Vol. III, pp. IX (29), (30).

4.139 Plesset was a National Research Fellow at the California Institute of Technology during the year 1932-33. He and Oppenheimer submitted a joint Letter to the Editor, "The Production of the Positive Electron," Phys. Rev. 44 (1933), 53-55, in which they correctly explained for the first time the mechanism of pair production by gamma rays, following the discovery of the positron by C.D. Anderson of the California Institute of Technology.

4.140 Birge, History of the Physics Department, Vol. IV, Appendix XVII, pp. XVII (a), (b), (c), "Ph.D. Degrees Awarded, up to July 1, 1932"; Ibid., pp. XI (19), (25).

The dissertation topics and related publications (where established) for these students were as follows:

M. Phillips, "Problems in the spectra of the alkalis: I. Photoionization probability of atomic potassium; II. Theoretical considerations on the inversion of doublets in alkali-like spectra." Phys. Rev. 44 (1933), 644-650.

G. Camp, "Is a relativistic, non-conservative mechanics possible?"

A. Nordsieck, "The Scattering of Radiation by an Electric Field";

H. Hall, "The Relativistic Theory of the Photoelectric Effect," phys. Rev. 38 (1931), 57-79;

J.F. Carlson, "The Energy Losses of Fast Particles," Phys. Rev. 41 (1932), 763-792 (with J.R. Oppenheimer);

L. Nedelsky, "Radiation from Slow Electrons," Phys. Rev. 42 (1932),

4.141 National Research Fellowships 1919-1944.

4.142 Biographies in AMWS, 12th edition and Birge, History of the Physics Department, Vol. IV, p. XI (25).

4.143 For examples see Oppenheimer bibliography in Physics Today (October, 1967), 52,63.

4.144 Crawley's thesis, "Characteristic functions and energy of a Li crystal," 1934, as listed in Marckworth, Dissertations in Physics, does not seem to have been published as such, but a paper "Electronic Functions for a Metallic Crystal" was presented by Houston and Crawley at the Berkeley APS meeting in June 1934 (Phys. Rev. 46 (1934), 329A).

Huff's thesis "The Motion of a Dirac Electron in a Magnetic Field" for which he received his degree in 1931, was published in Phys. Rev. 38 (1931), 501-512.

4.145 In addition to the usual sources of information relative to the granting of doctorates in physics by American institutions the reader's attention is called to two Harvard University publications dealing with physics doctorates earned at that institution:

Harvard University Department of Physics Doctors of Philosophy and Doctors of Science, 1873-1964 (Cambridge: Harvard Graduate Society for Advanced Study and Research, 1965) containing thesis titles arranged by year;

"A Directory of Harvard Physics Doctoral Alumni," a pamphlet issued by the Physics Department in the Summer 1975, which provides present addresses for living recipients of physics doctorates, together with information on the year of their doctorate and the identity of their thesis superviser.

The thesis titles and related publications for Kemble's students mentioned in the text are as follows:

Bartlett, "The Ionization of Multiatomic Molecules by Impact of Slow Electrons," Phys. Rev. 33 (1929), 169-174;

Zener, "Quantum Mechanics of the Formation of Certain Types of Diatomic Molecules";

Johnson, "The Theory of Complex Spectra," Proc. Nat. Acad. Sci. 19 (1933), 916-921;

Feenberg, "The Scattering of Slow Electrons by Neutral Atoms," Phys. Rev. 40 (1932), 17-32;

James, "The Quantitative Treatment of Molecules by Wave Mechanics," Jour. Chem. Phys. 1 (1933), 825-835; 2 (1934), 794-810; 3 (1935), 9-14;

Knipp, "Two Problems in the Quantum Mechanics of Diatomic Molecules," Phys. Rev. 47 (1935), 339A; 672-677.

4.146 Present, "I. Forbidden Transitions in Molecular Spectra; II Excited States of the Hydrogen Molecule," Phys. Rev. 48 (1935), 140-148.
Hebb, "I. On λ-type Doubling in II States of Diatomic Molecules Intermediate to Hund's Cases a and b; II. On the Paramagnetic Rotation of Tysonite," Phys. Rev. 46 (1934), 17-32.

4.147 MIT doctoral theses, publications and advisers:

Rosen, "Calculation of Energies of Diatomic Molecules," Phys. Rev. 38 (1931), 2099-2107, J.C. Slater;

Vinti, "Variational Calculation of Atomic Wave Functions," Phys. Rev. 43 (1933), 337-340, P.M. Morse;

Manning, "An Investigation of the Forms of the Potential Energy Function Permitting Exact Solutions of the One-dimensional Schrödinger Equation," Phys. Rev. 48 (1935), 161-164, Morse and J.A. Stratton;

Krutter, "The Thomas-Fermi Method in Metals and Energy Bands in the Copper Lattice," Phys. Rev. 47 (1935), 559-568, Slater;

Richtmyer, "Quantum Mechanical Study of Multiple Ionization Collisions of a Fast Electron with an Atom," Phys. Rev. 49 (1936), 1-8, Morse.

4.148 National Research Fellowships, 1919-1944.

4.149 University of Michigan doctoral theses, publications and advisors:

Adel, "The CO_2 Molecule: its Infra-red Spectrum and Mechanical Structure," Phys. Rev. 43 (1933), 716-723; 44 (1933), 99-194, Dennison;

Bacher, "Zeeman Effect of Hyperfine Structure," Phys. Rev. 34 (1929), 1501-1506, Goudsmit;

Baker, "The Application of Fermi Statistics to the Calculation of the Potential Distribution of Positive Ions," Phys. Rev. 36 (1930), 630-647, Laporte;

Fisher, "Hyperfine Structure in Ionized Bismuth," Phys. Rev. 37 (1931), 1057-1068, Goudsmit;

Gerhard, "The Infra-red Absorption Spectrum and the Molecular Structure of Ozone," Phys. Rev. 42 (1932), 622-631, Dennison;

Inglis, "Atomic Problems in the Perturbation Theory of Quantum Mechanics," Phys. Rev. 38 (1931), 862-872, Goudsmit and Uhlenbeck;

Koenig, "The Mathieu Functions and Associated Physical Problems," Phys. Rev. 44 (1933), 657-665, Laporte and Uhlenbeck;

Rose, "The Formation of Electron Pairs by Gamma Radiation," Phys. Rev. 48 (1935), 211-223, Uhlenbeck;

Uehling, "Transport Phenomena in Einstein-Bose and Fermi-Dirac Gases," Phys. Rev. 43 (1933), 522-561, Uhlenbeck;

Wu, "Thomas-Fermi Potential and Problems in Atomic Spectra," Phys. Rev. 43 (1933), 496L; 44 (1933), 727-731, Goudsmit;

Young, "The Wentzel-Brillouin-Kramers Approximate Solution of the Schrödinger Wave Equation," Phys. Rev. 36 (1930), 1154-1167, Uhlenbeck.

4.150 An example of such collaboration may be found in the case of an experimental thesis dealing with the spark spectra of rubidium and cesium, written in 1931 by George Miller which was published jointly with Laporte and R.A. Sawyer in Phys. Rev. 38 (1931), 843-853 and 39 (1932), 458-466.

4.151 National Research Fellowships 1919-1944.

4.152 Shortley, "The Theory of Complex Spectra," Phys. Rev. 40 (1932), 185-203;

Seitz, "A Matrix-algebraic Development of the Crystallographic Groups."

4.153 Phys. Rev. 43 (1933), 804-810; 46 (1934), 509-524.

4.154 Biography of John Bardeen, Nobel Lectures in Physics, 1942-1962, p. 342. Bardeen shared the Nobel Prize in Physics in 1956 with William Shockley and Walter Brattain "for their researches on semiconductors and their discovery of the transistor effect." Bardeen also shared in the 1972 Nobel Prize in Physics, the first person to have ever achieved the distinction of two Nobel awards in the same field, according to Physics Today, December 1972, p. 73. His award in 1972 was shared with Leon Cooper and J. Robert Schrieffer for their joint efforts in developing the theory of semiconductors.

4.155 Slater, Solid State and Molecular Theory, Chapter 21, entitled "MIT and Princeton, 1930-1940," describes the cordial relations that existed between the two institutions. The exchange visits of graduate students and staff are described on page 172.

4.156 National Research Fellowships, 1919-1944. It was during the period of this fellowship that the joint paper, Van Vleck and Sherman, "The Quantum Theory of Valence," Rev. Mod. Phys. 7 (1935), 167-228, was written.

Unfortunately, Sherman died a few years after this work was completed, as did another of Van Vleck's young students from this period, Amelia Frank who received her doctorate in 1934.

4.157 Van Vleck's doctoral students, their thesis topics and related publications are as follows:

Atanasoff, "The Dielectric Constant of Helium," Phys. Rev. 36 (1930), 1232-1242;

Amelia Frank, "The Effect of Crystalline Fields on the Magnetic Susceptibilities of Triply Charged Samarium and Triply Charged Europium and the Heat Capacity of Triply Charged Samarium," Phys. Rev. 39 (1932), 119-129;

Goble, "The Four-vector Problem in Quantum Mechanics and its Application to the Platimum-like Spectra," Phys. Rev. 48 (1935), 346-356;

Jordahl, "The Paramagnetic Susceptibility of Copper^{++}," Phys. Rev. 42 (1932), 901A.

Serber, "Some Optical Properties of Molecules," Phys. Rev. 41 (1932), 489; 43 (1933), 1003-1010, 1011-1021; 45 (1934), 461-467.

Whitelaw, "Theory of the Quantum Defect and Anomalous Multiplet Structure of Aluminum II," Phys. Rev. 44 (1933), 551-569.

Merrill's thesis on spectral multiplets was completed in 1935 and published in Phys. Rev. 46 (1934), 334A, 487-501. Stehn's work was not completed until 1937.

As a specific example of Van Vleck's influence on experimentalists we cite the work of R.B. Janes, a doctoral student of C.E. Mendenhall. Janes' thesis "Magnetic Susceptibilities of Inorganic Salts," published in Phys. Rev. 48 (1935), 78-83, contains thanks for Van Vleck's interest in his work and a statement to the effect that his study was designed as an experimental test of the theory set forth by Van Vleck.

4.158 Breit's doctoral students, their thesis topics and related publications were:

Brice, "The Band Spectrum of Silver Chloride," Phys. Rev. 35 (1930), 960-972 (begun with F.A. Jenkins);

Granath, "A Critical Study of the Hyperfine Structure of the 5485A line of the Lithium Ion," Phys. Rev. 42 (1932), 44-51;

Gray, "On the Calculation of the Magnetic Moment of the Lithium Nucleus from Hyperfine Structure Data," Phys. Rev. 44 (1933), 570-574;

Lowen, "Radiation Damping and Polarization of Fluorescent Radiation," Phys. Rev. 46 (1934), 590-597;

Willis, "Relativity Corrections in the Theory of Hyperfine Structure," Phys. Rev. 44 (1933), 470-490.

4.159 National Research Fellowships 1919-1944.

4.160 American Men and Women of Science, 12th edition.

4.161 "Remarks by I.I. Rabi" in Bromley and Hughes Facets of Physics, a Festschrift for Gregory Breit, pp. 181-185.

It appears that Rabi was hired by Columbia in 1929 primarily as a theoretician and he did in fact teach most of the modern theoretical courses there in the early 1930's. His own research, however, became increasingly experimental, focussing on magnetic moments and nuclear spins, a line of investigation that was recognized by the Nobel Prize in Physics in 1944.

Rabi and Breit collaborated in 1931 in developing a theoretical formula for the response to a magnetic field of a particle composed of spinning electrons as well as a spinning nucleus. (See Phys. Rev. 38 (1931), 2082, "Measurement of Nuclear Spin," a Letter to the Editor.)

4.162 Herzfeld came to Johns Hopkins from Munich in 1926; Dieke had studied at Leiden before coming to the University of California in 1925; Goeppert-Mayer arrived in 1930 soon after completing her doctoral studies at the University of Göttingen.

Source: American Men of Science, 7th edition.

4.163 National Research Fellowships 1919-1944.

4.164 Thesis topics, related publications and advisors as follows:

Lewis, "Coupled Vibrations with Applications to the Specific Heat and Infra-red Spectra of Crystals," Phys. Rev. 36 (1930), 68-86, Herzfeld;

Mauchly, "The Third Positive Group of Carbon Monoxide Brands," Phys. Rev. 40 (1932), 123A; 43 (1933), 12-30, experimental in part, Dieke;

Lee, "Forced Double Refraction in Cubic Crystals," Phys. Rev. 44 (1933), 625-631, Herzfeld;

Wheeler, "Theory of the Dispersion and Absorption of Helium," Phys. Rev. 43 (1933), 258-263, Herzfeld.

4.165 Same information for University of Chicago students:

Beardsley, "An Approximate Calculation of the Lower Energy Levels of the Carbon Atom," Phys. Rev. 39 (1932), 913-921, Eckart;

Dickinson, "The Specific Isotope Effect in the Hyperfine Spectrum of the Lead Atom," Phys. Rev. 46 (1934), 598-604, Eckart.

4.166 "Historical Background" in the Introduction, prepared by Mulliken, to Part II of Selected Papers of Robert S. Mulliken, J. Hinze, D. A. Ramsey, eds. (Chicago: University of Chicago Press, 1975).

Copies of these pages in manuscript form were kindly made available to me by Professor Mulliken in advance of publication.

4.167 Loomis recalled (Interview with Maynard Brichford, Archivist at the University of Illinois, 19 November 1965, p. 14) that Kusch was "the brightest graduate student I've ever had.... He was immensely bright, very, very energetic."

Kusch received the Nobel Prize in Physics in 1955 for his "precise determination of the magnetic moment of the electron." (Nobel Lectures in Physics 1942-1962.)

4.168 Bartlett's students, their thesis topics and related publications were as follows:

Furry, "Molecular Energies: the Lithium Molecule," Phys. Rev. 39 (1932), 210-225; 43 (1933), 316L;

Ireland, "Valence Forces in Beryllium Hydride," Phys. Rev. 43 (1933), 329-336.

Brown, "Atomic Wave Functions," Phys. Rev. 42 (1932), 914A;

Gibbons, "Theory of the Fine Structure (Neon Isotope Shift)," Phys. Rev. 44 (1933), 538-543;

Cook, "Bounds on Characteristic Energies of Atomic Systems."

4.169 I have been unable to uncover thus far much biographical information about Mott-Smith. His name never became listed in American Men of Science.

Wendell Furry recalled (private conversation with the author during Fall of 1975) that Mott-Smith impressed him in the 1930's as being a mature physicist of considerable competence. Furry was surprised to learn that Mott-Smith did not already have his Ph.D. at that time and noted that it was to Loomis' credit that he hired him anyway.

The Physical Review Index 1921-1950--Author Index lists Mott-Smith as author or co-author of some dozen papers between 1922 and 1930.

4.170 The title of Mott-Smith's thesis was "Certain Integral Transformations Connected with the Summation Problem." It is listed in American Doctoral Dissertations Printed in 1934, a Library of Congress Publication, as having been lithoprinted by Edwards Bros. Inc., Ann Arbor, Michigan.

Information on Mott-Smith's thesis supervisor was provided to me by J.H. Bartlett in private communication, 17 November 1975.

4.171 "The Quantum Mechanical Effective Cross-section for Ionization of Helium by Electron Impact," Phys. Rev. 44 (1933), 25-30.

4.172 White, "A Critique of the Indeterminacy Principle," Proc. N.A.S. 20 (1934), 525-529;

Donley, "The Calculation of Atomic Wave Functions for Two Stages of Ionization of Silicon by the Method of the Self-consistent Field," Phys. Rev. 50 (1936), 1012-1016.

4.173 This discussion of the writing of theoretical doctoral dissertations has been organized within an institutional framework, in contrast to what was done in our earlier chapters, due primarily to the large number of students involved, but also because the institutions and the faculty advisors are thereby highlighted.

4.174 Mulliken, "Molecular Scientists and Molecular Science: Some Reminiscences," Jour. Chem. Phys. 43 (1965), 33;

Slater, "Introduction to Professor J. H. Van Vleck," Internat. Jour. Qu. Chem. 5 (1971), 1-2.

4.175 I.I. Rabi has remarked (as quoted in The New Yorker, 13 October 1975, p. 88) on the lack of such a second tier in Germany, which he noted when he studied there in the 1920's. He believes this may have been influential during World War II in the German failure to succeed in their effort to make an atomic bomb and in their delay in developing radar--both of which were achieved in the United States through the combined efforts of a large number of scientists with varying degrees of skill.

4.176 I am indebted to Philip M. Morse for directing my attention to this topic during a private conversation on 2 October 1974.

An interesting example of the involvement by members of the American physics profession in the burgeoning field of communications occurred in the early decades of thd 20th century at the laboratories of the Bell Telephone Company. R.A. Millikan had close ties with members of the management of that corporation. He participated personally in some of their activity and, over the years, several of Millikan's most promising graduate students found productive employment there.

Souce: Millikan, Autobiography, pp. 116-123.

4.177 New York: John Wiley and Sons, Inc., 1931.

4.178 Private conversation with the author, 26 October 1974. On that occasion Lenzen related the following anecdote: Lenzen was very much intrigued with the notion of complementarity. Since he found a few fellow Americans among physicists or philosophers with whom to discuss the topic, he was especially interested to make the most of every opportunity to hear Bohr speak about his own views of it when visiting the United States in 1933. When Bohr spoke at Pasadena, Lenzen was in the front row of the audience. When Bohr spoke at Berkeley, Lenzen was in the front row of the audience. When Bohr spoke in Chicago, Lenzen was in the front row of the audience. Finally, Bohr said to him "Mr. Lenzen, I am curious about your wave equation. It seems to make it possible for you to materialize at so many locations where I am about to lecture."

4.179 New York: John Wiley and Sons, Inc., 1936.

4.180 Letter to the author, 14 January 1975. The Dover publication which Margenau mentions was in 1957.

4.181 In a letter to the author, 30 December 1974, Raymond Seeger stated "The Lindsay-Margenau book was planned in my house but I had to drop out after 2 years because of my heavy teaching load." Seeger was at the George Washington University in Washington, D.C. at that time. It is interesting to note that it was at Yale University that Lindsay, Margenau and Seeger (and also A.E. Ruark) had become acquainted in the late 1920's. All have written papers which examine some of the difficulties associated with the acceptance of quantum mechanics.

4.182 See, for example,

Lindsay, "The Broad Point of View in Physics," Sci. Mon. 34 (1932), 115-124; "Causality in the Physical World," Sci. Mon. 37 (1933),

330-337; "Where is Physics Going?", Sci. Mon. 38 (1934), 240-250.

Margenau, "Causality and Modern Physics," The Monist 41 (1931), 1-36; "Probability and Causality in Quantum Physics," The Monist 42 (1932), 161-188; "On the Application of Many-valued Systems of Logic to Physics," Phil. Sci. 1 (1934), 118-121; "Meaning and Scientific Status of Causality," Ibid., 133-148; "Methodology of Modern Physics," Phil. Sci. 2 (1935), 49-71, 164-187.

4.183 Phys. Rev. 47 (1935), 777-780.

4.184 The Philosophy of Quantum Mechanics (New York: John Wiley and Sons, Inc., 1974) pp. 160-251.
Jammer's footnotes include copious references to publications of other authors on this topic through 1972.

4.185 Podolsky took his Ph.D. at the California Institute of Technology in 1928 with Paul Epstein; Rosen took his at the Massachusetts Institute of Technology in 1932 with J.C. Slater.

4.186 "Knowledge of Past and Future in Quantum Mechanics," Phys. Rev. 37 (1931), 780-781.

4.187 "What Requirements Must the Schrodinger ψ-function Satisfy?", Phys. Rev. 37 (1931), 658

4.188 "The Correlation of Wave Functions with the States of Physical Systems," Phys. Rev. 47 (1935), 973-974.

4.189 "Note on the Quantum-mechanical Theory of Measurement," Phys. Rev. 49 (1936), 393-399; "Remarks on Measurements in Quantum Theory," Ibid., 476.

4.190 Letter to the author, 9 May 1972.

4.191 Proc. Roy. Soc. A 123 (1929), 714.

4.192 Condon "Reminiscences of a Life in and out of Quantum Mechanics," Internat. Jour. Qu. Chem. Symp. No. 7 (1973), 7-22; reference on page 17.

4.193 According to John R. Platt, in his article "1966 Nobel Laureate in Chemistry: Robert S. Mulliken," Science 154 (1966), 745-747, "Wits have said that the Journal of Chemical Physics was founded as an American Physical Society journal in 1932 just to have a place for Mulliken's papers; but the fact is that all the molecular theorists needed it, because the great Journal of the American Chemical Society did not, until the mid-1950's, publish any papers not containing experimental data."

4.194 An "orbital" in quantum mechanics, which is not to be confused with the old notion of orbits in the Bohr-Sommerfeld atomic theory, is "simply an abbreviation for *one-electron orbital wave function* or, preferably, for *one-electron orbital eigenfunction*. The last mentioned expression refers to any one of the so-called characteristic solutions or eigenfunctions of Schrödinger's quantum-mechanical wave equation for a single electron or an atom or molecule." The calculation of a molecular orbital involves consideration of "the attraction of two or more nuclei plus the averaged repulsion of the other electrons." Quotation taken from Mulliken "Spectroscopy, Molecular Orbitals and Chemical Bonding," Science 157 (1967), 13-24. This article is a printed version of Mulliken's lecture on the occasion of his reception of the Nobel Prize in Chemistry, on 12 December 1966. (Emphasis Mulliken's.)

Note: the term "orbital" did not come into use until 1932, according to Mulliken (Ibid., p. 17), although the initial idea of Hund and the subsequent development by Mulliken date back to 1928.

4.195 Ibid., p. 13.

4.196 Phys. Rev. 39 (1932), 938-947.

4.197 Phys. Rev. 41 (1932), 313-321.

4.198 Phys. Rev. 42 (1932), 210-217.

4.199 Cambridge (England): The University Press, 1935.

4.200 We note the existence of a third, proposed by Heitler and Rumer, which will not be discussed in this presentation because it did not involve any American participation in its development.

4.201 Zeit. f. Phys. 44 (1927), 455-472.

4.202 Phys. Rev. 37 (1931), 481-489; paper noted as received 22 January 1931.

4.203 Phys. Rev. 38 (1931), 1109-1144; paper noted as received 4 August 1931.

4.204 Jour. Amer. Chem. Soc. 53 (1931), 1367-1400; paper noted as received 17 February 1931.

Slater made note of this paper by Pauling in his publication refered to in Note 4.203.

4.205 Jour. Amer. Chem. Soc. 53 (1931), 3225-3237; 54 (1932), 988-1003; 3570-3582.

4.206 Jour. Chem. Phys. 1 (1933), 362-374; 606-617; 679-686. The first of these was written jointly with G.W. Wheland; the second and third with J. Sherman.

4.207 J.H. Van Vleck, Pure and Applied Chem. 24 (1970), 242.

4.208 R.S. Mulliken, Science 157 (1967), 17.

4.209 Jour. Chem. Phys. 1 (1933), 17-28.

4.210 Jour. Chem. Phys. 3 (1935), 80.

4.211 Science 157 (1967), 16.

4.212 Jour. Chem. Phys. 1 (1933), 177-182; 219-238; 2 (1934), 20-30.

4.213 Jour. Chem. Phys. 3 (1935), 803-806.

4.214 Rev. Mod. Phys. 7 (1935), 167-228.

4.215 The term "quantum chemistry" had come into use by the Spring of 1929 when Arthur Haas delivered a series of four lectures on the topic of the Physico-Chemical Society of Vienna, according to a statement in the English translation of those lectures by L.W. Codd, Quantum Chemistry (New York: Richard R. Smith Inc., 1930).

A list of contributors to quantum chemistry appeared in Nature 141 (1938), 667-668 as part of a book review of Einführung in die Quantenchemie by Hans Hellman (Leipzig: Franz Deuticke, 1937). The list contained nine names, the four Americans mentioned plus Walter Heitler, Friedrich Hund, J.E. Lennard-Jones, Fritz London and Michael Polanyi. This item came to my attention through a reference to it in Gray A History of the International Board, p. 101, where comment was made on the fact that seven of the nine had been holders of either International Education Board or National Research Council Fellows.

4.216 Letter to the author, 9 May 1972; it is planned that a copy of this letter will be deposited, with H.M. James' approval, at the Niels Bohr Library in New York City.

4.217 A table comparing the results achieved by various authors appears in an article by J.H. Van Vleck in Pure and Applied Chem. 24 (1970), 243.

4.218 A review of Van Vleck's book by Linus Pauling opened with the words "This excellent treatise on the theory of electric and magnetic susceptibilities is written by the world's leading authority in this field." Jour. Amer. Chem. Soc. 54 (1932), 4119.

4.219 Penney and Schlapp, Phys. Rev. 41 (1932), 194-207; Schlapp and Penney, Phys. Rev. 42 (1932), 666-686.

4.220 Phys. Rev. 41 (1932), 208-215.

4.221 "Reminiscences of the First Decade of Quantum Mechanics," Internat. Jour. Quan. Chem. 5 (1971), 15.

In this article, and in "Spin, the Great Indicator of Valence Behavior," Pure and Applied Chem. 24 (1970), 235-255, Van Vleck has commented in detail upon the work done during this period by himself and some of his younger associates. One of these, O. Jordahl, in the thesis that he wrote under Van Vleck's guidance, previously mentioned in Note 4.157, dealt with the paramagnetism of copper salts. It was his misfortune that much of the experimental data he was trying to fit were erroneous. He did, however, obtain some interesting results related to the Stark pattern for $CuSo_4 \cdot 5H_2O$. Although these results were initially greeted with suspicion by crystallographers and chemists, they were valid as later shown by x-ray studies made in England. In the course of this work Jordahl developed the degenerate perturbation theory that Van Vleck had introduced earlier. See Phys. Rev. 33 (1929), 467-506.

4.222 "Valence Strength and the Magnetism of Complex Salts," Jour. Chem. Phys. 3 (1935), 807-813.

4.223 "On the Constitution of Metallic Sodium," Phys. Rev. 43 (1933), 804-810; Part II, Phys. Rev. 46 (1934), 509-524.

4.224 "Electronic Energy Bands in Metals," Phys. Rev. 45 (1934), 794-801.

4.225 Ph.D. candidate Henry Krutter, for example, worked on copper, Phys. Rev. 48 (1935), 664-671, and National Research Fellow George Kimball on diamond, Jour. Chem. Phys. 3 (1935), 560-564. William Shockley, who received his Ph.D. in 1936, wrote his thesis on the energy band structure of sodium chloride.

4.226 "Probability of Radiative Transitions," Phys. Rev. 35 (1930), 939-947.

4.227 Phys. Rev. 45 (1934), 245-262; 343-344. Wendell Furry, whose name on these papers stood first, has recalled, in private conversation with the author 29 April 1976, that the classicist in Oppenheimer originally rebelled against accepting the word "positron" on etymological grounds; his resistance was relatively short lived, however, as shown by the title of another Letter to the Editor, Phys. Rev. 45 (1934), 903-904, "On the Limitation of the Theory of the Positron," written by Furry and Oppenheimer within six months of the items mentioned above.

4.228 Phys. Today (October, 1967), 36.

Conclusion

The account of the American response to quantum theory, as given in this thesis within the context of the overall development of physics in America, makes clear the interaction between the rise of quantum physics and the "coming of age" of physics in America. The period from 1920 to 1935, upon which this study has concentrated, was a crucial time for both of these accomplishments.

The quantum concept, first introduced by Planck in 1899, slowly invaded the discipline of physics in the ensuing two decades. By 1920, however, it was firmly linked to black body radiation, the photoelectric effect and atomic structure. The advent of quantum mechanics in 1925-26 provided the key event which enabled the new ideas to become fully integrated within the next ten years into every branch of physics and chemistry dealing with phenomena on the atomic scale.

Physics in America also achieved a milestone in 1899 with the founding of the American Physical Society and, in a sense, the birth of the American physics profession. Twenty years later, much growth had been accomplished by American physicists in terms of their professional society, research publication, laboratory facilities, undergraduate instruction in classical physics and respectable Ph.D. programs at several universities. But American physicists of the early 1920's worked in relative isolation from the mainstream of physics in Europe, and while the experimental achievements of a number of American physicists were internationally known and esteemed, there was in America a lack of an established tradition of theoretical activity, coupled with a weakness in the mathematical training of many physicists.

With regard to quantum theory, most established American physicists in 1920 were well aware of its existence and had come to accept the inevitability of its eventually attaining a permanent role in physics. For the most part, they did so with little enthusiasm and with no real sense of participation. There were, however, among the young American newcomers to the field of physics some who were attracted by the excitement of the new ideas involved in quantum theory and who would contribute to the rise of theoretical physics in America.

As we have just seen in Chapter 4, the state of physics in the United States in 1935 contrasted sharply with what had prevailed only fifteen years earlier. Despite external economic difficulties, physics had continued to grow and prosper in America. Noteworthy achievements had been made by the American physics profession as a whole in terms of number and diversity of membership in professional societies, of research publications, and of educational and research facilities. More remarkable, perhaps, is the story of what had taken place with respect to participation by Americans in the theoretical side of physics. Gone were the old deficiencies in American theoretical training and output. Diminishing American reliance on European physicists for theoretical advances was accompanied by increasing international respect for American theoretical competence. No longer were Americans isolated from the mainstream of physics. The center of gravity of international activity in physics had shifted westward toward the United States even before the major influx in the 1930's of physicists who were fleeing from Europe.

Yet the rise of theoretical physics in America in no way interfered with the continued appreciation for and preeminence of American experimental strength. In fact, in succeeding years both continued to grow in a symbiotic relationship. Physics in America achieved its mature status primarily through the addition of a strong theoretical component to the already well-established tradition of excellence in the experimental domain.

It was, however, the development of quantum physics within the discipline as a whole that provided the crucial element in the timing of the "coming of age" of physics in America, and it was the emerging group of American theoreticians who responded so positively and productively to quantum mechanics that provided the essential ingredient. Through their presence the character of the physics profession in America was altered. The necessary balance of emphasis between experiment and theory was reached that corresponded to the requirements of understanding and contributing to physics itself. Once this was accomplished, the United States could become--as it did--as propitious a place for doing physics as any in the world.

Postword

In developing the story of the interaction between the general growth of physics in America in roughly the first third of the twentieth century and the American response to quantum theory during those years, I have not, of course, by any means fully exhausted this intriguing topic. In the near future I propose to broaden my research to include the exploration of additional questions and the use of additional resources.

Among the additional topics which have aroused my interest, but to which I could not give adequate attention for reasons of space and time in the course of preparing this thesis, are the following:

1) a comparison of the American response to quantum theory with that of physics in other nations;

2) a deeper investigation of the specific objections to quantum theory, and the backgrounds of those American physicists who did not accept quantum theory or did so only with great reluctance;

3) the response of American experimental physicists to the rising importance of theoretical physics;

4) a search for possible correlating elements in the backgrounds of those Americans who did become quantum theorists;

5) evidence of the impact of quantum theory on the non-scientific American pouplation as, for example, occurred in the impact on religious thought owing to indeterminism;

6) American attitudes toward physical science in general on the part of influential members of the non-scientific community such as

philanthropists, politicians and social commentators;

7) the professional careers of American women physicists of the pre-World War II era;

8) theoretical areas other than quantum theory which attracted American physicists;

9) the theoretical "styles" of individual American physicists (as, for example, characterized by Dirac as "algebraic" or "geometric").

Among the presently available archival materials which I have not yet fully explored from the point of view of these questions, but which I believe would prove illuminating on several of the above topics are:

1) Records of the American Physical Society;

2) Correspondence involving American physicists as gathered in the Archive for the History of Quantum Physics;

3) Recent acquisitions of the Center for History of Physics at the American Institute of Physics;

4) Deposits of papers of individual American physicists at such sites as the Library of Congress, the California Institute of Technology, the Universities of California, Chicago and Michigan, Princeton University, and certain industrial archieve.

Also, I propose to make personal contact with additional individuals who were professionally active in physics before 1935 and who, I have reason to believe, would be able to provide me with further insights into the state of physics in America in the early decades of the twentieth century.

APPENDIX I

Selected Biographical Notes

Comments: No attempt has been made to provide here biographical information for every person whose name occurs in this dissertation. Those for whom we do provide the notes below include members of the following categories: 1) persons whose names occur repeatedly in various parts of this dissertation; 2) persons who have been quoted explicitly; 3) persons with the same or similar surnames who might become confused in the reader's mind; and 4) a few additional persons who seem to possess some intrinsic interest and about whom the reader can be expected to know little already.

European scientists have been included in these notes because it is assumed that most of those mentioned in this thesis are already well-known to all readers.

For each entry one or more sources of biographical information are provided; they should not be construed as being exhaustive. The author has chosen to list only those which have been most helpful in preparing these notes. Appropriate editions of *American Men of Science* have also been consulted, whether specifically cited or not.

In general, details of the professional careers beyond 1935 for the individuals listed are not included.

EDWIN PLIMPTON ADAMS (1878-1955) was born in Prague of American Missionary parents. He graduated from Beloit College in 1899. As a graduate student at Harvard University he received a Tyndall Fellowship which enabled him to study abroad at Berlin, Göttingen and Cambridge for two years before completing his doctoral degree program at Harvard in 1904. By that time he had become a member of the physics faculty at Princeton University where he remained until his retirement in 1943. He was a skilled experimentalist but gradually shifted to more theoretical work, primarily of a classical nature. He was also proficient in the old quantum theory.

Source: *Princeton Alumni Weekly* (5 July 1957) "Memorial".

ELLIOT QUINCY ADAMS was born in Medford, Massachusetts in 1888. Following undergraduate studies at the Massachusetts Institute of Technology he worked at the General Electric Research Laboratory from 1909 to 1912. Between 1912 and 1917 he was affiliated with the Chemistry Department of the University of California where he received his Ph.D. in 1914. He was subsequently employed by industrial organizations.

Source: *American Men of Science*, 7th edition.

JOSEPH SWEETMAN AMES (1864-1943) graduated from the Johns Hopkins University in 1886, following which he spent two years at Helmholtz' Laboratory in Berlin. He returned to Hopkins to take his Ph.D. with H.A. Rowland in 1890. His entire professional life was spent at that institution, beginning as a faculty member in 1890 and later (1901) serving as Director of the Physics Laboratory until becoming Provost (1926) and President (1929) of the University. He was elected President of the American Physical Society in 1919.

Source: Dict. Sci. Biog. Vol. I, 132-3.

CARL BARUS (1856-1935) took his Ph.D. at the University of Wurzburg in 1879 after having studied at Columbia University's School of Mines. He was employed by various governmental agencies, 1890-94, before becoming professor of physics and later dean of the graduate school at Brown University where he remained until his retirement in 1926. Barus' research focused on geophysics, pyrometry and condensation phenomena. He was a founder of the American Physical Society and served as its President in 1905-6.

Source: Dict. Biog. Vol. I, 490-1.

R(AYMOND) T(HAYER) BIRGE was born in Brooklyn in 1887. He studied physics at the University of Wisconsin where he received his Ph.D. in 1914 for a thesis dealing with the band spectra of nitrogen and written under C.E. Mendenhall. Following a five-year period of teaching at the University of Syracuse, Birge joined the faculty of the physics department at the University of California at Berkeley and has ever since been associated with that institution.

Source: American Men and Women of Science, 12th edition; Physics Today (May 1956), 20-8.

GREGORY BREIT was born in Russia in 1899. He studied at the Johns Hopkins University between 1914 and 1921 both as an undergraduate and graduate student. Following the reception of his Ph.D. he became a National Research Fellow first at the University of Leiden, then at Harvard University. He taught at the University of Minnesota for one year, 1923-4, before joining the Department of Terrestrial Magnetism of the Cargegie Institution as a mathematical physicist. In 1929 he became professor of physics at New York University. Five years later he accepted a similar position at the University of Wisconsin.

Source: Amer. Men Wom. Sci., 12th ed.

JAMES B. BRINSMADE (1884-1936) was born in Everett, Pennsylvania. He graduated from Yale College in 1906. After some years of teaching he resumed study at Harvard University in physics, receiving his M.A. in 1913, his Ph.D. in 1917. He remained there as an instructor until going into the army during World War I. In 1919 he joined the faculty of Williams College and remained there teaching in the physics department until the time of his death.

Source: Williams Alumni Review (October 1936), 2-3.

ARTHUR HOLLY COMPTON (1892-1962) was born in Wooster, Ohio, where he lived until the completion of his undergraduate study at the College of Wooster. He received his Ph.D. from Princeton University in 1916, then taught at the University of Minnesota for a year before working as a research engineer at the Westinghouse Co. He went to Cambridge University as a National Research Fellow, 1919-20, before becoming professor of physics at Washington University in St. Louis. Three years later he joined the faculty at the University of Chicago and remained there for many years.

Source: The Cosmos of Arthur Holly Compton, "Chronology," 449-457.

KARL TAYLOR COMPTON (1887-1954) was the older brother of Arthur Holly and preceded him through the College of Wooster and on to Princeton University. Karl received his Ph.D. in 1912 and remained one year as a teaching fellow. He was an instructor at Reed College, 1913-15 before returning to Princeton as an assistant professor. He subsequently rose to full professorship and chairman of the physics department there before leaving to become President of the Massachusettes Institute of Technology in 1930.

Source: Amer. Men Sci., 8th ed.

EDWARD UHLER CONDON (1902-1974) was born in Alamagordo, New Mexico. He studied at the University of California at Berkeley from 1920 to 1926, completing his undergraduate work in three years and receiving his Ph.D. in 1926 for work begun with R.T. Birge. He then traveled as an International Education Board Fellow to the quantum-mechanical centers of Göttingen and Munich for a year. In the Spring of 1928 he became a lecturer at Columbia University and in the fall of that year an assistant professor at Princeton University. He was made professor of theoretical physics at the University of Minnesota in 1929, but returned a year later to Princeton where he remained until 1937.

Source: Topics in Modern Physics: A Tribute to Edward U. Condon.

DAVID M. DENNISON (1900-1976) was born in Oberlin, Ohio. He received his bachelor's degree from Swarthmore College in 1921 and his Ph.D. from the University of Michigan in 1924. He then went abroad for further study as an International Education Board Fellow, principally in Copenhagen. He returned to the University of Michigan as an instructor in physics in 1927 and subsequently rose to full professorship in 1935. He remained associated with that institution until the time of his death.

Source: Amer. Men and Wom. Sci., 12th ed.; New York Times, 6 April 1976.

WILLIAM DUANE (1872-1935) was born in Philadelphia. Upon completion of his undergraduate studies at the University of Pennsylvania in 1892 he went to Harvard University where he obtained his A.M. in 1895. His doctoral studies were completed at the University of Berlin in 1897. He then became professor of physics at the University of Colorado until 1907 when he joined the Curie Radium Laboratory in Paris. He returned to Harvard University in 1913 as an assistant professor of physics and remained associated with that institution until the time of his death.

Source: Dict. Sci. Biog., Vol. II, 194-7.

CARK ECKART (1902-1973) was born in St. Louis, Missouri. He studied at Washington University through the master's degree level before going to Princeton University in 1923. Upon receiving his Ph.D. two years later, Eckart was awarded a National Research Fellowship to the California Institute of Technology. This was followed by a year of study abroad, 1927-28, as a Guggenheim Fellow. After returning to the United States, he became a member of the physics department at the University of Chicago.

Source: Amer. Men Sci., 11th ed.; Physics Today (January 1974), 87.

WENDELL HINKLE FURRY was born in Prairieton, Indiana, in 1907. After completing his undergraduate studies at de Pauw University in 1928, he did graduate work at the University of Illinois where he received his Ph.D. in 1932. For the next two years he studied with J. Robert Oppenheimer as a National Research Fellow in California. Since 1934 he has been a member of the physics department at Harvard University.

Source: Amer. Men and Wom. Sci., 12th ed.

EDWIN CRAWFORD KEMBLE was born in Delaware, Ohio, in 1889. His undergraduate study was done at the Case Institute where he received his B.S. in 1911. From 1913 to 1917 he pursued graduate study in physics at Harvard University, receiving his Ph.D. in 1917 for a thesis dealing with quantum theory and for which P.W. Bridgman was the advisor. In 1919, following a period of war-related work and a semester of teaching at Williams College, Kemble returned to the Harvard Physics Department as a member of the faculty. He has been associated with that institution ever since. In 1927 he spent seven months as a Guggenheim Fellow studying at the quantum-mechanical centers in Munich and Göttingen.

Source: Biographical materials at CHP-AIP.

F. WHEELER LOOMIS (1889-1976) was born in Parkersburg, West Virginia. Both his undergraduate and graduate studies were done at Harvard University where he received his Ph.D. in 1917. For two years he worked as a research physicist at the Westinghouse Lamp Co. He then joined the faculty of the physics department at New York University. During 1928-29, as a Guggenheim Fellow, he studied in Zurich and Göttingen. While abroad he received an invitation to become head of the physics department at the University of Illinois, a position which he retained until the time of his retirement in 1959.

Sources: Physics Today (May 1966), 83; private communication to the author.

HENRY MARGENAU, born in Bielefeld, Germany in 1901, came to America as a young man. His early scientific studies were done in Nebraska, first at Midland College and then at the University of Nebraska where he received an M.S. degree in 1926. He remained there as an instructor for a year before going to Yale University. His doctoral studies at Yale on an experimental topic with Louis McKeehan were completed in 1929. He was then granted a travelling fellowship by Yale to go to Europe for quantum-mechanical study in Munich and Berlin. He returned to Yale as a member of the faculty in the physics department and has been associated with that institution ever since.

Source: Amer. Men and Wom. Sci., 12th ed.; private communication to the author.

MORTON MASIUS was born in Egg Harbor, New Jersey, in 1883. His doctoral studies in physical chemistry were completed at the University of Leipzig in 1908. Following a year-long fellowship at Harvard University he became a member of the faculty at Worcester Polytechnic Institute in 1909 and remained associated with that institution until his retirement in 1954.
Source: Amer. Men Sci., 9th ed.

ALBERT ABRAHAM MICHELSON (1852-1931) emigrated with his parents from Poland at the age of 4. His early education took place in Virginia City, Nevada and in San Francisco. In 1869 he was appointed to the U.S. Naval Academy from which he graduated four years later and where he later returned as an instructor. His studies on the speed of light, begun at Annapolis, were continued in Europe between 1880 and 1882. Upon returning to the United States he joined the faculty of the new Case School of Applied Physics where he became associated with Edward Morley. Between 1889 and 1893 Michelson was affiliated with Clark University. In 1803 he moved to the University of Chicago where he remained for the rest of his life.

Source: Dict. Sci. Biog., Vol. IX, 371-4.

ROBERT ANDREWS MILLIKAN (1868-1953), born in Morrison, Illinois, grew up in rural Iowa. He studied at Oberlin College between 1886 and 1803 before going to Columbia University where he received his Ph.D. in 1895 for work done with Michael Pupin. He then enjoyed a year of postdoctoral study at Paris, Berlin and Göttingen. From 1896 to 1921 Millikan was a faculty member of the University of Chicago. The rest of his professional life centered about the California Institute of Technology which he guided through its great period of growth.

Source: Dict. Sci. Biog., Vol. IX, 395-400.

PHILIP McCORD MORSE, born in Shreveport, Louisiana in 1903, grew up in Cleveland. Following the completion of his undergraduate studies at the Case Institute Morse embarked on graduate study in physics with K.T. Compton at Princeton University where he received his Ph.D. in 1929. He remained at Princeton as an instructor for one year before going abroad as a Rockefeller Foundation Fellow for study at Munich and Cambridge in 1930-31. Upon returning to the United States, he joined the faculty of the physics department at the Massachusetts Institute of Technology and has ever since been associated with that institution.

Source: Autobiographical notes provided to the author.

ROBERT SANDERSON MULLIKEN was born in Newburyport, Massachusetts in 1896. He received his bachelor's degree from M.I.T. in 1917, his Ph.D. in physical chemistry from the University of Chicago in 1921. He stayed at Chicago as a National Research Fellow for a year before coming to Harvard where he remained until 1926. He was an assistant professor at New York University from 1926 to 1928 when he was made professor of physics at the University of Chicago. He studied in Europe as a Guggenheim Fellow in 1930 and 1932.

Source: Amer. Men and Wom. Sci., 12th ed.

ALFRED L. PARSON was born in India in 1889. His bachelor's degree in chemistry was awarded by Oxford University in 1911. He studied chemistry in the United States first at Harvard University, 1913-14, then at the University of California, 1914-15. It was during this period that he developed his theory of chemical bonding.

Source: Kohler, "The Origin of G.N. Lewis's Theory of the Shared Pair Bond," Hist. Stud. Phys. Sci., Vol. 3.

LINUS CARL PAULING was born in Portland, Oregon in 1901. Following the completion of his undergraduate studies at Oregon State College in 1922 he went to the California Institute of Technology as a graduate student in Chemistry, receiving his Ph.D. there in 1925 for x-ray crystal studies with Roscoe Dickinson. Upon receipt of his doctoral degree Pauling remained at Cal Tech as a National Research Fellow before going to Europe as a Guggenheim Fellow in March of 1926. During the next year and a half he studed at the quantum-mechanical centers in Munich, Zurich and Copenhagen. He then returned to the United States to become assistant professor of theoretical chemistry at the California Institute of Technology in 1927. He became full professor there in 1931. Pauling was a strong motivating force for the incorporation of quantum mechanics into the study of chemistry at American universities.

Source: Private communication to the author; Daedalus (Fall 1970), 988.

BENJAMIN OSGOOD PEIRCE (1854-1914) was born in Beverly, Massachusetts. While still an undergraduate student at Harvard College, Peirce published his first research paper. His bachelor's degree was awarded in 1876. From 1877 to 1880 Peirce studied in Europe receiving his Ph.D. in 1879 at the University of Leipzig where he studied with E. Wiedemann. The following year he spent with Helmholtz in Berlin before returning to Harvard to teach mathematics and physics until the time of his death. He was a founder of the American Physical Society and was elected its President in 1913.

Source: N.A.S. Biog. Mem., Vol. 8, 437-64.

GEORGE WASHINGTON PIERCE (1872-1956) was born in Webberville, Texas. He graduated from the University of Texas in 1894, following which he held teaching and other jobs before winning a fellowship to Harvard University in 1898. His Ph.D. thesis was completed in 1900 on the measurement of short radio waves. He then studied for a year at Boltzmann's Laboratory in Leipzig before returning to Harvard where he remained throughout his entire professional career as a member of the faculty of the physics department. His principal scientific contributions lay in the area of electrical communication.

Source: Dict. Sci. Biog., Vol. X, 604-5.

MICHAEL PUPIN (1858-1935), born in Idvor, located in what is now Yugoslavia, emigrated alone to the United States in 1874. Five years later, following hard work and serious study, he won a scholarship to Columbia University. His undergraduate studies, completed with distinction in 1883, were followed by European study. His doctoral work was done with Helmholtz and Kirchhoff in Berlin. After receiving his Ph.D. in 1899 for a thesis on osmotic pressure he returned to Columbia University to teach mathematical physics in the new department of electrical engineering and continued to do so until his retirement in 1931.

Source: Dict. Sci. Biog., Vol. XI, 213.

I(SIDOR) I(SAAC) RABI, born in Raymanov, Austria in 1898, came to the United States as a child. As an undergraduate at Cornell University he studied chemistry, receiving his bachelor's degree in 1919. He later did graduate work in physics at Cornell but his doctoral studies were completed at Columbia University in 1927. He then spent two years at the European centers for the study of quantum mechanics. Since 1929 he has been a member of the faculty of the physics department at Columbia University.

Source: Amer. Men and Wom. Sci., 12th ed.

JOHN CLARKE SLATER (1900-1976) was born in Oak Park, Illinois. Following undergraduate work in physics at the University of Rochester he came to Harvard in 1920. He completed his doctoral work under P.W. Bridgman in 1923 and was awarded a travelling fellowship by Harvard University for the next year which he spent in Cambridge, England and in Copenhagen. From 1924 to 1930 Slater was a member of the faculty of the physics department at Harvard University. He went to Leipzig as a Guggenheim Fellow in 1929 before becoming Head of the Physics Department at the Massachusetts Institute of Technology in 1930. He remained associated with that institution until he retired in 1966. At the time of his death he was a Graduate Research Professor at the University of Florida.

Source: Amer. Men and Wom. Sci., 12th ed.; Phys.Today(Oct.1976)68.

JOHN TORRENCE TATE (1889-1950) was born in Adams County, Iowa. His early scientific studies in electrical engineering were done at the University of Nebraska from which he received his B.S. and M.A. degrees in 1910 and 1912, respectively. He then went to study physics with James Franck at the University of Berlin where he was awarded a Ph.D. in 1914. Following his return to the United States he became an instructor of physics, first at the University of Nebraska, then at the University of Minnesota where he rose to full professor in 1919. In 1926 he became Managing Editor of the Physical Review.
Source: N.A.S. Biog. Mem., Vol. XLVII, 461-84.

RICHARD CHACE TOLMAN (1881-1948), born in Newton, Massachusetts, received his B.S. degree in chemical engineering at the Massachusetts Institute of Technology in 1903. He then spent a year studying in Germany before returning to M.I.T. for graduate work in physical chemistry with Arthur Noyes. His Ph.D. was awarded in 1910. For the next eight years he taught successively at the University of Michigan, Cincinnati, California and Illinois. He was a member of the Fixed Nitrogen Research Laboratory of the Ordinance Department in Washington, D.C. from 1919 until 1922 when he was invited by R.A. Millikan to become professor of theoretical chemistry at the California Institute of Technology. In 1935 he was made dean of the graduate school at that institution.

Source: N.A.S. Biog. Mem., Vol. XXVII.

JOHN HASBROUCK VAN VLECK was born in Middleton, Connecticut in 1899, the son of a mathematician and the grandson of an astronomer, both distinguished in their fields. After undergraduate study at the University of Wisconsin Van Vleck came to study physics at Harvard University in 1920. His doctoral dissertation, written under the guidance of E.C. Kemble and completed in 1922, was the first entirely theoretical thesis on quantum theory to be accepted by an American university. He remained as an instructor at Harvard University for a year before accepting an assistant professorship at the University of Minnesota where he taught and did research until 1928. Between 1928 and 1934 he was professor of physics at the University of Wisconsin. He then returned to Harvard University in a similar capacity and has been associated with that institution ever since. In 1930 he held a Guggenheim Fellowship for study in Europe.

Source: Autobiographical notes at CHP-AIP.

HAROLD WEBB was born in Ithaca, New York in 1884. His undergraduate and graduate studies were carried on at Columbia University where his Ph.D. was awarded in 1908 for work done with Michael Pupin. The following year, as a Tyndall Fellow, he spent at the Cavendish Laboratory and at the University of Berlin. He returned to Columbia as a member of the faculty of the physics department and remained associated with that institution until his retirement in 1953. At two different times, 1923-28 and 1939-41, Webb served as Secretary of the American Physical Society.

Source: CHP-AIP Autobiography and Inverview; private communication to the author.

ARTHUR GORDON WEBSTER (1863-1923), born in Brookline, Massachusetts, received his bachelor's degree from Harvard College in 1885. A travelling fellowship awarded by Harvard allowed him to go abroad for further study. His Ph.D. was earned at the University of Berlin for studies done with Helmholtz and Kundt. Upon returning to the United States in 1890 Webster joined the faculty at the recently established Clark University in Worcester, Mass. He remained there until his death by suicide in 1923. His research was primarily experimental but he wrote several successful textbooks on topics of a mathematical-theoretical nature. He was a prime mover in the formation of the American Physical Society and served as its President in 1903-4.

Source: N.A.S. Biog. Mem., Vol. XVIII, 337-47.

DAVID LOCKE WEBSTER (1888-1977) was born in Boston. He did both his undergraduate and graduate studies at Harvard University where his Ph.D. was awarded in 1913. He remained at Harvard as an instructor in the physics department until 1917. Then followed an interruption due to war work and a brief period of teaching at the Massachusetts Institute of Technology. In 1920 Webster became professor of physics at Stanford University where he continued to work beyond his retirement in 1954 until shortly before his death.

Source: Amer. Men. Sci., 11th ed.; private communication to the author; Physics Today (May 1977), 98.

APPENDIX II

European Scientists who Visited and Lectured in the United States between 1872 and 1935

Comments: 1) This list is not claimed to be all inclusive. It is, rather, the chronological compilation of information uncovered in the course of the general research which formed the background for this thesis. It focuses primarily on physicists.

2) The meetings of the British Association for the Advancement of Science have been included since they provided additional opportunity for contact with European physicists for those Americans who attended the meetings and, in some instances, the Europeans made visits to the United States before returning home.

3) The items in { } at the end of the individual entries are the principal sources of the information summarized therein.

1872-73 JOHN TYNDALL delivered popular lectures on light in Boston, Philadelphia, Baltimore, Washington, New Haven, Brooklyn and New York City between October 1872 and January 1873. The lectures were subsequently published as Six Lectures on Light (New York: Appelton and Co., 1873).

{K. J. Sopka "An Apostle of Science Visits America: John Tyndall's Journey of 1872-73" The Physics Teacher 10 (1972) 369-375.}

1884 The British Association for the Advancement of Science held its 54th annual meeting in Montreal, Canada during the late summer. Among the British physicists present at this meeting were Lord Rayleigh, the President of the Association; William Thomson (later Lord Kelvin) President of Section A; O. L. Lodge, and G. F. Fitzgerald. American physicists A. A. Michelson and H. A. Rowland attended the meeting as did the American chemist, Woolcott Gibbs.

{Report of the 54th Meeting of the British Association for the Advancement of Science (London: John Murray, 1885).}

1884 LORD RAYLEIGH visited the United States after presiding at the BAAS meeting in Montreal making stops in Baltimore, Philadelphia and Boston. He visited Harvard University where he met with E. C. Pickering and John Trowbridge and inspected the recently completed Jefferson Physical Laboratory.

{R. J. Strutt John William Strutt, Third Baron Rayleigh (London: Edward Arnold and Co., 1924).}

1884 WILLIAM THOMSON (later LORD KELVIN) came to the United States after the BAAS meeting in Montreal. During his visit Thomson delivered a series of 20 lectures at the Johns Hopkins University which attracted members of the American physics community from other institutions as well. Twenty years later these lectures, somewhat revised and updated, were published as Baltimore Lectures on Molecular Dynamics and the Wave Theory of Light (London: C. J. Clay and Sons, 1904).

{S. P. Thompson The Life of William Thomson, Baron Kelvin of Largs (London: Macmillan and Co. Ltd., 1910).}

1893 HERMANN VON HELMHOLTZ participated in the International Congress of Electricians, held in Chicago during August as part of the World's Fair, and was elected Honorary President of the Congress which attracted other European delegates as well. Posthumous honor was paid to Joseph Henry during the Congress with the naming of the "henry" as the unit of induction, upon the motion of E. Mascart of France and W. E. Ayrton of Great Britain.

{Phys. Rev. 1 (1893-94) 226-229.}

1896 J(OHN) J(OSEPH) THOMSON attended the sesquicentenary celebration of Princeton University, held in the autumn, and delivered a course of lectures, later published as The Discharge of Electricity Through Gases (New York: Charles Scribner's Sons, 1898) Thomson also visited the Johns Hopkins University during this trip.

{J. J. Thomson Recollections and Reflections (London: G. Bell and Sons, Ltd. 1936).}

1897 The British Association for the Advancement of Science held its 67th annual meeting in Toronto during August. Among the European attending were: O. Hahn, Lord Kelvin, J. Larmor, O. L. Lodge, F. Paschen, C. Runge, and S. P. Thompson. American participants were: B. B. Brackett, A. W. Duff, G. E. Hale, G. F. Hull, W. J. Humphreys, P. Lowell, A. A. Michelson, E. B. Rosa, S. W. Stratton, A. G. Webster and A. P. Wills.

{Report of the 67th Meeting of the British Association for the Advancement of Science (London: John Murray, 1898)}

1898-1907 ERNEST RUTHERFORD, during the period when he was a faculty member of McGill University in Montreal, enjoyed frequent contact with the members of the physics profession in the United States. He participated in several meetings of the American Physical Society and in the International Congress of Arts and Sciences, held in St. Louis in 1904. He was a lecturer at the Summer School of the University of California at Berkeley in 1906 and also at Yale University in 1905 where he delivered the Silliman Lectures which were published as Radioactive Transformations (New Haven: Yale University Press, 1906). He was awarded honorary degrees by the Universities of Pennsylvania and Wisconsin.

{A. S. Eve Rutherford (New York: Macmillan Co., 1939)}

1902 LORD KELVIN (WILLIAM THOMSON) toured eastern United States for three weeks in April and May, making stops at the Eastman Photographic Works, Rochester and at several universities such as Columbia, Cornell and Yale where he was given an honorary degree. Kelvin was very much interested in furthering the industrial development that was taking place in America especially in the area of harnessing water power as electrical energy.

{Thompson Life of Lord Kelvin}

1903 J. J. THOMSON delivered the Silliman Lectures at Yale University. They were subsequently published as Electricity and Matter (New York: Charles Scribner's Sons, 1904).

{Thomson Recollections and Reflections}

1904 SVANTE ARRHENIUS, LUDWIG BOLTZMANN, PAUL LANGEVIN, WILHELM OSTWALD, HENRI POINCARÉ, ERNEST RUTHERFORD, and J. H. VAN'T HOFF participated in the International Congress of Arts and Sciences (part of the Universal Exposition honoring the 100th anniversary of the Louisiana Purchase) in St. Louis in September.

{Proceedings of the International Congress of Arts and Sciences, eight volumes (Boston: Houghton Mifflin and Co., 1905)}

1905 LUDWIG BOLTZMANN lectured at the summer session of the University of California at Berkeley on thermodynamics and statistical mechanics. Boltzmann described some of his American experiences in "Reise eines deutschen Professors ins Eldorado" which was published in his Populäre Schriften (Leipzig: Barth, 1905).

{R. T. Birge History of the Physics Department, volume I}

1905-1909 JAMES H. JEANS served as Professor of Mathematical Physics at Princeton University. During this period he wrote two books based on the courses he taught: An Elementary Treatise on Theoretical Mechanics (Boston: Ginn and Co., 1907) and The Mathematical Theory of Electricity and Magnetism (Cambridge: The University Press, 1908).

{Prefaces of works cited above and Allen Shenstone "Sixty Years of Palmer Laboratory" address delivered at the dedication of the Jadwin Physics Building.}

1906 HENDRIK A. LORENTZ, during March and April, delivered lectures in mathematical physics at Columbia University which were later published as The Theory of Electrons and its Application to the Phenomena of Light and Radiant Heat (Leipzig: B. Teubner, 1909).

{H. A. Lorentz: Impressions of his Life and Work , G. L. deHass-Lorentz, ed. (Amsterdam: North Holland Publishing Co., 1957)}

1906 WALTHER NERNST delivered the Silliman Lectures at Yale University during the Fall semester. His lectures included a presentation of his "Third Law of Thermodynamics." They were subsequently published as Experimental and Theoretical Applications of Thermodynamics to Chemistry (New York: Charles Scribner's Sons, 1907).

{Text cited above}

1906-1913 OSWALD W. RICHARDSON was Professor of Physics at Princeton University. His book The Electron Theory of Matter (Cambridge, England: The University Press, 1914) was based on a course he gave there.

{Preface to work cited above and Shenstone "Sixty Years of Palmer Laboratory".}

1909 MAX PLANCK delivered lectures at Columbia University which were published as Acht Vorlesungen über Theoretische Physik (Leipzig: S. Hirzel, 1910).

{Text cited above}

1909 ERNEST RUTHERFORD and VITO VOLTERRA participated in the celebration of the 20th anniversary of the founding of Clark University.

{Lectures Delivered at the Celebration of the Twentieth Anniversary of the Founding of Clark University (Worcester: Clark University, 1912.}

1909 The British Association for the Advancement of Science held its 79th annual meeting in Winnepeg, Canada in late August. Among the British physicists in attendance were: J. J. Thomson, the President of the Association, E. Rutherford, President of Section A, J. H. Poynting and A. S. Eddington. American physicists participating in the meeting included: W. J. Humphreys, T. Lyman, L. A. McKeehan, E. Merritt, D. C. Miller, E. L. Nichols, and J. Zeleny.

{Report of the 79th meeting of the British Association for the Advancement of Science (London: John Murray, 1910)}

1912 MAX BORN lectured on relativity during the summer session at the University of Chicago.

{University of Chicago Catalogue and D. M. Livingston, The Master of Light (New York: Charles Scribners' Sons, 1973)}

1913 WILHELM WIEN delivered lectures at Columbia University which were subsequently published as Neure Probleme der Theoretische Physik (Leipzig: B. Teubner, 1913).

{Text cited}

1913 FREDERICK A. LINDEMANN came to the University of Chicago as guest lecturer during the summer session.

{Sources for the History of Quantum Physics and University of Chicago Catalogue}

1914 WILLIAM H. BRAGG visited Brown and Harvard Universities. His lectures at Harvard were delivered during November on the topic "X-Rays and Crystals."

{Correspondence between Bragg and T. Lyman in the Lyman Papers, 1914 Box, Harvard University Physics Department.}

1914 ERNEST RUTHERFORD delivered two lectures at the National Academy of Sciences, inaugurating a series of such lectures. He also visited and lectured at Columbia, Harvard, Princeton and Yale Universities.

{Eve Rutherford}

1917 ERNEST RUTHERFORD came to the United States to discuss the submarine problem with Hale, Millikan and others. He also travelled to New York, Boston and New Haven where he was awarded an honorary Doctor of Science degree by Yale University.

{Eve Rutherford}

1921 ALBERT EINSTEIN visited the United States under Zionist auspices in April and May. He was awarded an honorary degree by Princeton University where he delivered four lectures on relativity that were later published as The Meaning of Relativity, E. P. Adams, translator (Princeton: University Press, 1923). Einstein also visted Columbia University, the City College of New York, the University of Chicago and Harvard University. In addition, he addressed the annual meeting of the National Academy of Sciences.

{P. Frank Einstein: His Life and Times (New York: Alfred Knopf Inc., 1947)}

1921 MARIE CURIE arrived in New York City on May 11 and spent about six weeks touring in the United States amid great acclaim. She was presented with a gram of radium, valued at $120,000, "from the women of America" and personally welcomed by President Harding. She visited Niagara Falls and the Grand Canyon before returning home.

{Science 53 (1921) 327, 382, 483, 497, 509, 513, 517}

1922 FRANCIS W. ASTON came to Philadelphia in March as Franklin Institute Lecturer. In addition, he delivered two lectures at Harvard University on Atomic Structure and Isotopes and also lectured in Washington, North Carolina and New York City.

{Journal Franklin Institute 193 (1922) 581-608, 712, Harvard University Gazette,March 18, 1922; and, Science 55 (1922) 395}

1922 HENDRIK A. LORENTZ served as Visiting Professor at the California Institute of Technology between January and March. A symposium in his honor was held at the University of Wisconsin where he had delivered four lectures. He also lectured at Harvard University before returning to Europe in April. Problems of Modern Physics: A Course of Lectures delivered in the California Institute of Technology was published some years later based on the 1922 lectures by H. A. Lorentz, H. Bateman, ed. (Boston: Ginn and Co., 1927).

{Science 54 (1921) 572; 55 (1922) 312, 426; preface of text cited}

1922-1923 CHARLES G. DARWIN served as Visiting Professor at the California Institute of Technology.

{R. A. Millikan Autobiography (New York: Prentice Hall, 1950)}

1922-1923 ARNOLD SOMMERFELD came to the University of Wisconsin as Carl Schurz Exchange Professor. During his six month stay, October to April, he also lectured at the California Institute of Technology, the University of California at Berkeley, Harvard University and the National Bureau of Standards.

{U. Benz unpublished biography of Sommerfeld; Birge History of the Physics Department; and, Science 57 (1923) 20, 202, 323}

1923 J(OHN) J(OSEPH) THOMSON delivered five lectures at the Franklin Institute that were subsequently published as The Electron in Chemistry (Philadelphia: J. B. Lippincott, 1923). Also, he visited The Johns Hopkins University, Princeton and Yale Universities, and the General Electric and Bell Telephone Laboratories during March and April.

{Thomson Recollections and Reflections}

1923 NIELS BOHR delivered endowed lectures at Amherst College and at Yale University (Silliman Lectures) during October and November. He gave two colloquia at Harvard University and addressed the American Physical Society meeting held in Chicago.

{The Amherst Student September 24, October 15, 18, 22, 25, 1923; Harvard University Gazette October 20, 1923; Science 58 (1923) 429; and, Physical Review 23 (1924) 104.}

1923-1925 OSKAR KLEIN was a member of the physics department faculty at the University of Michigan where he supervised the thesis of D. M. Dennison.

{Sources for the History of Quantum Physics; unpublished reminiscences of D. M. Dennison as listed in the Bibliography}

1924 H. ABRAHAM, W. L. BRAGG, C. FABRY, E. RUTHERFORD, and P. ZEEMAN took part in the centenary celebration of the Franklin Institute, September 17-19.

{Journal Franklin Institute 198 (1924) 407-419}

1924 PAUL EHRENFEST came to the California Institute of Technology in early 1924 as Research Associate. He travelled widely, lecturing at other institutions such as the University of Minnesota, Harvard University, Columbia University and the University of California at Berkeley.

{Cal. Inst. Tech. Annual Catalogue 1923: Harvard University Gazette April 18, 1924; R. Kronig "The Turning Point" (see Bibliography); recollections of J. H. VanVleck, shared privately with the author; and, Birge History of the Physics Department, Volume II, p.VIII (15)}

1924 ARTHUR S. EDDINGTON was Visiting Professor at the University of California at Berkeley during the Fall semester. He lectured on astronomy and relativity to academic and public audiences.

{Birge History of the Physics Department, Volume 2, pp. VIII (19) and (20)}

1924 HENDRIK A. LORENTZ visited the California Institute of Technology.

{Sources for the History of Quantum Physics}

1925 MANNE SIEGBAHN gave two lectures at Harvard University on X-Rays during March.

{Harvard University Gazette, March 7, 1925}

1925 STANISLAW LORIA, of the University of Lwow, Poland, lectured during the summer session of the University of California at Berkeley on atomic theories and the experimental foundation of quantum theory.

{Birge History of the Physics Department, Volume 2, p. VIII (25)}

1925 PETER DEBYE was a visiting lecturer at the Massachusetts Institute of Technology during February and March. He delivered three lectures on magnetism at Harvard University during this period.

{Mass. Inst. Tech. Presidential Report, 1925, and Harvard University Gazette, February 14, 28, 1925}

1925-1926 MAX BORN delivered thirty lectures at the Massachusetts Institute of Technology and five at Harvard University between the middle of November 1925 and the end of January 1926. Subsequently, he travelled widely in the United States and lectured at the California Institute of Technology, the University of California at Berkeley, the University of Wisconsin, the University of Chicago and Columbia University. The lectures he delivered at M.I.T. were published as Problems of Atomic Dynamics (Cambridge: M.I.T. Press, 1926). Born's lectures aroused considerable excitement since they contained the earliest presentations in America of the new matrix mechanics.

{Preface to Problems of Atomic Dynamics; Harvard University Gazette January 2, 9, 16, 1926; Birge History of the Physics Department; W. Weaver Scene of Change (New York: Scribner's Sons, 1970); and Born-VanVleck correspondence, AHQP Microfilm 49}

1926 KARL HERZFELD, then of the University of Munich, lectured at the University of Michigan Summer School.

{Goudsmit "The Michigan Symposium"}

1926-1927 HENDRIK A. LORENTZ delivered thirty lectures at Cornell University during the Fall Semester, afterward went to the California Institute of Technology as Visiting Professor for several months. On March 31, 1927 he delivered an address at the Franklin Institute entitled "How Can Atoms Radiate?" later printed in Journal Franklin Institute 205 (1928) 449-471. He sailed for Europe on April 7.

{The Cornell Daily Sun, September 30, October 13, 1926; Sources for the History of Quantum Physics; and Science 65 (1927) 351}

1926-1927 ARTHUR HAAS, of the University of Vienna, lectured at the Massachusetts Institute of Technology during the academic year. He addressed the Philosophical Society of Washington in February.

{Mass. Inst. Tech. Presidental Report, 1926-27, and Science 65 (1927) 180}

1927 PETER DEBYE was acting professor of mathematical physics during the Spring semester at the University of Wisconsin.

{Science 65 (1927) 58}

1927 VICTOR HENRI, a physical chemist from Zurich, lectured to the physics departments of Harvard University and the Massachusetts Institute of Technology on the structure of molecules during the Spring semester.

{Mass. Inst. Tech. Presidential Report, 1926-27; Harvard University Gazette, April 2, 1927; and, Science 65 (1927) 33}

1927 ABRAM F. JOFFE, President of the Russian Physical Society, lectured at Harvard University, the Massachusetts Institute of Technology and the University of Pittsburgh before travelling west to the University of California at Berkeley where he was Guest Lecturer during the Spring semester. His lectures at Berkeley were published as The Physics of Crystals L. B. Loeb, ed. (New York: McGraw-Hill, 1928).

{Science 65 (1927) 33, 180; Report of the Jefferson Laboratory of Harvard University, 1926-27; Mass. Inst. Tech. Presidential Report 1926-27; and, Birge History of the Physics Department Volume 2, p. VIII (37)}

1927 ERWIN SCHÖDINGER lectured at a number of American universities during the Spring. Among them were the California Institute of Technology, Harvard University, the University of Iowa, the Massachusetts Institute of Technology and the University of Wisconsin.

{Sources for the History of Quantum Physics; Harvard University Gazett, March 5, 12, 1927; Mass. Inst. Tech. Presidential Report, 1926-27; Science 65 (1927) 252; and, Van Vleck "Reminiscences of the First Decade of Quantum Mechanics"}

1927 EDWARD A. MILNE, Manchester University, lectured at the University of Michigan Summer School.

{Goudsmit "The Michigan Symposium"}

1927-1928 WILLIAM L. BRAGG lectured at the Massachusetts Institute of Technology for several months and gave one lecture at Harvard University in March.

{Mass. Inst. Tech. Presidential Report 1927-28, and Harvard University Gazette, March 24, 1928}

1928 LEON BRILLOUIN was visiting professor at the University of Wisconsin. Also, he delivered a colloquium at Harvard University during January on the statistics of quantum mechanics.

{Sources for the History of Quantum Physics and Harvard University Gazette, January 28, 1928

1928 JAMES FRANCK spent several months in the United States during which time he visited many different institutions. A symposium in his honor was held soon after his arrival in January at the Loomis Laboratory in Tuxedo Park, New York. Some ninety physicists were in attendance as guests of Alfred Loomis. Franck gave three lectures at Harvard University during January before travelling west to the University of California at Berkeley where he spent two months as guest lecturer and was awarded an honorary LL.D. degree.

{Journ. Frank. Inst. 205 (1928) 473; Science 67 (1928) 63; Harvard University Gazette, January 7, 1928; and Birge History of the Physics Department, Volume 2, pp. VIII (36)-(40)}

1928 HANS A. KRAMERS was guest lecturer at the University of Michigan Summer Symposium in Theoretical Physics.

{Goudsmit "The Michigan Symposium"}

1928-29 ARNOLD SOMMERFELD came to the California Institute of Technology as Visiting Professor as part of a year of world travel which also included India and Japan. Sommerfeld paid a visit to the University of Minnesota where he delivered a lecture on "Electrons in Metals" and participated in the meeting of the American Physical Society that was held in Washington, D. C. in April.

{Sources for the History of Quantum Physics; Sommerfeld "Autobiographische Skizze"; H. A. Erikson "History of the Minnesota Physics Department"; and, Physical Review 33 (1929) 1067.}

1928-1929 HERMANN WEYL was Visiting Professor at Princeton University and also delivered a series of lectures at Harvard University under the joint auspices of the Departments of Mathematics and Physics.

{Sources for the History of Quantum Physics and Harvard University Gazette, April 27, 1929.}

1929 WERNER HEISENBERG delivered a series of lectures at the University of Chicago that were subsequently published in an English translation prepared by C. Eckart and F. C. Hoyt as The Physical Principles of Quantum Theory (Chicago: The University of Chicago, 1930; reprinted by Dover Publications,Inc.). In addition, Heisenberg travelled widely and lectured at many American institutions such as the University of California at Berkeley, the Massachusetts Institute of Technology, Oberlin College and Ohio State University.

{Text cited; Birge History of the Physics Department, Volume 3, p.IX (40); Mass. Inst. Tech. Presidential Report 1928-29; and Science 69 (1929) 667.}

1929 FRIEDRICH HUND was Visiting Professor at Harvard University during the Spring semester. He lectured on Band Spectra and Molecular Structure and joined E. C. Kemble and J. C. Slater in offering a Seminar in Theoretical Physics.

{Harvard University Catalogue, 1928-29}

1929 LEONARD S. ORNSTEIN, of Utrecht, was guest lecturer at the Massachusetts Institute of Technology. He delivered a lecture at the University of Minnesota.

{Mass. Inst. Tech. Presidential Report, 1928-29 and Erikson "History of the Minnesota Physics Department" }

1929 LEON BRILLOUIN was Guest Lecturer at the University of Michigan Summer Symposium in Theoretical Physics.

{Goudsmit "The Michigan Symposium"}

1929 P(AUL) A. M. DIRAC came to the University of Wisconsin as Visiting Professor during the spring semester. He delivered a lecture at the University of Minnesota and was one of the guest lecturers at the University of Michigan Summer Symposium in Theoretical Physics.

{VanVleck "Travels with Dirac in the Rockies"; Erikson "History of the Minnesota Physics Department"; and, Goudsmit "The Michigan Symposium" }

1929 EDWARD A. MILNE was a guest lecturer at the University of Michigan Summer Symposium in Theoretical Physics.

{Goudsmit "The Michigan Symposium"}

1929, 1930 ALFRED LANDÉ came to the Ohio State University as guest lecturer. He visited and lectured at the University of Minnesota. (In 1931 Landé accepted a permanent position at Ohio State University.)

{Landé letter to the author, 18 February 1972 and Erikson "History of the Minnesota Physics Department".
Vorlesungen über Wellenmechanik gehalten an der Staatsuniversität zu Columbus U.S.A. (Leipzig: Akademische Verlagsgesellschaft M.B.H 1930)}

1930 ENRICO FERMI was a lecturer during the University of Michigan Summer Symposium.

{Goudsmit "The Michigan Summer Symposium"}

1930 MAX VON LAUE and RUDOLF LADENBURG came to the United States in the Fall at the expense of the Rockefeller Foundation which was funding an Institute of Physics at the Kaiser Wilhelm Institute at Dahlem-Berlin with Einstein as director and von Laue, acting director. Von Laue and Ladenburg toured such American institutions as might be expected to contribute ideas for the new institute. Among their stops were the University of California at Berkeley, Harvard University, the National Bureau of Standards and the laboratories of the Westinghouse and General Electric Companies.

{Birge, Vol. III, p. X7; <u>SHQP</u>, p. 149; <u>Report of Jefferson Physical Laboratory, 1930-31</u>.}

1930 OTTO STERN was a Visiting Lecturer at the University of California at Berkeley.

{Birge, Vol. III, pp. IX39, 40; <u>SHQP</u>.}

1930 GREGOR WENTZEL served as Visiting Professor at the University of Wisconsin.

{Van Vleck, "First Decade;" <u>SHQP</u>.}

1930-31 P.A.M. DIRAC was a visiting lecturer at Princeton University.

{<u>SHQP</u>.}

1930 PAUL EHRENFEST lectured during the Summer Symposium at the University of Michigan. He also gave two colloquia at Harvard University and lectured at the California Institute of Technology.

{Goudsmit, "The Michigan Summer Symposium;" <u>SHQP</u>; <u>Report of the Jefferson Physical Laboratory 1930-31</u>.}

1930-31 ALBERT EINSTEIN was Visiting Professor at the California Institute of Technology. He returned there during the following two winters. (He settled in the United States permanently in the Fall 1933.)

{SHQP; Frank, Einstein. pp. 224, 226, 270.}

1930-31 JAKOV FRENKEL was Guest Professor of Theoretical Physics at the University of Minnesota. He also visited Harvard University where he delivered a colloquium.

{Erikson, "History of the Minnesota Physics Department;" SHQP; Report of the Jefferson Physical Laboratory 1930-31.}

1930-31 HANS A. KRAMERS was a guest lecturer at Purdue University and at the University of California at Berkeley during the academic year. He then served as one of the lecturers at the University of Michigan Summer Symposium.

{Birge, Vol. III, p. X7; Goudsmit, "The Michigan Summer Symposium;" SHQP.}

1931 R.H. FOWLER was Visiting Professor during the Spring Semester at the University of Wisconsin and also lectured at the University of Minnesota during May.

{Van Vleck, "First Decade;" Erikson, "History of the Minnesota Physics Department."}

1931 ARTHUR HAAS lectured at the University of Minnesota during April.

{Erikson, "History of the Minnesota Physics Department."}

1931 WOLFGANG PAULI and ARNOLD SOMMERFELD lectured during the Summer Symposium at the University of Michigan.

{Goudsmit, "The Michigan Summer Symposium."}

1931-32 PAUL SCHERRER was Visiting Lecturer at the Massachusetts Institute of Technology.

{MIT Presidential Report 1931-32; G.R. Harrison, "Karl Compton and American Physics" Phys. Today (Nov. 1957) 19-22.}

1932 WERNER HEISENBERG lectured during the Summer Symposium at the University of Michigan.

{Goudsmit, "The Michigan Summer Symposium."}

1933 NIELS BOHR lectured at the California Institute of Technology and at the annual Summer Symposium at the University of Michigan. He also participated in the APS meeting held in connection with the World's Fair in Chicago.

{SHQP; Goudsmit, "The Michigan Summer Symposium;" Phys. Rev. 44 (1933) 313.}

1933 F.W. ASTON, J. BJERKNES, J.D. COCKCROFT, L. FEJER, E. FERMI and T. LEVI-CIVITA took part in the APS meeting held in connection with the World's Fair in Chicago, June 19-24.

{Phys. Rev. 44 (1933) 313-315.}

1933 ENRICO FERMI was one of the guest lecturers at the University of Michigan Summer Symposium.

{Goudsmit, "The Michigan Summer Symposium."}

1933-34 JAMES FRANCK was Speyer Professor at the Johns Hopkins University and visited at the Massachusetts Institute of Technology.

{SHQP, p. 149.}

1934 GEORGE GAMOW lectured at the University of Michigan during the annual Summer Symposium in Theoretical Physics.

{Goudsmit, "The Michigan Summer Symposium."}

1934-35 P.A.M. DIRAC was a member of the Insitute for Advanced Study at Princeton.

{SHQP.}

1935 ENRICO FERMI lectured at the University of Michigan Summer Symposium.

{Goudsmit, "The Michigan Summer Symposium."}

ADDENDUM

1924 The British Association for the Advancement of Science held its 92nd annual meeting in Toronto during August. Among the British physicists in attendance were: W.H. and W.L. Bragg, A.S. Eddington, C.V. Raman and E. Rutherford. There were many American physicists who participated in the program which included a debate between A.H. Compton and Wm. Duane on the validity of the "Compton Effect."

{Report of the 92nd Meeting of the BAAS; Compton "The Scattering of X-Rays as Particles" Amer. Jour. Phys. 29 (1961) 817-820.}

APPENDIX III

Some European Scientific Emigres to the United States Prior to 1936

Year	Name	American Institution Joined
1908	Jakob Kunz	Univ. of Michigan (1915, Univ. of Illinois)
1913	W.F.G. Swann	Carnegie Institution
1920	L. Silberstein	Eastman Kodak
1921	P. Epstein	California Institute of Technology
1923	A.L. Hughes	Washington University, St. Louis
1925	H. Mueller	Massachusetts Institute of Technology
	F. Zwicky	California Institute of Technology
1926	K.F. Herzfeld	Johns Hopkins University
1927	S.A. Goudsmit	University of Michigan
	G.E. Uhlenbeck	University of Michigan
1928	K. Lark-Horovitz	Purdue University
1929	L.H. Thomas	Ohio State University
1930	G.H. Dieke	Johns Hopkins University
	Maria Goeppert-Mayer	Johns Hopkins University
	J. von Neumann	Princeton University
	O. Oldenberg	Harvard University
	E.P. Wigner	Princeton University
1931	R. Ladenburg	Princeton University
	C. Lanczos	Purdue University
	A. Landé	Ohio State University
1933	A. Einstein	Institute for Advanced Study
	O. Stern	Carnegie Institute of Technology
	H. Weyl	Institute for Advanced Study
1934	F. Bloch	Stanford University
	G. Gamow	George Washington University
1935	H.A. Bethe	Cornell University
	J. Franck	Johns Hopkins University
	L.W. Nordheim	Purdue University

Sources: Sources for History of Quantum Physics, for all except Gamow, Kunz, Lark-Horovitz, Mueller, Swann and Zwicky. In these instances American Men of Science was consulted; 5th edition for Kunz, 7th for Gamow, Lark-Horovitz, Meuller, Swann and Zwicky.

Note: AHQP erroneously placed Gamow at Washington University, St. Louis.

APPENDIX IV

The American Physical Society 1899-1935

Part I: Letter of Call to the Initial Meeting

Dear Sir:

The American Mathematical Society has been in existence for over ten years, and has been markedly successful in bringing together mathematicians and influencing the growth of mathematics in this country. Many other professional societies might be mentioned with similar aims. It seems to us that the time has now arrived for the organization of physicists into a national society, which shall be for this country what the Physical Society is for England, and the Deutsche Physikalische Gesellschaft for Germany. We propose a society meeting, four or more times yearly, in New York for the reading and discussion of papers. Of the advantages of such a society there is small need to speak. Few things inure more to the advantage of scientists than frequent meetings for the purpose of interchange of ideas and learning of one another's work. Such meetings have so far been too infrequent. Persons who have attended meetings of the American Association for the Advancement of Science must have wished that such opportunities for social and scientific intercourse were more frequent, as well as possible at other times of the year than in the summer. An organization like the one proposed could not fail to have an important influence in all matters affecting the interests of physicists, whether in connection with work done under Government auspices or otherwise.

The proposed society will conflict with the interests of no other organization, and will represent no institution or clique, but will be devoted to the advancement of our science. We invite you to attend a meeting to be held at Columbia University, New York City, on Saturday, May 13, at ten A.M., in Fayerweather Hall, for the purpose of discussing this matter, and if possible, organizing a Physical Society. If you are unable to attend, will you not communicate with one of the undersigned, making any suggestions that you may think important, and signifying your desire to join and support such a society.

A. G. WEBSTER, Worcester, Mass.
J. S. AMES, Baltimore, Md.
E. L. NICHOLS, Ithaca, N.Y.
C. BARUS, Providence, R.I.
M. I. PUPIN, New York, N.Y.
B. O. PEIRCE, Cambridge, Mass.
W. F. MAGIE, Princeton, N.J.

May 1, 1899

[Note: an 1899 copy of this letter is among the G. Stanley Hall Papers in the Archives of Clark University, Worcester, Massachusetts.]

APPENDIX IV

Part 2: Names and Institutional Affiliations of Physicists who attened the organizational meeting of the American Physical Society, May 20, 1899*

Abbe, Cleveland, U. S. Weather Bureau
Ames, Joseph S., Johns Hopkins University
Andrews, William C., Stanley Instrument Company
Bancroft, Wilder D., Cornell University
Barus, Carl, Brown University
Beach, Frederick E., Yale University
Bedell, Frederick, Cornell University
Bumstead, Henry A., Yale University
Carhart, Henry S., University of Michigan
Cooley, LeRoy C., Vassar College
Crawford, Morris Barker, Wesleyan University
Davis, Bergen, Columbia University
Day, William Schofield, Columbia University
Edmondson, Thomas Williams, New York University
Hallock, William, Columbia University
Hastings, Charles S., Yale University
Hering, Daniel, W., New York University
Hoadley, George A., Swarthmore College
Kieth, Marcia A., Lake Erie College
Kimball, Arthur L., Amherst College
Mackenzie, A. Stanely, Dalhousie University
Magie, William Francis, Princeton University
McClenahan, Howard, Princeton University
Merritt, Ernest, Cornell University
Michelson. A. A., University of Chicago
Nichols, E. L., Cornell University
Parker, Herschel C., Columbia University
Peirce, B. Osgood, Harvard University
Pupin, M. I., Columbia University
Rowland, Henry A., Johns Hopkins University
Shearer, John Sanford, Cornell University
Slocum, Allison W., University of Vermont
Stone, Isabelle, Vassar College
Thomson, Elihu, General Electric Company
Trowbridge, Charles C., Columbia University
Tufts, Frank L., Columbia University
Waterman, Frank A., Smith College
Webster, Arthur Gordon, Clark University

From E. Merritt "Early Days of the Physical Society" Rev. Sci. Inst. 5 (1934) 143-148.

*One week later than date originally planned in Letter of Call.

APPENDIX IV

Part 3: American Physical Society Membership 1900 - 1935

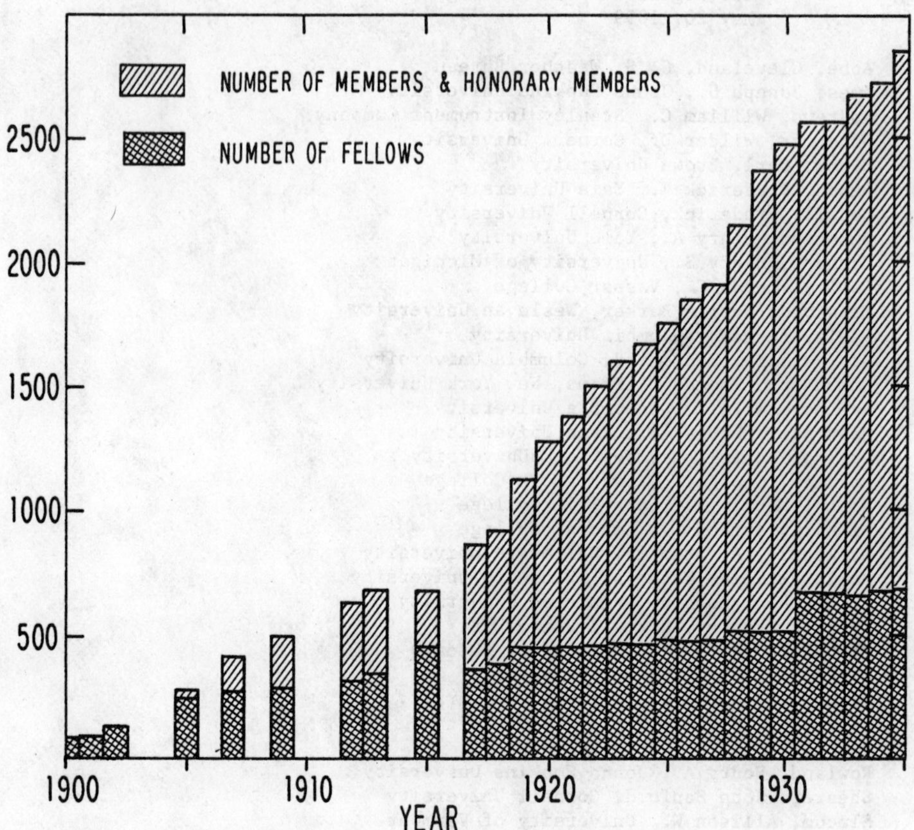

Note: Honorary Membership is reserved for distinguished physicists who are not residents of the United States, Canada, or Mexico.

Source: Bull. Amer. Phys. Soc. 13 (1968) 1050, 1061.

APPENDIX IV

Part 4: American Physical Society Presidents 1899-1935, their terms of office and titles of their Presidential Addresses.

Note: The journal reference given with each entry provided the information presented. Those references which are * contain the texts of the Addresses.

1899-1900	H.A. Rowland	The Highest Aim of the Physicist * Science 10 (1899) 825-833
1901-1902	A.A. Michelson	(none delivered according to A.G. Webster)
1903-1904	A.G. Webster	Some Practical Aspects of the Relation Between Mathematics and Physics * Phys. Rev. 18 (1904) 297-318
1905-1906	Carl Barus	Condensation Nuclei * Phys. Rev. 22 (1906) 82-110
1907-1908	E.L. Nichols	Theories of the Color of the Sky * Phys. Rev. 26 (1908) 497-511
1909-1910	Henry Crew	The Debt of Physics to Metaphysics * Phys. Rev. 31 (1910) 79-92
1911-1912	W.F. Magie	The Primary Concepts of Physics * Phys. Rev. 34 (1912) 125-138
1913	B.O. Pierce	(died in office)
1914-1915	Ernest Merritt	Luminescence * Phys. Rev. 5 n.s. (1915) 319-334
1916-1917	R.A. Millikan	Atomic Structure and Radiation * Phys. Rev. 10 (1917) 194-205
1918	H.A. Bumstead	(resigned due to overseas assignment)
1919-1920	J.S. Ames	Einstein's Law of Gravitation * Phys. Rev. 15 (1920) 206-216
1921-1922	Theodore Lyman	The Spectroscopy of the Extreme Ultra-violet * Science 55 (1922) 161-166
1923-1924	C.E. Mendenhall	Some Recent Developments in the Study of the Solid State * Science 59 (1924) 219-230

1925-1926	D.C. Miller	The Michelson-Morley Etherdrift Experiment - Its History and Significance * Science 63 (1926) 433-443
1927-1928	K.T. Compton	Recent Studies of the Electrical Discharges in Gases Phys. Rev. 31 (1928) 300
1929-1930	H.G. Gale	The Interplay between Theory and Experiment in Modern Spectroscopy Phys. Rev. 35 (1930) 289
1931-1932	W.F.G. Swann	Reality in Physics Phys. Rev. 39 (1932) 855
1933	P.D. Foote	Industrial Physics * Rev. Sci. Inst. 5 (1934) 57-66
1934	A.H. Compton	(none found)
1935	R.W. Wood	Optical and Physical Effects of High Explosives Phys.Rev.49(1936)408

APPENDIX IV

Part 5: American Physical Society Symposia Topics and Panelists

Comments: No attempt has been made to insure that we have included <u>every</u> APS symposium held prior to 1936. However, those listed are sufficient to indicate the extent of this activity, the topics that were considered noteworthy, and the identities of the members of the American physics profession who were considered to be knowledgeable about them.
The symposia listed as having been held in December were part of the annual APS - AAAS Section B meeting. Occasionally, there was also participation in these by the American Chemical and/or Mathematical Societies.
The * journal references contain published reports of the texts of the presentations.

Date / Location	Topic / Participants	Reference
1911 December, Washington, D.C.	Ether Theories A. Michelson, E. Morley, A.G. Webster, W.S. Franklin, D.F. Comstock, G.N. Lewis	<u>Phys. Rev</u>. <u>34</u> (1912) 63
1913 November, Chicago	Quantum Theory C.E. Mendenhall, R.A. Millikan, Max Mason, J. Kunz, A.C. Lunn	<u>Phys. Rev</u>. <u>3</u> (1914) 57
1914 November, Chicago	Spectroscopic Evidence Regarding Atomic Structure H.B. Lemon, H. Gale, G. Fulcher, G.W. Stewart, K.K. Darrow	<u>Phys. Rev</u>. <u>5</u> (1915) 72
1914 December, Philadelphia	The Use of Dimensional Equations E. Buckingham, A.C. Lunn, A.G. Webster, W.S. Franklin	<u>Phys. Rev</u>. <u>5</u> (1915) 176
1916 December, New York	The Structure of Matter (with Amer. Chem. Soc.) R.A. Millikan, G.N. Lewis, R.W. Wood, W.J. Humphreys, A.P. Wills, W.D. Harkins, I. Langmuir, L.W. Jones	*See Note 1.144
1917 December, Pittsburgh	The Relationship of Physics to the War A.L. Day, C.E. Mendenhall, G.P. Thomson (Eng.), G. Abeti (Italy)	<u>Phys. Rev</u>. <u>13</u> (1919) 136

1919	October Philadelphia	The Present State of Theories of Atomic Structure Saul Dushman, I. Langmuir, F.A. Saunders, P.W. Bridgman Phys. Rev. 14 (1919) 530
1919	December St. Louis	Phenomena in the Ultra-violet Spectrum, Including X-Rays D.L. Webster, R.A. Millikan, Wm. Duane, A.W. Hull *Science 51 (1920) 504-408
1920	December Chicago	Recent Progress in Magnetism S.J. Barnett, S.R. Williams, A.H. Compton Phys. Rev. 17 (1921) 368
1921	December Toronto	The Quantum Theory Saul Dushman (APS), H.B. Phillips (AMS), R.C. Tolman (ACS) *J.O.S.A. 6 (1922) 211-253
1922	December Cambridge	Ionization Potentials and Atomic Radiation K.T. Compton, P.D. Foote, H.N. Russell Phys. Rev. 21 (1923) 367
1923	December Cincinnati	The Reflection and Scattering of X-Rays A.H. Compton, Bergen Davis, Wm. Duane Phys. Rev. 23 (1924) 287
1925	December Kansas City	Relativity H.G. Gale, C.E. St. John, A.C. Lunn Phys. Rev. 27 (1926) 240
1928	November Minneapolis	Electron Impact Phenomena and Quantum Theory H. Weyl, K.T. Compton, E.U. Condon, J.R. Oppenheimer, F.L. Mohler, I. Langmuir, L. Tonks Phys. Rev. 33 (1929) 111
1928	December New York	Quantum Mechanics (with Amer. Math. Soc.) J.C. Slater, J.H. Van Vleck, H. Weyl, N. Wiener Phys. Rev. 33 (1929) 275 (See also p. 3.64 and Note 3.161)
1929	April Washington	Dielectrics J.B. Whitehead, F. Zwicky, L. Tonks, H.L. Curtis, K.F. Herzfeld, V. Karapetoff, G. Mackay, C.P. Smyth Phys. Rev. 33 (1929) 1067
1930	February New York	Photoelectric and Thermionic Phenomena F.L. Mohler, F.K. Richtmyer, W.V. Houston, C.E. Mendenhall, G. Wentzel (Zurich) Phys. Rev. 35 (1930) 656

1930 December Cleveland	Acoustics P.E. Sabine, H. Fletcher, C.W. Hewlett <u>Phys. Rev</u>. <u>37</u> (1931) 453
1931 June Pasadena	The Physics of Crystals F. Zwicky, P.W. Bridgman, C.J. Davisson, A. Goetz Present Status of the Problem of Nuclear Structure R.H. Fowler (Eng.), W. Pauli (Zurich), S.A. Goudsmit, W.M. Latimer, W.F.G. Swann Production of High Energy Electrical Particles W.D. Coolidge, M.A. Tuve, E.O. Lawrence, C.C. Lauritsen, A. Brasch (Germany), F. Lange (Germany) <u>Phys. Rev</u>. <u>38</u> (1931) 579-80
1931 September Schenectady	Ferromagnetism F. Bitter, S.L. Quimby, R.M. Bozorth *<u>Phys. Rev</u>. <u>39</u> (1932) 337-378
1931 December New Orleans	Mathematical Symposium with Amer. Math. Soc. G.D. Birkhoff, W.F.G. Swann, G.A. Bliss, P.W. Bridgman <u>Phys. Rev</u>. <u>39</u> (1932) 854
1932 February Cambridge	Properties of Matter P.Debye (Zurich), B.E. Warren, C.P. Smyth, C.J. Davisson, H.H. Race. F. Bitter <u>Phys. Rev</u>. <u>40</u> (1932) 118
1932 December Atlantic City	Cosmic Rays A.H. Compton, G.L. Locher, R.A. Millikan, H.V. Neher <u>Phys. Rev</u>. <u>43</u> (1933) 370
1933 February New York	Electron Optics C.J. Davisson, H.E. Farnsworth, W.F.G. Swann, W.E. Danforth, V.K. Zworykin <u>Phys. Rev</u>. <u>43</u> (1933) 773
1933 June Chicago	Application of Quantum Mechanics in Chemistry J.C. Slater, L. Pauling, H. Eyring, R.S. Mulliken Mathematical Aspects of Quantum Mechanics T. Levi-Civita (Italy), G.D. Birkhoff Isotopes W.D. Harkins, F.W. Aston (Eng.), K.T. Bainbridge, E.W. Washburn, H.C. Urey, F. Allison Nuclear Disintegrations E.O. Lawrence, J.D. Cockcroft (Eng.), M.A. Tuve, W.D. Harkins, N. Bohr (Denmark) Additional topics and speakers in <u>Phys. Rev</u>. <u>44</u> (1933) 313-5.

1933 December Boston — Nuclear Physics
G.E. Uhlenbeck, C.D. Anderson, J.R. Oppenheimer — Phys. Rev. 45 (1934) 284

1934 June Berkeley — Spectroscopy in Astrophysics
R.W. Wood, P.W. Merrill, J. Kaplan, R.J. Trumpler, I.S. Bowen
Nuclear Structure
R.J. Van de Graaf, M.A. Tuve, E.O. Lawrence, C.C. Lauritsen, C.D. Anderson, T.H. Johnson, F. Bloch
Fundamental Constants
R.A. Millikan, P.A. Ross, P. Kirkpatrick, J. DuMond, W.V. Houston, C.D. Shane, R.T. Birge — Phys. Rev. 46 (1934) 321,2

1934 June Ann Arbor — Nuclear Moment
G. Breit, R.S. Mulliken, R. Bacher, I.I. Rabi, A. Ellett — Phys. Rev. 46 (1934) 333

1934 December Pittsburgh — Group Theory and Quantum Mechanics
E. Wigner, J.H. Van Vleck, G. Breit
Heavy Hydrogen and its Compounds
R.C. Gibbs, G.H. Dieke, O. Stern, M.A. Tuve, H.L. Johnston, H.S. Taylor, J.R. Bates, F.G. Brickwedde — Phys. Rev. 47 (1935) 323

1935 February New York — Atmospheric Optics
B. O'Brien, E.W. Woolard, J.A. Anderson, R. Ladenburg, O.R. Wulf — Phys. Rev. 47 (1935) 637

1935 June Minneapolis — Nuclear Physics
L.R. Hafstad, G. Breit, C.C. Lauritsen, H. Bethe — Phys. Rev. 48 (1935) 472

1935 December St. Louis — Photoelectricity
A.A. Knowlton, A.L. Hughes, L.O. Grondahl — Phys. Rev. 49 (1936) 407

Addendum

Although it was not designated a "Symposium", it is noteworthy that at the APS - AAAS Section B joint meeting in Philadelphia in December 1926 W. F. G. Swann delivered an address entitled "The New Quantum Dynamics" with additional discussion of the topic provided by Gregory Breit and John H. Van Vleck. Phys. Rev. 29 (1927) 350

Selected Bibliographic and Other Sources

Comment: This list of Sources is organized according to the outline presented below. For this kind of study the divisions "materials Produced Prior to 1936" and "Materials Produced Since 1935" are more meaningful than the usual "Primary" and Secondary Sources", since emphasis is then placed on the materials contemporary to the period of focus with those of a retrospective nature.

I A 1 a: Books, General - Histories, Biographies, Commentaries

Boltzmann, Ludwig. _Populäre Schriften_ . Leipzig: Barth, 1905

Flexner, Abraham. _Universities: American, English, German_. New York: Oxford University Press, 1930

Fosdick, R.B. _The Old Savage in the New Civilization_. Garden City, N.Y.: Doubleday, Doran and Co.,Inc.,1931

A History of the Cavendish Laboratory 1871 - 1910. London: Longmans, Green and Co. Ltd., 1910

Keppel, Frederick P. ,ed. _A History of Columbia University 1754-1904_ New York: Columbia University Press, The Macmillan Co. Agents,1904.

Merz, John T. _A History of European Scientific Thought in the Nineteenth Century_. London: Wm.Blackwood and Sons, 1904 - 1912. Reprinted, New York: Dover Publications Inc., 1965

Morison, S.E., ed. _Development of Harvard University 1869 - 1929._ Cambridge: Harvard University Press, 1930

Palmer, A.M. and Garvan, F.P. _Aims and Purposes of The Chemical Foundation and the Reasons for its Organization._ New York: 1919

Poincare, Henri. _Dernieres Pensees_. Paris: Flammarion,1913. English Translation: _Mathematics and Science: Last Essays._ John W. Bolduc, translator. New York: Dover Publications Inc., 1963

Pupin, Michael. _From Immigrant to Inventor_. New York: Charles Scribner's Sons, 1923

Slosson, Edwin E. _Creative Chemistry: Descriptive of Recent Achievements in the Chemical Industries_. Garden City, N.Y.: Garden City Publishing Co., Inc. 1919; 2nd ed. revised by Harrison E. Howe. New York: The Century Co., 1930

Strutt. Robert J. _John William Strutt, Third Baron Rayleigh._ London: Edward Arnold and Co.Ltd., 1924

Thompson, Silvanus P. _The Life of William Thomson, Baron Kelvin of Largs_ London: Macmillan and Co.Ltd., 1910

Woodruff, L.L., ed. _The Development of the Sciences_. New Haven: Yale University Press, 1923

Yerkes, Robert M., ed. _The New World of Science: Its Development during the War_. New York: The Century Co., 1920

I A 1 b: Books, Technical - Monographs, Textbooks, Lecture Series

Bacher, R.F. and Goudsmit, S.A., compilers. Atomic Energy States as derived from the Analyses of Optical Spectra. New York: McGraw-Hill Book Co., 1932.

Birtwistle, George. The New Quantum Mechanics. Cambridge, Eng.: University Press, 1928.

Bohr, Niels. On the Quantum Theory of Line Spectra. Kobenhavn: A.F. Host and Son, 1918 - 1922.

________. A.D.Udden, trans. The Theory of Spectra and Atomic Constitution. Cambridge, Eng.: University Press, 1922 and 1924.

Born, Max. Atomtheorie des festen Zustandes. Leipzig: B.Teubner, 1923

________. Problems in Atomic Dynamics. Cambridge, Mass.: The MIT Press, 1926; German edition: Probleme der Atomdynamik. Berlin: Julius Springer, 1926.

________. Vorlesungen über Atommechanik. Berlin: Julis Springer, 1925

Bridgman, P.W. The Logic of Modern Physics. New York: The Macmillan Co., 1927.

Condon, E.U. and Morse, P.M. Quantum Mechanics. New York: McGraw-Hill Book Co., 1929

Condon, E.U. and Shortley, George. The Theory of Atomic Spectra. Cambridge, Eng.: University Press, 1935

Darrow, K.K. Elementare Einführung in die Wellenmechanik. Leipzig: Verlag von S. Hirzel, 1929 and 1932.

________. Elementare Einführung in die Physikalische Statistik. Leipzig: Verlag von S. Hirzel, 1931.

________. Elementare Einführung in die Quantenmechanik. Leipzig: Verlag von S. Hirzel, 1933.

________. Introduction to Contemporary Physics. New York: D. Van Nostrand, 1926 and 1939

Einstein, Albert. E.P.Adams, trans. The Meaning of Relativity: Four Lectures delivered at Princeton University, May 1921. Princeton: Princeton University Press, 1923.

Foote, P.D. and Mohler, F.L. The Origin of Spectra. New York: The Chemical Catalog Co., Inc., 1922.

Fowler, R.H. Statistical Mechanics: The Theory of the Properties of Matter in Equilibrium. Cambridge, Eng.: University Press, 1929.

Frenkel, J. Wave Mechanics: Elementary Theory. Oxford: The Clarendon Press, 1932.

______ Wave Mechanics: Advanced General Theory. Oxford: The Clarendon Press, 1934.

Gibbs, J.W. Elementary Principles in Statistical Mechanics. New York: Charles Scribner's Sons, 1902.

______ Vector Analysis. E.B.Wilson, ed. New York: Charles Scribner's Sons, 1901

Haas, Arthur. L.W.Codd, trans. Quantum Chemistry New York: R.R.Smith, Inc.,1930.

______ L.W.Codd, trans. Wave Mechanics and the New Quantum Theory London: Constable and Co.Ltd., 1928.

Heisenberg,W. C.Eckart and F.Hoyt, trans. The Physical Principles of the Quantum Theory. Chicago: University of Chicago Press, 1930. reprinted, New York: Dover Publications Inc., 19?.

Jeans, J.H. An Elementary Treatise on Theoretical Mechanics. Boston: Ginn and Co.,1906

______ The Mathematical Theory of Electricity and Magnetism. Cambridge, Eng.: University Press, 1907.

______ Report on Radiation and the Quantum Theory. London: "The Electricians" Printing and Publishing Co. Ltd., 1914. 2nd ed. Fleetway Press, Ltd., 1924

Kelvin, Lord. Baltimore Lectures on Molecular Dynamics and the Wave Theory of Light. London:C.J.Clay and Sons,Ltd.,1904

Krammers, H.A. and Holst, H. R.B. and Rachel Lindsay, trans. The Atom and the Bohr Theory of its Structure. London: Glyndendal, 1923.

Kronig, Ralph. The Optical Basis of Chemical Valence. Cambridge,Eng.: University Press, 1935.

Lande, Alfred. Die Neure Entwicklung der Quantentheorie. Dresden: Verlag von T. Steinkopff, 1926.

Vorlesungen über Wellenmechanik. Leipzig: Akademische Verlagsgesellschaft, 1930.

Lenzen, V.F. The Nature of Physical Theory: A Study in the Theory of Knowledge. New York: John Wiley and Sons, Inc., 1931.

Lewis, G.N. Valence and the Structure of Atoms and Molecules. New York: The Chemical Catalog Co., Inc., 1923. Reprinted, Dover Publications, Inc., 1966.

Lindsay, R.B. and Margenau, H. Foundations of Physics. New York: John Wiley and Sons, Inc., 1936.

Lorentz, H.A. The Theory of Electrons and its Applications to the Phenomena of Light and Radiant Heat. Leipzig: B.G.Teubner, 1909.

Millikan, R.A. The Electron. Chicago: University of Chicago Press, 1968. Facsimile of the original 1917 edition.

National Research Council Bulletins

No. 5 Quantum Theory E.P.Adams.
6 Data Relating to X-Ray Spectra. Wm.Duane.
7 Intensity of Emissin of X-Rays and the Reflection from Crystals. B.Davis.
14 Atomic Structure. L. Page and D.L.Webster.
18 Theories of Magnetism. A.P.Wills et al.
20 Secondary Radiation Produced by X-Rays. A.H.Compton.
48 Critical Potentials. K.T.Compton and F.L.Mohler.
54 Quantum Principles and Line Spectra. J H.Van Vleck.
57 Molecular Spectra in Gases. E.C.Kemble et al.

Page, Leigh. Introduction to Theoretical Physics. New York: D. Van Nostrand Co., Inc., 1928, 1935.

Pauling, L. and Goudsmit, S.A. The Structure of Line Spectra. New York: McGraw-Hill Book Co., Inc., 1930.

Pauling, L. and Wilson, E.B. Introduction to Quantum Mechanics with Applications to Chemistry. New York: McGraw-Hill Book Co., Inc., 1935.

Peirce, B.O. Elements of the Theory of the Newtonian Potential Function. Boston: Ginn and Co., 1886.

__________. Short Table of Integrals. Boston: Ginn and Co., 1899.

Planck, Max. Acht Vorlesungen über Theoretische Physik. Leipzig: Verlag von S. Hirzel, 1910. English Translation by A.P.Wills. New York: Columbia University Press, 1915.

__________. The Theory of Heat Radiation. Morton Masius, trans. Philadelphia: P.Blakiston's Son and Co., 1914. Reprinted by Dover Publications, Inc., 1959.

Planck, Max. Vorlesungen über die Theorie der Wärmestrahlung. Leipzig: Barth, 1906. 2nd. ed. 1913.

Richtmyer, F.K. Introduction to Modern Physics. New York: McGraw-Hill Book Co., 1928. 2nd ed. 1934.

Ruark, A.E. and Urey, H.C. Atoms, Molecules and Quanta. New York: McGraw Hill Book Co., 1930.

Slater, J.C. and Frank, N.H. Introduction to Theoretical Physics. New York: McGraw-Hill Book Co., 1933.

Sommerfeld, Arnold. Atombau und Spektrallinien. Braunschweig: Friedr. Vieweg und Sohn, 1919, 1920, 1922 and 1924. Wellenmechanik Ergänzungsband. 1929, translated by H.L.Brose as Wave Mechanics. New York: E.P.Dutton and Co., Inc., 1930.

Stoner, E.C. Magnetism and Atomic Structure. London: Methuen and Co.Ltd. 1926.

Thomson, J.J. The Discharge of Electricity through Gases. New York: Charles Scribner's Sons, Inc., 1898.

Tolman, R.C. Relativity, Thermodynamics and Cosmology. Oxford: The Clarendon Press, 1934.

__________. The Theory of the Relativity of Motion. Berkeley:University of California Press, 1917.

Tyndall, John. Six Lectures on Light delivered in America 1872 - 1873. New York: D.Appleton and Co., 1873.

Van Vleck, J.H. The Theory of Electric and Magnetic Susceptibilities. Oxford: The Clarendon Press, 1932.

Webster, A.G. Dynamics of Particles and of Rigid, Elastic and Fluid Bodies. New York:G.E.Stechert, 1904, 1912, and 1922. Leipzig: B.G.Teubner, 1925.

__________. Partial Differential Equations of Mathematical Physics. S.J.Plimpton,ed. New York:G.E.Stechert, 1927, 1933 and 1950. Reprinted by Dover Publications, Inc. 1955.

__________. The Theory of Electricity and Magnetism. London: Macmillan and Co.,Ltd., 1897.

Weyl, Hermann. Gruppentheorie und Quantenmechanik. Leipzig: Verlag von s!Hirzel, 1928.

Weyl, Hermann. H.P.Robertson,trans. The Theory of Groups and Quantum Mechanics. London: Methuen and Co.,Ltd.,1931.

Wien, W. Neure Probleme der Theoretischen Physik. Leipzig: B.G.Teubner, 1913.

I A 1 c: Collected Works of Individual Scientists and Festschriften.

The Scientific Papers of J. Willard Gibbs. H.A.Bumstead and R.G. Van Name. eds. London: Longmans, Green and Co.Ltd.,1906.

Recueil de Travaux offer par les auteus a H. A. Lorentz. La Haye: M. Nijhoff, 1900.

Benjamin Osgood Peirce: Mathematical and Physical Papers, 1903 - 1913. Cambridge, Mass.: Harvard University Press, 1926.

The Physical Papers of Henry Augustus Rowland. Baltimore: The Johns Hopkins University Press, 1902.

I A 1 d: Conference Proceedings.

Reports of the Meetings of the British Association for the Advancement of Science. London: John Murray
67th Meeting, Toronto, Canada, August 1897.
54th Meeting, Montreal, Canada, August 1884.
79th Meeting, Winnepeg, Canada, August 1909.
82nd Meeting, Dundee, Scotland, September, 1912.
83rd Meeting, Birmingham, England, September, 1913
92nd Meeting, Toronto, Canada, August 1924.

Rapports Presentes au Congres International de Physique, 3 Tomes. Paris: Gauthier-Villars, 1900.

Atti del Congresso Internazionale dei Fisici, 11-20 Settembre 1927, Como, Pavia, Roma: Onoranze ad Alessandros Volta nel primo centenario della morte. Vol. I, II. Bologna: Nicola Zanichelli,1928.

Solvay Congress Reports. Paris: Gauthier-Villars
1911 La Theorie du Rayonnement et les Quanta
1913 La Structure de la Matiere
1921 Atomes et Electrons
1924 Conductibilite Electrique
1927 Electrons et Photons
1930 Le Magnetisme
1933 Structure et Proprietes des Noyaux Atomiques

I A 1 e: Reference Works - Foundation Reports and Doctorate Listings

American Scandinavian Foundation Annual Reports, 1912 - 1936.

The General Education Board Annual Reports, 1925 - 1930.

The Rockefeller Foundation Annual Reports, 1929 - 1935.

The International Education Board Annual Reports, 1923 - 1928.

National Research Council Reprint and Circular Series, Nos. 12, 26, 42, 75, 80, 86, 91, 95, 101, 104, 105. Doctorates Conferred in the Sciences in the years 1920 to 1933 by American Universities.

Library of Congress, A List of the American Doctoral Dissertations Printed in 1922, 1923, 1924, and 1925.

Gilchrist, Donald, ed., Doctoral Dissertations Accepted by American Universities, No.1(1933-34), No.2(1934-35) New York: H.W.Wilson Co.,1934 and 1935.

I A 2 a: Journal Articles, General - Addresses, Biographical Memoirs etc.

Ames, J.S. "Einstein's Law of Gravitation". Phys.Rev.15(1920)206-216.

Barton, H.A. "Support for Science" Rev.Sci.Inst.5(1934)339-340.

"Science in the Public Press" Ibid. 386.

Bridgman, P.W. "Permanent Elements in the Flux of Present-Day Physics" Science 71(1930)19-23.

"William Duane". N.A.S.Biog.Mem. Vol.XVIII, pp23 -41.

Brown, F.C. "The Predicament of Scholarship in America and One Solution" Science 39(1914)587-595.

Bumstead, H.A. "Physics" in The Development of the Sciences, L.L.Woodruff,ed. New Haven: Yale University Press, 1923.

Compton, K.T. "The American Institute of Physics" Rev.Sci.Inst.4(1933) 57-58.

Clarke, F.W. "American Colleges versus American Science" Pop.Sci.Mon. 9(1876)467-479.

Crew, Henry "The Debt of Physics to Metaphysics". Phys.Rev.31(1910) 79-92.

Diederich, Henry W. "American and German Universities". Science 20 (1904)157.

Foote, P.D. "Industrial Physics". Rev.Sci.Inst.5(1934)57-66.

Hall, E.H. "Physics" in Development of Harvard University 1869-1929. S.E.Morison, ed.

Humphreys, W.J. "The Relation of Magnetism to the Structure of the Atom" Science 46(1917)273-279.

Jones, L.W. "Electromerism". Ibid.493-502.

Lewis, G.N. "The Static Atom". Ibid. 297-302.

Lyman, Theodore "The Future of the Department of Physics". Harvard Alumni Bulletin (13 June1929)1056-1058.

Merritt, Ernest "Early Days of the Physical Society" Rev.Sci.Inst.5 (1934)143-148.

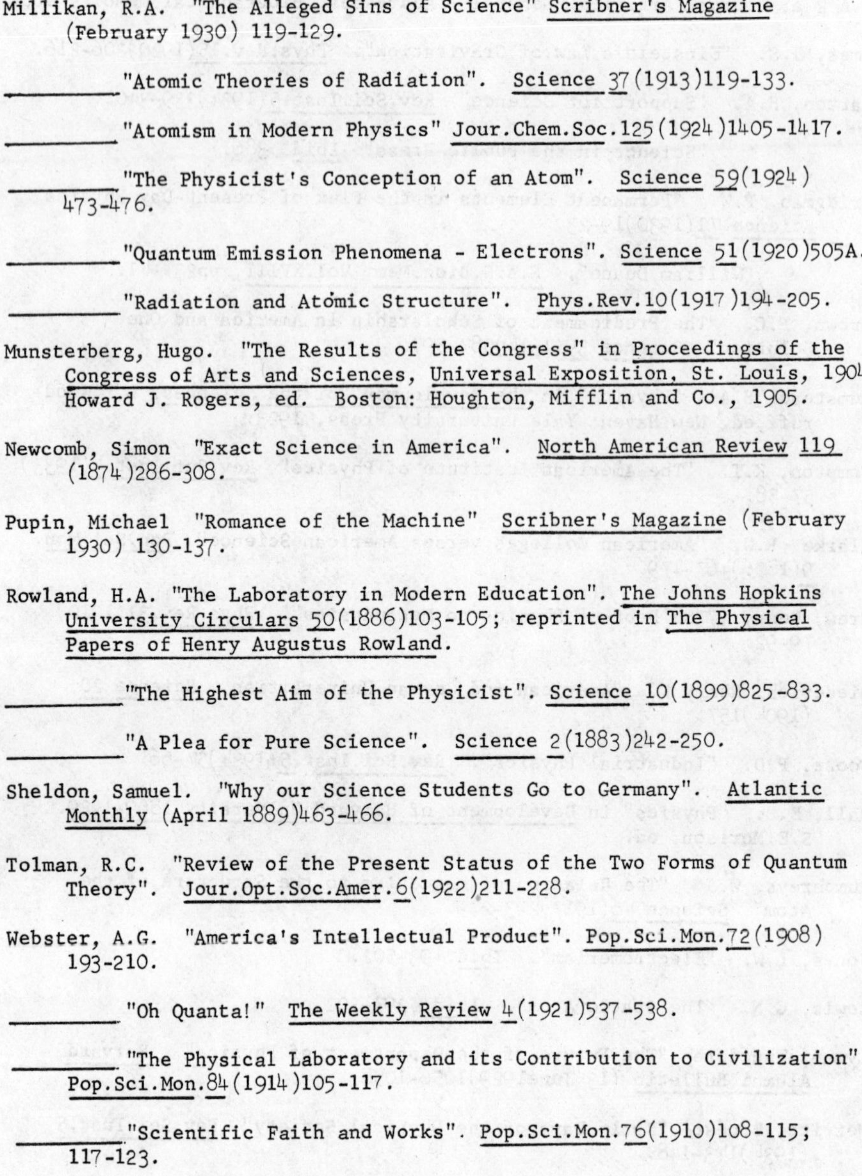

Millikan, R.A. "The Alleged Sins of Science" Scribner's Magazine (February 1930) 119-129.

________ "Atomic Theories of Radiation". Science 37(1913)119-133.

________ "Atomism in Modern Physics" Jour.Chem.Soc.125(1924)1405-1417.

________ "The Physicist's Conception of an Atom". Science 59(1924) 473-476.

________ "Quantum Emission Phenomena - Electrons". Science 51(1920)505A.

________ "Radiation and Atomic Structure". Phys.Rev.10(1917)194-205.

Munsterberg, Hugo. "The Results of the Congress" in Proceedings of the Congress of Arts and Sciences, Universal Exposition, St. Louis, 1904 Howard J. Rogers, ed. Boston: Houghton, Mifflin and Co., 1905.

Newcomb, Simon "Exact Science in America". North American Review 119 (1874)286-308.

Pupin, Michael "Romance of the Machine" Scribner's Magazine (February 1930) 130-137.

Rowland, H.A. "The Laboratory in Modern Education". The Johns Hopkins University Circulars 50(1886)103-105; reprinted in The Physical Papers of Henry Augustus Rowland.

________ "The Highest Aim of the Physicist". Science 10(1899)825-833.

________ "A Plea for Pure Science". Science 2(1883)242-250.

Sheldon, Samuel. "Why our Science Students Go to Germany". Atlantic Monthly (April 1889)463-466.

Tolman, R.C. "Review of the Present Status of the Two Forms of Quantum Theory". Jour.Opt.Soc.Amer.6(1922)211-228.

Webster, A.G. "America's Intellectual Product". Pop.Sci.Mon.72(1908) 193-210.

________ "Oh Quanta!" The Weekly Review 4(1921)537-538.

________ "The Physical Laboratory and its Contribution to Civilization" Pop.Sci.Mon.84(1914)105-117.

________ "Scientific Faith and Works". Pop.Sci.Mon.76(1910)108-115; 117-123.

Webster, A.G. "Some Practical Aspects of the Relations between Physics and Mathematics". Phys.Rev. 18(1904)297-318.

A.G.Webster - In Memoriam. Publication of the Clark University Library, Vol.7, March 1924.

Williams, S.R. "Centers for Research" Science 68(1928)61.

Wills, A.P. "The Relation of Magnetism to Molecular Structure". Science 46(1917)349-351.

I A 2 b: Journal Articles: Technical and Research

Adel, A. Phys.Rev.43(1933)716-723; 44(1933)99-114.

Atanasoff, J., Phys.Rev.36(1930)1232-1242.

Baker, E., Phys.Rev.36(1930)630-647.

Bacher, R.F., Phys.Rev.34(1929)1501-1506.

Barnett, S.J., Phys.Rev.10(1917)7-21.

Bartlett, J.H., Phys.Rev.33(1929)169-174.

Beardsley, N., Phys.Rev.39(1932)913-921.

Bichowsky, F.R. and Urey, H.C., Proc.N.A.S.12(1926)80-85.

Blackburn, C.M., Proc.N.A.S.11(1925)28-34

______ Astrophys.Jour.62(1925)61-64

Bohr,N., Zeit.f.Phys.9(1922)1-67.

______, Kramers,H.A., and Slater,J.C. Phil.Mag.47(1924)785-802.

Born,M. and Jordan,P. Zeit.f.Phys.34(1925)858-888.

Born,M. and Wiener,N., M.I.T.Jour.Math.Phys.2(1925-26)84-98; Zeit.f. Phys.36(1926)174-187.

Bourgin,D.G., Phys.Rev.29(1927)794-816.

Breit,G. Phys.Rev.17(1921)649-677.

______ Proc.N.A.S.9(1923)230-243.

______ and Rabi,I., Phys.Rev.38(1931)2082.

Brice,B., Phys.Rev.35(1930)960-972.

Brinsmade, J.B., Kemble,E.C., Proc.N.A.S.3(1917)420-425.

Brown, F.W., Phys.Rev.42(1932)914A.

Bubb,F.W., Phys.Rev.24(1924)177-189.

______ Phil.Mag.49(1925)824-838.

Buckingham, E. and Dellinger, J.H., Bull.Nat.Bur.Stds.7(1911)393-406.

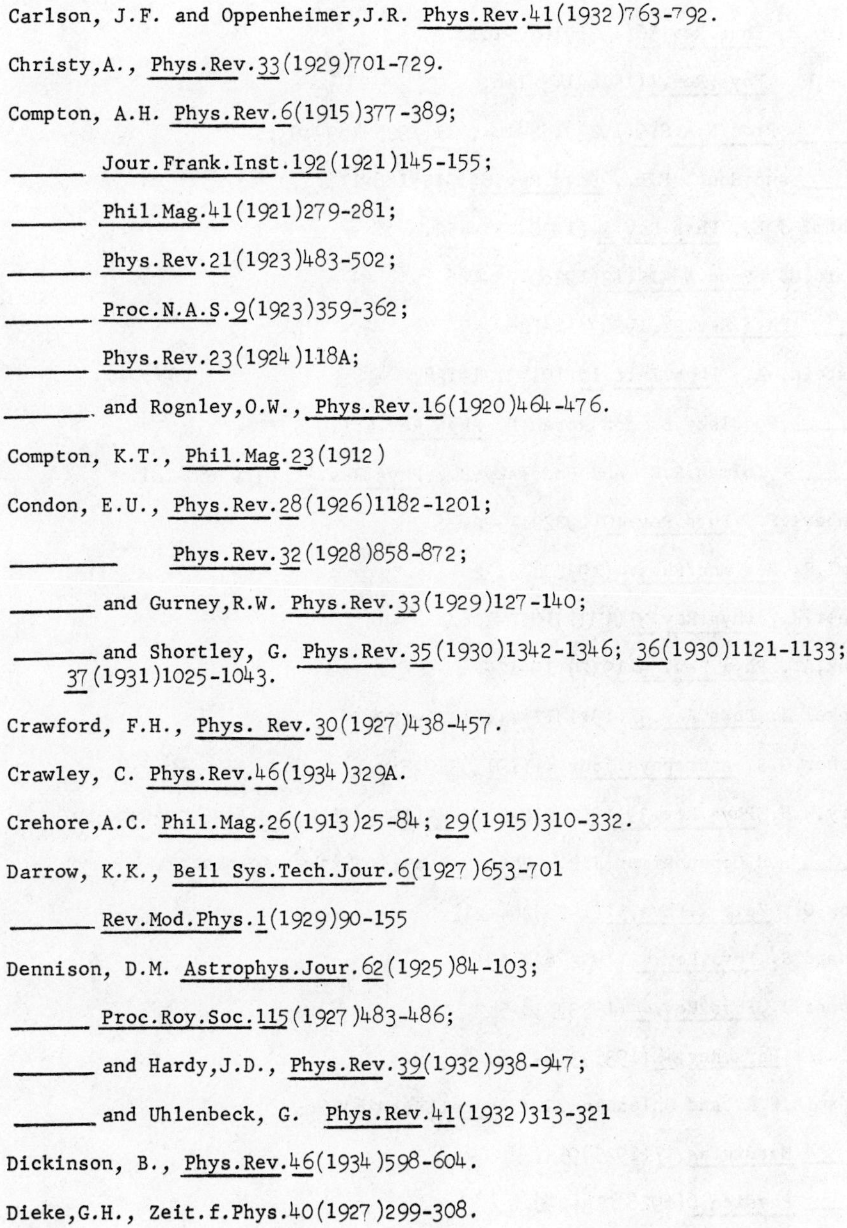

Carlson, J.F. and Oppenheimer, J.R. Phys.Rev.41(1932)763-792.

Christy, A., Phys.Rev.33(1929)701-729.

Compton, A.H. Phys.Rev.6(1915)377-389;

______ Jour.Frank.Inst.192(1921)145-155;

______ Phil.Mag.41(1921)279-281;

______ Phys.Rev.21(1923)483-502;

______ Proc.N.A.S.9(1923)359-362;

______ Phys.Rev.23(1924)118A;

______ and Rognley, O.W., Phys.Rev.16(1920)464-476.

Compton, K.T., Phil.Mag.23(1912)

Condon, E.U., Phys.Rev.28(1926)1182-1201;

______ Phys.Rev.32(1928)858-872;

______ and Gurney, R.W. Phys.Rev.33(1929)127-140;

______ and Shortley, G. Phys.Rev.35(1930)1342-1346; 36(1930)1121-1133; 37(1931)1025-1043.

Crawford, F.H., Phys. Rev.30(1927)438-457.

Crawley, C. Phys.Rev.46(1934)329A.

Crehore, A.C. Phil.Mag.26(1913)25-84; 29(1915)310-332.

Darrow, K.K., Bell Sys.Tech.Jour.6(1927)653-701

______ Rev.Mod.Phys.1(1929)90-155

Dennison, D.M. Astrophys.Jour.62(1925)84-103;

______ Proc.Roy.Soc.115(1927)483-486;

______ and Hardy, J.D., Phys.Rev.39(1932)938-947;

______ and Uhlenbeck, G. Phys.Rev.41(1932)313-321

Dickinson, B., Phys.Rev.46(1934)598-604.

Dieke, G.H., Zeit.f.Phys.40(1927)299-308.

Donley,H. Phys.Rev.50(1936)1012-1016.

Duane,W., Phys.Rev.7(1916)143-146;

________ Proc.N.A.S.9(1923)158-164; 11(1925)489-493;

________ and Hunt, F.L., Phys.Rev.6(1915)166-171.

Dunham, J.L., Phys.Rev.34(1930)438-452.

Eckart,C., Proc.N.A.S.12(1926)473-476;

________ Phys.Rev.28(1926)711-726.

Einstein, A., Phys.Zeit.18(1918)121-128;

________,Podolsky,B. and Rosen,N. Phys.Rev.47(1935)777-780;

________, Tolman,R.C. and Podolsky,B., Phys.Rev.37(1931)780-781.

Feenberg,E., Phys.Rev.40(1932)17-32.

Fermi,E.,Rev.Mod.Phys.4(1932)87-132.

Fisher,R., Phys.Rev.37(1931)1057-1068.

Frank,A., Phys.Rev.39(1932)119-129.

Frenkel,J.,Phys.Rev.37(1931)17-44; 1276-1294.

Fulcher,G.S. Astrophys.Jour.41(1915)359-372.

Furry,W.H. Phys.Rev.39(1932)210-225; 43(1933)316L; 49(1936)393-399;476.

________and Oppenheimer,J.R., Phys.Rev.45(1934)245-262;343-344;903-904.

Gamow,G., Zeit.f.Phys.51(1928)204-212.

Gerhard,S.,Phys.Rev.42(1932)622-631.

Gibbons,J.,Phys.Rev.44(1933)538-543.

Goble,A.,Phys.Rev.48(1935)346-356.

Goudsmit,S.A. and Uhlenbeck,G. Nature117(1926)264;

________ Naturwiss.47(1925)953;

________ Physica 6(1926)273-290.

Granath, L., Phys.Rev.42(1932)44-51.

Gray,N.,Phys.Rev.44(1933)570-574.

Hall,H.,Phys.Rev.38(1931)57-79.

Ham, L.B., Phys.Rev.25(1925)762-767.

Harkins, W.D., Science 46(1917)419-427;443-448.

Hebb,M.,Phys.Rev.46(1934)17-32.

Heisenberg,W. Zeit.f.Phys. 32(1925)841-860; 33(1925)879-893.

Heitler,W. and London,F., Zeit.f.Phys.44(1927)455-472.

Hill.E.L., Phys.Rev.32(1928)250-272.

Houston, W.V., Zeit.f.Phys.48(1928)449-468.

Huff,L., Phys.Rev.38(1931)501-512.

Hutchisson,E.,Phys.Rev.28(1926)1022-1029; 29(1927)270-284.

Inglis,D., Phys.Rev.38(1931)862-872.

Ireland,C., Phys.Rev.43(1932)329-336.

James,H.M. Jour.Chem.Phys.1(1933)825-835; 2(1934)794-810;3(1935)9-14.

Janes,R., Phys.Rev.48(1935)78-83.

Johnston,M.H., Proc.N.A.S.19(1933)916-921.

Jordahl,O., Phys.Rev.42(1932)901A.

Kemble,E.C., Phys.Rev.8(1916)691-700; 701-714; 15(1920)95-109;

______ Phil.Mag.42(1921)123-133;

______ Phys.Rev.47(1935)973-974.

Kennard,E.H., Phys.Rev.19(1922)420.

Knipp,J., Phys.Rev.47(1935)672-677.

Koenig,H., Phys.Rev.44(1933)657-665.

Krammers,H.A., Zeit.f.Phys.13(1923)312-341;

Krammers, H.A. and Ittmann, G.P., Zeit.f.Phys.53(1929)553-565; 58(1929) 217-231; 60(1930)663-681.

Kratzer, A. Zeit.f.Phys.3(1920)460.

Kronig, R., Nature 117(1926)550.

Krutter, H.M., Phys.Rev.47(1935)559-568.

Lamgmuir, I., Phys.Rev.17(1921)339-353.

Lee, R., Phys.Rev.44(1933)624-631.

Lewis, A., Phys.Rev.36(1930)68-86.

Lewis, G.N., Jour.Amer.Chem.Soc.38(1916)762-785;

______ Jour.Chem.Phys.1(1933)17-28;

______ and Adams, E.Q., Phys.Rev.3(1914)92-102; 4(1914)331-343.

Lindsay, R.B., M.I.T.Jour.Math.Phys.3(1924)191-236

______ Sci.Mon.34(1932)115-124; 37(1933)330-337; 38(1934)240-250

Loomis, F.W., Nature 106(1920)179;

______ Astrophys.Jour.52(1920)248.

Lowen, I. Phys.Rev.46(1934)590-597.

Manning, M.F., Phys.Rev.48(1935)161-164.

Lyman, T., Nature 93(1914)241.

Margenau, H., The Monist 41(1931)1-36; 42(1932)161-188;

______ Phil.Sci.1(1934)118-121;133-148; 2(1935)49-71;164-187.

Mauchly, J., Phys. Rev. 40(1932)123A; 43(1933)12-30.

Merrill, R., Phys.Rev.46(1934)334A;487-501.

Millikan, R.A. Phys.Rev.2(1913)109-143;

______ Proc.N.A.S. 2(1916)78-83;

______ and Winchester, G. Phil.Mag.14(1907)188-210.

Morse, P.M. and Steuckelberg, E.C., Phys.Rev.33(1929)932-947; 34(1929) 57-64;
Helv.Phys.Acta.2(1929)304-306;

_______ and Rosen, N. Phys.Rev.42(1932)210-217; 37(1931)658.

Mulliken, R.S., Nature 113(1924)744, 785;

_______ Phys.Rev.32(1928)186-222;761-772; 33(1929)730-747.

Muskat, M., Proc.N.A.S.15(1929)405

Oppenheimer, J.R., Proc.Camb.Phil.Soc.23(1926)327-335;422-431;

_______ Ann.d.Phys.84(1927)457-484;
_______ Proc.N.A.S.14(1928)363-365;

_______ Phys.Rev.35(1930)939-947;

_______ and Born, M. Zeit.f.Phys.41(1927)268-294;

_______ and Plesset, M., Phys.Rev.44(1933)53-55.

Page, L., Phys.Rev.20(1922)18-25.

Parson, A.L., Smithsonian Misc. Coll.65 (nov.29,1915) Publication 2371.

Pauling, L. Chem.Rev.5(1928)173-213;

_______ Proc.N.A.S.14(1928)359-362;

_______ Jour.Amer.ChemSoc.53(1931)1367-1400; 3225-3237; 54(1932)988-1003; 3570-3582;

_______ Jour.Chem.Phys.1(1933)362-374; 606-617; 679-686.

Penney, W.G. and Schlapp, R., Phys.Rev.41(1932)194-207.

Phillips, M., Phys.Rev.44(1933)644-650.

Podolsky, B., Proc.N.A.S.14(1928)253-258.

Pomeroy, W.C., Phys.Rev.29(1927)59-78.

Present, R.D., Phys.Rev.48(1935)140-148.

Richardson, O.W. and Compton, K.T., Phil.Mag.24(1912)575-594; 26(1913) 549-567

Richter, C.F., Proc.N.A.S.13(1927)476-479.

Richtmyer, R.D., Phys.Rev.49(1936)1-8.

Rojansky,V., Phys.Rev.33(1929)1-15.

Rose,M., Phys.Rev.48(1935)211-223.

Rosen,N., Phys.Rev.38(1931)2099-2107.

Rosenthal,J.E., Phys.Rev.33(1929)163-168;

______ Proc.N.A.S.15(1929)381-387.

Russell, H.N. and Saunders,F.A., Astrophys.Jour.61(1925)38-69.

Schlapp,R. and Penney, W.G., Phys.Rev.42(1932)666-686.

Schrödinger,E., Phys.Rev.28(1926)1049-1070.

Seeger, R.J., Proc.N.A.S.17(1931)301-310; 18(1932)303-310.

Serber,R., Phys.Rev.41(1932)489; 43(1933)1003-1010;1011-1021; 45(1934) 461-467.

Shea, J.D., Phys.Rev.27(1926)245A.

Slater, J.C., Nature 113(1924)307-308; 116(1925)278;

______ Proc.N.A.S.11(1925)732-738;

______ Phys.Rev.34(1929)1293-1322; 37(1931)481-489; 38(1931)1109-1144; 45(1934)794-801.

Sommerfeld,A., Jour.Opt.Soc.Amer.7(1923)509-516;

______, and Frank, N.H., Rev.Mod.Phys.3(1931)1-42.

Stratton, J.A., Helv.Phys.Acta 1(1928)47-74.

Tolman,R.C., Phys.Rev.4(1914)145-153.

Turner, L.A., Astrophys.Jour.58(1923)176-194.

Uehling,E., Phys.Rev.43(1933)522-561.

Urey,H.C., Astrophys.Jour.59(1924)1-10;

______ Jour.Amer.ChemSoc.45(1923)1445-1455;

______ Zeit.f.Phys.29(1924)86-90.

Vinti,J.P., Phys.Rev.43(1933)337-340.

Vallarta,M.S., M.I.T.Jour.Math.Phys.4(1925)65-83.

Van Vleck,J.H., Phil.Mag.44(1922)842-869;

______ Phys.Rev.21(1923)372; 24(1924)330-365;

______ Jour.Opt.Soc.Amer.9(1924)27-30; 14(1927)108-112; 16(1928)301-306;

______ Jour.Chem.Phys.1(1933)117-182; 219-238; 2(1934)20-30;3(1935)803-6;

______ Phys.Rev.33(1929)467-506; 41(1932)208-215; 46(1934)509-524;

______ and Sherman, A., Rev.Mod.Phys.7(1935)168-228.

Wang,S.C., Phys.Rev.31(1928)579-586; 34(1929)243-252.

Wetzel,W.,Phys.Rev.44(1933)25-30.

Wheeler,J.A.,Phys.Rev.43(1933)258-263.

White,F.,Proc.N.A.S.20(1934)525-529.

Whitelaw,N.,Phys.Rev.44(1933)551-569.

Wigner,E. and Seitz,F.,Phys.Rev.43(1933)804-810;46(1934)509-524.

Willis,L.,Phys.Rev.44(1933)470-490.

Wu,Ta-You, Phys.Rev.43(1933)496L; 44(1933)727-731.

Young,L., Phys.Rev.36(1930)1154-1167.

Part II B: Unpublished Materials.

II B 1: Lecture Notes

E.C.Kemble's Quantum Theory Courses at Harvard University, 1919 - 1928, loaned to the author by E.C.Kemble, now in his files.

J.H.Van Vleck's Quantum Mechanics Course, 1927 - 1928, at the University of Minnesota; notes taken by R.B.Whitney, Xerox copy provided to the author by R.B.Whitney.

II B 2: Letters (Arranged by Present Location)

Center for History of Physics - American Institute of Physics: E.U. Condon to J.T.Tate, 2 October 1928; G.H.Dieke to S.A.Goudsmit, 15 August 1925, 15 October 1925, 20 January 1927.

Theodore Lyman Papers, Physics Department, Harvard University: J.R. Oppenheimer to E.C.Kemble, 24 May 1923; T.Lyman to J.R.Oppenheimer 8 June 1923; E.C.Kemble to T.Lyman, 9 June 1927; T.Lyman to E.C.Kemble, 28 June 1927.

Massachusetts Institute of Technology Archives: Max Born to S.W. Stratton, 13 February 1927.

E.C.Kemble's Files: J.H.Van Vleck to E.C.Kemble, 30 March 1927.

Part II: Materials Published Since 1935

II A 1 a: Books, General - Histories and Social Commentaries

Brickman, W.W. and Lehrer, S. eds. A Century of Higher Education: Classical Citadel to Collegiate Colossus New York: Society for the Advancement of Education, 1962.

Brubacher, John S. and Rudy, Willis. Higher Education in Transition - An American History, 1636 - 1956. New York: Harper and Row, 1958.

Burnham, John C. ed. Science in America: Historical Selections. New York: Holt, Rinehart and Winston, 1971.

Cline, Barbara. The Questioners: Physicists and the Quantum Theory. New York: Thomas Y. Crowell Co., 1965.

Cochrane, Rexmond C. Measures for Progress: A History of the National Bureau of Standards. Washington, D.C.: U.S.Department of Commerce, 1966.

Dupree, A.H. Science in the Federal Government New York: Belknap Press, 1964.

Daniels, George H. Science in American Society: A Social History. New York: Alfred A. Knopf, 1971.

__________, ed. Nineteenth Century American Science: A Reappraisal Evanston: Northwestern University Press, 1972.

Fermi, Laura. Illustrious Immigrants: The Intellectual Migration from Europe 1930-1941. Chicago: University of Chicago Press, 1968.

Fleming, D. and Bailyn, B. The Intellectual Migration: Europe and America 1930 - 1960. Cambridge, Mass: Harvard University Press, 1969.

Fosdick, Raymond B. Adventure in Giving: The Story of the General Education Board. New York: Harper and Row, 1962.

The Story of the Rockefeller Foundation. New York: Harper and Brothers, 1952.

Friis, Erik J. The American Scandinavian Foundation: A Brief History. New York: American Scandinavian Foundation, 1961.

Gamow, George. Thirty Years that Shook Physics: The Story of the Quantum. Garden City, N.Y.: Doubleday and Co, Inc., 1966.

Gray, George W. Education on an International Scale: A History of the International Education Board. New York: Harcourt,Brace Co., 1941.

Hermann, Armin. The Genesis of Quantum Theory 1899-1913. Claude W.Nash, trans. Cambridge, Mass.: The MIT Press,1971.

Hund, Friedrich. Geschichte der Quantentheorie. Mannheim: Bibliographisches Institut AG, 1967.

Jammer, Max. The Conceptual Development of Quantum Mechanics. New York: McGraw-Hill Book Co., Inc.,1966.

The Philosophy of Quantum Mechanics. New York: John Wiley and' Sons, Inc., 1974.

Jungk, Robert. James Cleugh,trans. Brighter than a Thousand Suns New York: Harcourt, Brace and Co., 1958. From Heller als tausend Sonnen. Bern: Alfred Scherz Verlag, 1956.

Kargon, Robert, ed. The Maturing of American Science. Washington,D.C.: The American Association for the Advancement of Science, 1974.

Morison, Samuel E. The Oxford History of the American People. New York: Oxford University Press, 1965.

Reingold, Nathan, ed. Science in Nineteenth - Century America. New York: Hill and Wang, 1964.

Stuewer, Roger H. The Compton Effect: Turning Point in Physics. New York: Science History Publications, 1975.

Van der Waerden, B.L., ed. Sources of Quantum Mechanics. New York: Dover Publications, Inc., 1967

Wilson, Edwin B. History of the Proceedings of the National Academy of Sciences. Washington, D.C.: National Academy of Sciences, 1966.

II A 1 b: Books: Biographies, Autobiographies, Correspondence

Born, Irene, trans. The Born - Einstein Letters New York: Walker and Co., 1971.

Born, Max. My Life and My Views. New York: Charles Scribner's Sons, 1968.

Childs, Herbert. An American Genius: The Life of Ernest Orlando Lawrence. New York: E.P.Dutton and Co., Inc., 1968.

Coulson, Thomas. Joseph Henry: His Life and Work. Princeton: The Princeton University Press, 1950.

Davis, N.P. Lawrence and Oppenheimer. New York: Simon and Schuster Co. 1968.

Einstein, Albert. Sonja Bargmann, trans. Ideas and Opinions. New York: Crown Publishers, Inc., 1954.

Fermi, Laura. Atoms in the Family. Chicago: The University of Chicago Press, 1954.

Jaffe, Bernard. Michelson and the Speed of Light. Garden City,N.Y.: Doubleday and Co., Inc. 1960.

Frank, Philip. Einstein: His Life and Times. New York: Alfred Knopf Co. 1967.

Heisenberg, Werner. A.J.Pomerans, trans. Physics and Beyond: Encounters and Conversations. New York: Harper and Row Publishers, Inc., 1971.

Hermann, Armin, compiler. Albert Einstein/ Arnold Sommerfeld Brief-Wechsel. Basel: Schwabe and Co.Verlag, 1968.

Johnston, Marjorie, ed. The Cosmos of Arthur Holly Compton. New York: Alfred A. Knopf, 1967.

Livingston, Dorothy Michelson The Master of Light: A Biography of Albert A. Michelson. New York: Charles Scribner's Sons, 1973.

Lurie, Edward. Louis Agassiz: A Life in Science. Chicago: The University of Chicago Press, 1960.

Michelmore, Peter. The Swift Years. New York: Dodd, Mead and Co., 1969.

Michelson Museum Albert Abraham Michelson: The Man who taught a World to Measure. China Lake, Cal.: Michelson Museum Publication, 1970

Millikan, R.A. The Autobiography of Robert A. Millikan. New York: Prentice-Hall, Inc., 1950.

Rabi,I., Serber, R., Weisskopf, V.F., Pais,A. and Seaborg, G.T. Oppenheimer. New York: Charles Scribner's Sons, Inc. 19691

Rayleigh, Lord. The Life of Sir J. J. Thomson. Cambridge,Eng.: The University Press, 1942.

Seabrook, William. Doctor Wood - Modern Wizard of the Laboratory. New York: Harcourt, Brace and Co., 1941.

Slater, J.C. Solid State and Molecular Theory: A Scientific Autobiography. New York: John Wiley and Sons, Inc.,1973.

Wheeler, L.P. Josiah Willard Gibbs: The History of a Great Mind. New Haven: Yale University Press, 1951, rev.1952. Reprinted by Archon Books, 1970.

Wiener, Norbert I am a Mathematician. Garden City, N.Y.:Doubleday and Co., 1956.

Weisskopf, V.F. Physics in the Twentieth Century: Selected Essays. Cambridge,Mass.: The MIT Press, 1972.

II A 1 c: Festschriften, Memorial Volumes

Note: listed alphabetically by name of person honored.

de Broglie: Price,W.C., Chisswick,S.S., Ravensdale,T.,eds. Wave Mechanics: The First Fifty Years. New York: Halsted, 1973.

Breit,G.: Bromley,D.A., Hughes,V.W.,eds. Facets of Physics. New York: Academic Press, 1970.

Condon,E.U.: Brittin,W.E. and Odabasi,H.,eds. Topics in Modern Physics Boulder: Colorado Associated University Press,1971.

Dirac, P.A.M.: Salam,A. and Wigner,E.P. Aspects of Quantum Theory Cambridge,Eng.: University Press,1972.

Lorentz, H.A.: de Haas-Lorentz,G.L.,ed. H.A.Lorentz: Impressions of his Life and Work. Amsterdam: North Holland Publishing Co.,1957.

Morse, P.M.: Feshback,H. and Ingard,K.U.,eds. In Honor of Philip M. Morse. Cambridge,Mass.: The MIT Press,1969.

Pauli, W.: Fierz,M. and Weisskopf,V.F.,eds. Theoretical Physics in the Twentieth Century. New York: Wiley Interscience Publishers,Inc., 1960.

Pauling,L.: Rich,A. and Davidson,N.,eds. Structural Chemistry and Molecular Biology. SanFrancisco:W.H.Freeman and Co., 1968.

II A 1 d: Collected Works of Individual Scientists

Birkhoff, G.D. Collected Mathematical Papers Vol.I-III New York: The American Mathematical Society, 1950.

Suits,C.G. and Way,H.E.,eds. The Collected Works of Irving Langmuir New York: Pergamon Press,1962.

Hinze,J. and Ramsey,D.A.,eds. Selected Papers of Robert S.Mulliken Chicago: University of Chicago Press,1975.

Kronig,R. and Weisskopf, eds. Collected Scientific Papers of Wolfgang Pauli. New York: John Wiley and Sons,Inc.1964.

Sommerfeld, Arnold Gesammelte Schriften Vol.I-IV. Braunschweig:Friedr. Vieweg und Sohn,1968.

II A 1 e: Information Sources - Doctorates, Fellowships, Nobel Prizes, Biographies and General Resources

Dissertation in Physics, 1861 - 1959, M. Lois Marckworth, compiler. Stanford: Stanford University Press, 1961.

Harvard University Department of Physics Doctors of Philosophy and Doctors of Science, 1873 - 1964, Cambridge: The Graduate Society for Advanced Study and Research, 1965.

"A Directory of Harvard Physics Doctoral Alumni", pamphlet issued by the Harvard University Physics Department, 1975.

Commonwealth Fund Fellows, 1925 - 1937. New York: The Commonwealth Fund, 1938.

Directory of the Fellows of the John Simon Guggenheim Memorial Foundation 1925 - 1967. New York: The John Simon Guggenheim Memorial Foundation, 1968.

National Research Fellowships, 1919 - 1944, Neva E. Reynolds, compiler. Washington, D.C.: National Research Council, 1944.

The Rockefeller FoundationDirectory of Fellowship Awards, 1917 - 1950. New York: The Rockefeller Foundation, 1951.

Nobel Lectures in Physics, 1901 - 1921; 1922 - 1941; 1942 - 1962. Amsterdam: Elsevier Publishing Co., 1967, 1965, 1964

Dictionary of Scientific Biography, Charles C. Gillespie, Ed. in Chief. New York: Charles Scribner's Sons, Vol.I, 1970 - Vol.XII, 1975.

Sources for History of Quantum Physics, T.S. Kuhn, J.L. Heilbron, P.L. Forman and L. Allen. Philadelphia: The American Philosophical Society, 1967.

II A 2 a: Articles: Reminiscences and Commentaries

Anderson, Philip, "Van Vleck and Magnetism", Physics Today (October 1968)23-26.

Barton, H. A., "The Story of the American Institute of Physics", Physics Today (January 1956)55-66.

Bedell, Frederick, "What led to the Founding of the American Physical Society", Phys.Rev.75(1949)1601-1604.

Birge, R.T., "Physics and Physicists of the Past Fifty Years", Physics Today(May 1956)20-28.

Birkhoff, G.D., "Fifty Years of American Mathematics", American Mathematical Society, Semicentennial Adresses 2(1938)270-315.

Buwalda, Imra, "The Roots of the California Institute of Technology", Engineering and Science (October1966)8-11; (November 1966)20-26, (December 1966)18-23.

Condon, E.U., "The Development of American Physics", Amer.Jor.Phys.17 1949)404-408.

________ "The Franck-Condon Principle and Related Topics", Amer. Jour. Phys.15(1947)365-408.

________ "Reminiscences of a Life in and out of Quantum Mechanics", Internat.Jour.Qu.Chem. Symp.No.7(1973)7-22.

________ "60 Years of Quantum Physics", Physics Today (October 1962) 37-49.

Dennison, D.M., "Physics and the Department of Physics since 1900" in Research - Definitions and Reflections: Essays on the Occasion of the University of Michigan Sesquicentennial, Ann Arbor: University of Michigan Press, 1967.

________ "Recollections of Physics and of Physicists during the 1920's", Amer.Jour.Phys.42(1974)1051-1056.

Germer,Lester, "Low Energy Electron Diffraction", Physics Today (July 1964) 19-23.

Goudsmit, S.A.,"The Michigan Symposium in Theoretical Physics", Michigan Alumni Quarterly Review (May 20,1961)178-182.

Harrison, G.R., "Karl Compton and American Physics", Physics Today (November 1957)19-22.

Hull, G.F., "Experimental Discoveries Announced at the Meeting of the American Physical Society Fifty Years Ago", Sci.Mon.77(1953)13-18.

"Reminiscences of a Scientific Comradeship", American Physics Teacher 4(1936)61-65.

Hylleraas, E.A.,"Reminiscences from Early Quantum Mechanics of Two-electron Atoms", Rev.Mod.Phys.35(1963)421-431.

Kronig,R.,"The Turning Point" in Fierz and Weisskopf Theoretical Physics in the Twentieth Century: A Memorial Volume to Wolfgang Pauli.

Lande,A.,"Quantum Fact and Fiction, I,II,III", Amer.Jour.Phys.33(1965) 123-127; 34(1966)1160-1163; 37(1969)541-548.

__________"Unity in Quantum Theory", Foundations of Physics 1(1971) 191-202.

__________"The Decline and Fall of Quantum Dualism", Phil.Sci.38(1971) 221-223.

Mulliken, R.S., "Molecular Scientists and Molecular Science: Some Reminiscences", Jour.Chem.Phys.43(1965)S2-S11.

__________"The Path to Molecular Orbital Theory", Jour.Pure and Appl. Chem.24(1970)203-215.

__________ "Spectroscopy, Molecular Orbitals and Chemical Bonding", Science 157(1967)13-24.

Pauling, L.,"Fifty Years of Physical Chemistry in the California Institute of Technology", Annual Rev.Phys.Chem.(1965)1-14.

__________ "Fifty Years of Progress in Structural Chemistry and Molecular Biology", Daedalus(Fall 1970)988-1014.

Platt, J.R., "1966 Nobel Laureate in Chemistry, Robert S.Mulliken", Science 154(1966)745-747.

Shenstone, A., "Princeton Physics: A Remarkable History", Princeton Alumni Weekly (February 24, 1961)6-12,20.

__________"Sixty Years of Palmer Laboratory" , address delivered at the dedication of the Jadwin Physics Building, Princeton University.

Slater, J.C., "Quantum Physics in America Between the Wars", Internat. Jour.Qu.Chem. 1S(1967)1-23.

__________ "The Development of Quantum Mechanics in the Period 1924-1926" in Price, Chisswick and Ravensdale Wave Mechanics: The First Fifty Years.

Uhlenbeck, G. "Reminiscences of Professor Paul Ehrenfest", Amer.Jour. Phys.24(1956)431-433.

Van Vleck, J.H.,"American Physics Comes of Age", Physics Today(June 1964) 21-26.

________ Inaugural Address as Lorentz Professor at the University of Leiden, March 4, 1960.

________ "My Swiss Visits of 1906, 1926 and 1930", Helv.Phys.Acta 41 (1968)1234-1237.

________ "Reminiscences of the First Decade of Quantum Mechanics" Internat.Jour.Qu.Chem.5(1971)3-20.

________ "Spin, the Great Indicator of Valence Behavior", Pure and Appl.Chem.24(1970)235-255.

________ "Travels with Dirac in the Rockies" in Salam and Wigner Aspects of Quantum Theory.

II A 2 b: Articles: Historical Studies

Bartlett, A.A., "Compton Effect: Historical Background", Amer.Jour.Phys. 32(1964)120-127.

Beach,M., "Was there a Scientific Lazzaroni?" in Nineteenth Century American Science, G.H.Daniels,ed.

Bridgman,P.W., "P.W.Bridgman's The Logic of Modern Physics after Thirty Years", Daedalus 88(Summer 1959)518-526.

Coben,S., "The Scientific Establishment and the Transmission of Quantum Mechanics to the United States, 1919-1932", Amer.Hist.Rev.76(1971) 442-466.

Cohen,I.B., "American Physicists at War: from the First World War to 1942", Amer.Jour.Phys.13(1945)333-346.

________ "Science in America: The Nineteenth Century" in Paths of American Thought, A.M.Schlesinger,Jr. and M.White, eds. Boston: Houghton Mifflin Co.,1963.

Compton, A.H., "The Scattering of X-Rays as Particles", Amer.Jour.Phys. 29(1961)817-820.

Forman,P., "Weimar Culture, Causality and Quantum Theory, 1918-1927" Hist.Stud.Phys.Sci. Vol.3(1971)1-115.

______,Heilbron,J.L., Weart,S., "Physics circa 1900: Personnel, Funding and Productivity. Hist.Stud.Phys.Sci. Vol.5(1975)1-200.

Haberer,J.,"Politicalization in Science", Science 178(1972)713-724.

Kevles,D.J., "On the Flaws of American Physics" in Science in Nineteenth Century America, G.H.Daniels,ed.

______ "George Ellery Hale, The First World War, and the Advancement of Science in America", Isis 59(Winter 1968)427-437.

Klein,M.J., "Einstein, Specific Heats and the Early Quantum Theory", Science 148(1965)173-180.

Kohler,R.E.,Jr., "Irving Langmuir and the 'Octet' Theory of Valence" Hist.Stud.Phys.Sci. Vol4(1973)39-87;

______"The Origin of G.N.Lewis's Theory of the Shared Pair Bond", Hist.Stud.Phys.Sci. Vol.3(1971) 343-376.

McCormmach,R., "Henri Poincare and the Quantum Theory", Isis 58 (Spring 1967)37-55.

Mendelsohn,E. "The Emergence of Science as a Profession in Nineteenth Century Europe" in The Management of Science, Karl Hill,ed. Boston:Beacon Press,1964.

______"Science in America: The Twentieth Century" in Paths of American Thought, A.M.SchlesingerJr. and M.White,eds.

Rand,M.J., "The National Research Fellowships", Sci.Mon.73(1951)71-80.

Shryock,R.H., "American Indifference to Basic Science during the Nineteenth Century" in The Sociology of Science, B.Barber and W.Hirsch, eds., New York:The Free Pressof Glencoe, a division of Macmillan Co.,1962.

Sopka,K.J., "An Apostle of Science Visits America: John Tyndall's Journey of 1872-1873", The Physics Teacher 10(1972)369-375.

Weiner,C. "How the Transistor Emerged", IEEE Spectrum 10(1973)24-33;

______"Joseph Henry and the Relations between Teaching and Research" Amer.Jour.Phys.34(1966)1093-1100;

______"A New Site for the Seminar: The Refugees and American Physics in the Thirties" in The Intellectual Migration,D.Fleming and B.Bailyn, eds.

Weiner, C., "1932 - Moving into the New Physics", Physics Today (May 1972)40-49;

__________"Physics in the Great Depression", Physics Today (October 1970) 31-37;

__________"Science and Higher Education" in Science and Society in the United States, D.D.vanTassell and M.G. Hall, eds. Homewood,Ill: The Dorsey Press, 1966.

II A 2 c: Biographical Articles.

Bernstein,J., "I.I. Rabi", The New Yorker (13October1975)47-110 and (20October1975)47-101.

Lang, D., "A Farewell to String and Sealing Wax" (S.A.Goudsmit) The New Yorker (7November1953)47-64 and (14November1953)46-67.

II B: Unpublished Materials

II B 1: Interviews and Biographical Notes

Archive for History of Quantum Physics: D.M.Dennison, C.Eckart, P. Epstein, S.A.Goudsmit, K.Herzfeld, F.C.Hoyt, R.Kronig, E.C.Kemble, O.Laporte, L.Loeb, I.I.Rabi, J.C.Slater, J.H.Van Vleck.

Center for History of Physics, American Institute of Physics: R.F.Bacher, H.A.Barton, E.U.Condon, K.K.Darrow, H.Dodge, W.H.Furry, W.V.Houston, E.Hutchisson, R.B.Lindsay, R.Serber, D.L.Webster, H.Webb.

University of Illinois: F.W.Loomis.

II B 2: Departmental Histories

University of California at Berkeley: R.T.Birge History of the Physics Department, issued privately in mimeographed form by the University of California.

University of Michigan:"Physics" by Meyer, Lindsay, Rich, Barker and Dennison in The University of Michigan, available at CHP-AIP.

University of Minnesota: History of the Minnesota Physics Department, manuscript by H.A.Erikson deposited at CHP-AIP.

II B 3: Doctoral Dissertations

Benz, Ulrich, "Arnold Sommerfeld", selected pages from work in progress at the University of Stuttgart.

Gehrenbeck, Richard, "C.J.Davisson, L.H.Germer and the Discovery of Electron Diffraction", U. of Minn. Ph.D.thesis, 1973.

Goldberg, Stanley, "The Early Responses to Einstein's Special Theory of Relativity, 1905 - 1911", Harvard University Ph.D. in Education thesis, 1968.

Kevles, D.J., "The Study of Physics in America, 1865 - 1916", Princeton University Ph.D. thesis, 1964.

Small, Henry, "The Helium Atom in the Old Quantum Theory", University of Wisconsin Ph.D. thesis, 1971.

Stuewer, Roger, "The Compton Effect", University of Wisconsin Ph.D. thesis, 1968.

II B 4: Private Communications received by the Author.

Comment: The author is indebted to the following persons who have graciously shared with her their recollections of events and experiences that occurred up to 1935 in oral or written form. The author's files contain all of the correspondence involved and notes on conversations that took place.

K. T. Bainbridge, H. A. Bethe, R. T. Birge, Mrs. G. Breit, L. de Broglie, A. S. Coolidge, K. K. Darrow, D. M. Dennison, Jane Dewey, N. H. Frank, W. H. Furry, S. A. Goudsmit, V. Guillemin, G. R. Harrison, W. Heinsenberg, F. C. Hoyt, Charlotte Houtermanns, G. F. Hull, Jr., F. Hund, E. Hutchisson, H. M. James, J. M. Jauch, E. C. Kemble, R. Kronig, A. Landé, V. F. Lenzen, R. B. Lindsay, F. W. Loomis, H. Margenau, P. M. Morse, R. S. Mulliken, M. Muskat, O. O. Oldenberg, L. Pauling, Mrs. H. B. Phillips, I. I. Rabi, Jenny Rosenthal, A. E. Ruark, R. J. Seeger, A. G. Shenstone, J. C. Slater, J. A. Stratton, D. Struik, M. A. Tuve, H. C. Urey, M. S. Vallarta, J. H. Van Vleck, S. C. Wang, H. Webb, D. L. Webster, R. B. Whitney, E. Bright Wilson, C. Zenner.

At this time I wish to express my deep appreciation for their kind interest and encouragement in my research and for the assistance they provided.

Index

THREE CENTURIES OF SCIENCE IN AMERICA

An Arno Press Collection

Adams, John Quincy. **Report of the Secretary of State upon Weights and Measures.** 1821.

Archibald, Raymond Clare. **A Semicentennial History of the American Mathematical Society: 1888-1938** *and* **Semicentennial Addresses of the American Mathematical Society.** 2 vols. 1938.

Bond, William Cranch. **History and Description of the Astronomical Observatory of Harvard College** *and* **Results of Astronomical Observations Made at the Observatory of Harvard College.** 1856.

Bowditch, Henry Pickering. **The Life and Writings of Henry Pickering Bowditch.** 2 vols. 1980.

Bridgman, Percy Williams. **The Logic of Modern Physics.** 1927.

Bridgman, Percy Williams. **Philosophical Writings of Percy Williams Bridgman.** 1980.

Bridgman, Percy Williams. **Reflections of a Physicist.** 1955.

Bush, Vannevar. **Science the Endless Frontier.** 1955.

Cajori, Florian. **The Chequered Career of Ferdinand Rudolph Hassler.** 1929.

Cohen, I. Bernard, editor. **The Career of William Beaumont and the Reception of His Discovery.** 1980.

Cohen, I. Bernard, editor. **Benjamin Peirce: "Father of Pure Mathematics" in America.** 1980.

Cohen, I. Bernard, editor. **Aspects of Astronomy in America in the Nineteenth Century.** 1980.

Cohen, I. Bernard, editor. **Cotton Mather and American Science and Medicine: With Studies and Documents Concerning the Introduction of Inoculation or Variolation.** 2 vols. 1980.

Cohen, I. Bernard, editor. **The Life and Scientific Work of Othniel Charles Marsh.** 1980.

Cohen, I. Bernard, editor. **The Life and the Scientific and Medical Career of Benjamin Waterhouse: With Some Account of the Introduction of Vaccination in America.** 2 vols. 1980.

Cohen, I. Bernard, editor. **Research and Technology.** 1980.

Cohen, I. Bernard, editor. **Thomas Jefferson and the Sciences.** 1980.

Cooper, Thomas. **Introductory Lecture** *and* **A Discourse on the Connexion Between Chemistry and Medicine.** 2 vols. in one. 1812/1818.

Dalton, John Call. **John Call Dalton on Experimental Method.** 1980.

Darton, Nelson Horatio. **Catalogue and Index of Contributions to North American Geology:** 1732-1891. 1896.

Donnan, F[rederick] G[eorge] and Arthur Haas, editors. **A Commentary on the Scientific Writings of J. Willard Gibbs** *and* Duhem, Pierre. **Josiah-Willard Gibbs: A Propos de la Publication de ses Mémoires Scientifiques.** 3 vols. in two. 1936/1908.

Dupree, A[nderson] Hunter. **Science in the Federal Government: A History of Policies and Activities to 1940.** 1957.

Ellicott, Andrew. **The Journal of Andrew Ellicott.** 1803.

Fulton, John F. **Harvey Cushing: A Biography.** 1946.

Getman, Frederick H. **The Life of Ira Remsen.** 1940.

Goode, George Brown. **The Smithsonian Institution 1846-1896: The History of its First Half Century.** 1897.

Hale, George Ellery. **National Academies and the Progress of Research.** 1915.

Harding, T. Swann. **Two Blades of Grass: A History of Scientific Development in the U.S. Department of Agriculture.** 1947.

Hindle, Brooke. **David Rittenhouse.** 1964.

Hindle, Brooke, editor. **The Scientific Writings of David Rittenhouse.** 1980.

Holden, Edward S[ingleton]. **Memorials of William Cranch Bond, Director of the Harvard College Observatory, 1840-1859, and of his Son, George Phillips Bond, Director of the Harvard College Observatory, 1859-1865.** 1897.

Howard, L[eland] O[sslan]. **Fighting the Insects: The Story of an Entomologist, Telling the Life and Experiences of the Writer.** 1933.

Jaffe, Bernard. **Men of Science in America.** 1958.

Karpinski, Louis C. **Bibliography of Mathematical Works Printed in America through 1850.** Reprinted with **Supplement and Second Supplement.** 1940/1945.

Loomis, Elias. **The Recent Progress of Astronomy: Especially in the United States.** 1851.

Merrill, Elmer D. **Index Rafinesquianus: The Plant Names Published by C.S. Rafinesque with Reductions, and a Consideration of his Methods, Objectives, and Attainments.** 1949.

Millikan, Robert A[ndrews]. **The Autobiography of Robert A. Millikan.** 1950.

Mitchel, O[rmsby] M[acKnight]. **The Planetary and Stellar Worlds: A Popular Exposition of the Great Discoveries and Theories of Modern Astronomy.** 1848.

Organisation for Economic Co-operation and Development. **Reviews of National Science Policy: United States.** 1968.

Packard, Alpheus S. **Lamarck: The Founder of Evolution; His Life and Work.** 1901.

Pupin, Michael. **From Immigrant to Inventor.** 1930.

Rhees, William J. **An Account of the Smithsonian Institution.** 1859.

Rhees, William J. **The Smithsonian Institution: Documents Relative to its History.** 2 vols. 1901.

Rhees, William J. **William J. Rhees on James Smithson.** 2 vols. in one. 1980.

Scott, William Berryman. **Some Memories of a Palaeontologist.** 1939.

Shryock, Richard H. **American Medical Research Past and Present.** 1947.

Shute, Michael, editor. **The Scientific Work of John Winthrop.** 1980.

Silliman, Benjamin. **A Journal of Travels in England, Holland, and Scotland, and of Two Passages over the Atlantic in the Years 1805 and 1806.** 2 vols. 1812.

Silliman, Benjamin. **A Visit to Europe in 1851.** 2 vols. 1856

Silliman, Benjamin, Jr. **First Principles of Chemistry.** 1864.

Smith, David Eugene and Jekuthiel Ginsburg. **A History of Mathematics in America before 1900.** 1934.

Smith, Edgar Fahs. **James Cutbush: An American Chemist.** 1919.

Smith, Edgar Fahs. **James Woodhouse: A Pioneer in Chemistry, 1770-1809.** 1918.

Smith, Edgar Fahs. **The Life of Robert Hare: An American Chemist (1781-1858).** 1917.

Smith, Edgar Fahs. **Priestley in America: 1794-1804.** 1920.

Sopka, Katherine. **Quantum Physics in America: 1920-1935** (Doctoral Dissertation, Harvard University, 1976). 1980.

Steelman, John R[ay]. **Science and Public Policy: A Report to the President.** 1947.

Stewart, Irvin. **Organizing Scientific Research for War: The Administrative History of the Office of Scientifc Research and Development.** 1948.

Stigler, Stephen M., editor. **American Contributions to Mathematical Statistics in the Nineteenth Century.** 2 vols. 1980.

Trowbridge, John. **What is Electricity?** 1899.

True. Alfred. **Alfred True on Agricultural Experimentation and Research.** 1980.

True, F[rederick] W., editor. **The Semi-Centennial Anniversary of the National Academy of Sciences: 1863-1913** *and* **A History of the First Half-Century of the National Academy of Sciences: 1863-1913.** 2 vols. 1913.

Tyndall, John. **Lectures on Light: Delivered in the United States in 1872-73.** 1873.

U.S. House of Representatives. **Annual Report of the Board of Regents of the Smithsonian Institution...A Memorial of George Brown Goode together with a selection of his Papers on Museums and on the History of Science in America.** 1901.

U.S. National Resources Committee. **Research: A National Resource.** 3 vols. in one. 1938-1941.

U.S. Senate. **Testimony Before the Joint Commission to Consider the Present Organizations of the Signal Service, Geological Survey, Coast and Geodetic Survey, and the Hydrographic Office of the Navy Department.** 2 vols. 1866.